Y0-BYK-831

THE ALKALOIDS

Antitumor Bisindole Alkaloids from *Catharanthus roseus* (L.)

VOLUME 37

[Manske, R.H.F. (Richard Helmuth Fred)

THE ALKALOIDS

Antitumor Bisindole Alkaloids from *Catharanthus roseus* (L.)

Edited by

Arnold Brossi
National Institutes of Health
Bethesda, Maryland

Matthew Suffness
National Cancer Institute
National Institutes of Health
Bethesda, Maryland

VOLUME 37

QD421
M28
v.37
1990

Academic Press, Inc.
Harcourt Brace Jovanovich, Publishers
San Diego New York Berkeley Boston
London Sydney Tokyo Toronto

This book is printed on acid-free paper. ∞

COPYRIGHT © 1990 BY ACADEMIC PRESS, INC.
All Rights Reserved.
No part of this publication may be reproduced or transmitted in any form or by any means, electronic or mechanical, including photocopy, recording, or any information storage and retrieval system, without permission in writing from the publisher.

ACADEMIC PRESS, INC.
San Diego, California 92101

United Kingdom Edition published by
ACADEMIC PRESS LIMITED
24-28 Oval Road, London NW1 7DX

LIBRARY OF CONGRESS CATALOG CARD NUMBER: 50-5522

ISBN 0-12-469537-X (alk. paper)

PRINTED IN THE UNITED STATES OF AMERICA
90 91 92 93 9 8 7 6 5 4 3 2 1

CONTENTS

Chapter 4. Medicinal Chemistry of Bisindole Alkaloids From *Catharanthus*

H. L. PEARCE

Chapter 5. Pharmacology of Antitumor Bisindole Alkaloids from *Catharanthus*

JOHN J. MCCORMACK

Chapter 6. The Therapeutic Use of Bisindole Alkaloids from *Catharanthus*

NORBERT NEUSS AND MICHAEL N. NEUSS

CONTRIBUTORS

Numbers in parentheses indicate the pages on which the authors' contributions begin.

GÁBOR BLASKÓ (1), Central Research Institute for Chemistry, Hungarian Academy of Sciences, Budapest H-1525, Hungary

LINDA S. BORMAN (133), Department of Pharmacology and Vermont Regional Cancer Center, University of Vermont, Burlington, Vermont 05405

GEOFFREY A. CORDELL (1), Program for Collaborative Research in the Pharmaceutical Sciences, College of Pharmacy, University of Illinois at Chicago, Chicago, Illinois 60612

MARTIN E. KUEHNE (77, 133), Department of Chemistry and Vermont Regional Cancer Center, University of Vermont, Burlington, Vermont 05405

ISTVÁN MARKÓ (77), Department of Chemistry, The University of Sheffield, Sheffield S37HP, England

JOHN J. MCCORMACK (205), Department of Pharmacology, University of Vermont, Burlington, Vermont 05405

MICHAEL N. NEUSS (229), Oncology and Hematology, Cincinnati, Ohio 25219

NORBERT NEUSS (229), formerly of Lilly Research Laboratories, Eli Lilly and Company, Indianapolis, Indiana 46285

H. L. PEARCE (145), Lilly Research Laboratories, Eli Lilly and Company, Indianapolis, Indiana 46285

PREFACE

This volume entitled "Bisindole Alkaloids from *Catharanthus roseus* (L.)," commonly called Vinca alkaloids, *Catharanthus* alkaloids, or dimeric Vinca alkaloids, was written by a group of internationally known experts. It provides for the first time, a comprehensive view of the basic chemistry, biogenesis, synthesis, medicinal chemistry, pharmacology, and medical uses of these important antitumor agents. Because of the complex structures of the alkaloids, the transformations within the basic series, such as the medicinal chemistry, are best presented using the Chemical Abstracts numbering system, while the transformations between series, as in the total synthesis, are presented using the biosynthetic numbering system. We apologize if the reader finds the exposure to two numbering systems confusing.

This treatise is a companion work to Volume 25 of this series entitled "Antitumor Alkaloids," which covers all classes except the bisindole alkaloids presented here.

Matthew Suffness
Arnold Brossi
National Institutes of Health

—CHAPTER 1—

ISOLATION, STRUCTURE ELUCIDATION, AND BIOSYNTHESIS OF THE BISINDOLE ALKALOIDS OF *CATHARANTHUS*

GÁBOR BLASKÓ AND GEOFFREY A. CORDELL

Program for Collaborative Research in the Pharmaceutical Sciences
College of Pharmacy
University of Illinois at Chicago
Chicago, Illinois 60612

I. Introduction

The isolation of the antitumor agents vincaleukoblastine (**1**) and leurocristine (**2**) from *Catharanthus roseus* (L.) G. Don has proved to be one of the most important developments in both natural product chemistry and the clinical treatment of cancer during the 1960s to 1980s. More alkaloids (over 90) have been isolated from *C. roseus* than from any other plant, and because of the complexity of the alkaloid mixture this work has required the most advanced isolation and structure determination techniques. The exceptional interest in the broad spectrum of antitumor activity of these compounds has resulted in numerous achievements in the pharmaceutical, clinical pharmacologic, and therapeutical sciences. Simultaneously, strenuous efforts have been made in three areas of the natural product chemistry: (i) elaboration of a practical semisynthesis of

THE ALKALOIDS, VOL. 37
Copyright © 1990 by Academic Press, Inc.
All rights of reproduction in any form reserved.

1 and **2** from the two major building blocks vindoline (**3**) and catharanthine (**4**) which are found as major alkaloids in *C. roseus*, (ii) total synthesis of these alkaloids in order to develop analogs for structure–activity relationship studies and to find potentially more active and less toxic derivatives, and (iii) biosynthetic studies aimed at the large-scale production of vinblastine* (**1**) and vincristine (**2**) in cell-free systems or using recombinant DNA technology.

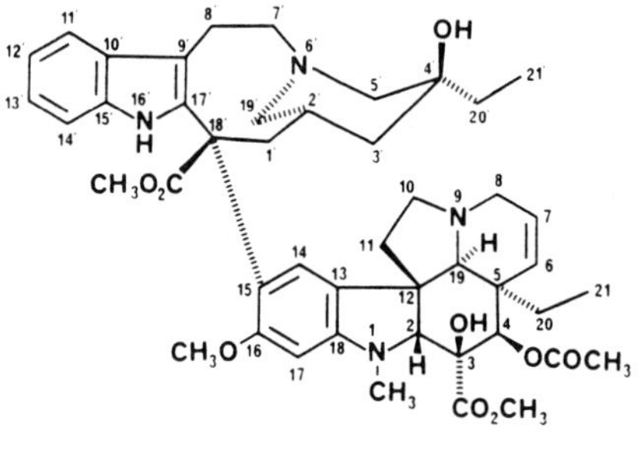

1

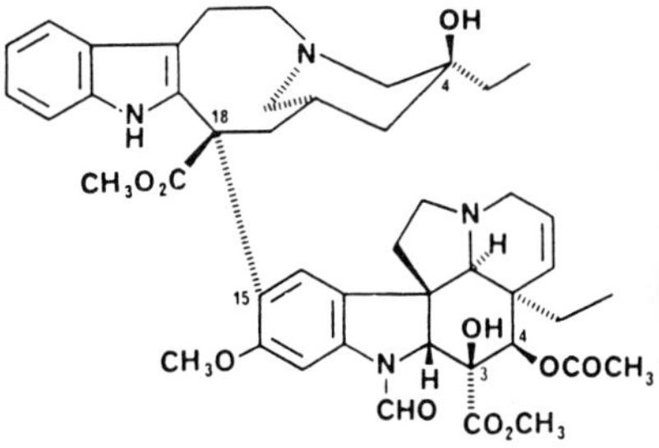

2

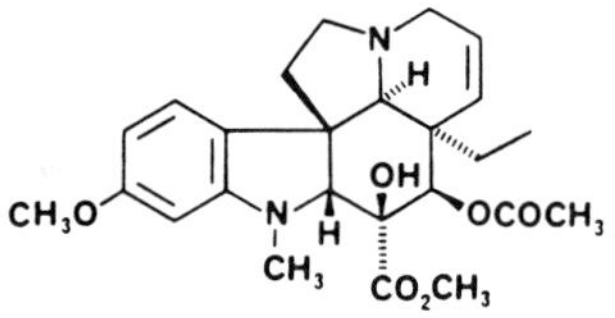

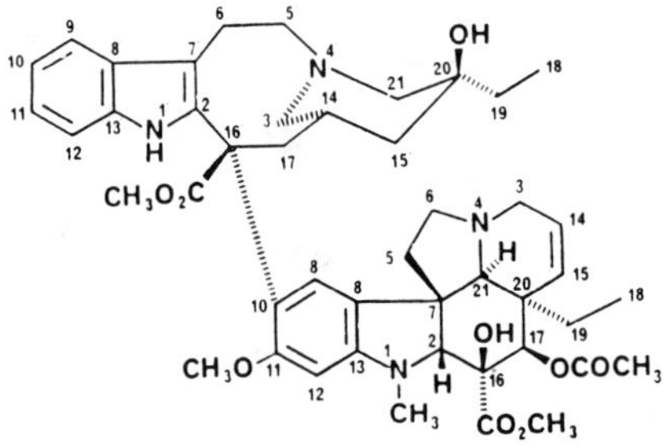

1A

3

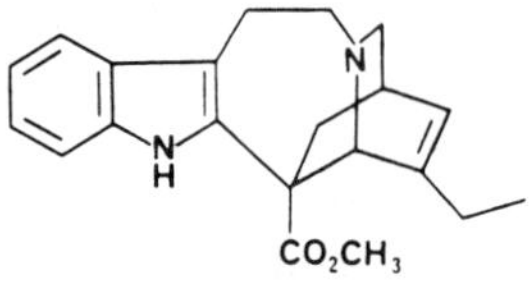

4

* Instead of the originally given alkaloid names vincaleukoblastine and leurocristine we adopt the more commonly used names vinblastine (VLB) and vincristine (VCR) for alkaloids **1** and **2**, respectively.

Since the pioneering work initiated by Dr. Gordon Svoboda at the Eli Lilly & Co. on the commercial development of the antitumor alkaloids vinblastine (**1**) and vincristine (**2**), these compounds have become widely utilized as antitumor agents in clinical practice. During this time a large number of publications have covered the improving knowledge on the antitumor alkaloids of *Catharanthus*. The first comprehensive book covering the phytochemical, chemical, and pharmacological research on the *Catharanthus* alkaloids was published in 1975 (*1*). Since that time several reviews have been published (*2–25*) focusing on different aspects of these scientific efforts. In this chapter we review the literature since 1975 on the occurrence, isolation, structure elucidation, spectroscopy, and biosynthetic studies of the antitumor alkaloids of *Catharanthus*.

1. Botanical Identification

The Madagascan periwinkle is a pantropical everblooming pubescent herb or subshrub. Previously, the botanical nomenclature of this plant was somewhat confused, and it was named as *Ammocallis rosea, Catharanthus roseus, Lochnera rosea,* and *Vinca rosea.* After recognizing the morphological characteristics, the currently accepted genus name is *Catharanthus*. The genus *Catharanthus* is chemotaxonomically very distinct from the genus *Vinca,* and the correct botanical name for the Madagascan periwinkle is *Catharanthus roseus* (L.) G. Don (*26*).

2. Nomenclature, Numbering System

Because of the previous inaccurate botanical determination of the Madagascan periwinkle, the alkaloids of this plant were formerly considered as "Vinca" alkaloids, an erroneous subclassification for alkaloids isolated from a plant belonging to the genus *Catharanthus*. It also should be noted that the alkaloids of *C. roseus* containing two different (most commonly indole and dihydroindole) alkaloid building blocks were, and sometimes still are, referred to as "dimeric" indole alkaloids. It is more accurate to use the term "binary" or "bisindole" alkaloids, since chemically these alkaloids are not dimers of two equal subunits, but rather comprised of two (bis) different alkaloid building blocks.

There are two generally accepted numbering systems for the bisindole-type alkaloids. The system used throughout the chapter is shown around formula **1**; however, the biogenetic numbering system presented for vinblastine as formula **1A** is still in wide use.

3. Occurrence

The occurrence of structurally established bisindole alkaloids in *Catharanthus roseus* (L.) G. Don and in the minor species of *Catharanthus*, namely, *C. lanceus* (Boj. ex DC.) Pich., *C. longifolius* Pich., *C. ovalis* Mgf., *C. pusillus* (Murr.) G. Don, and *C. trichophyllus* (Bak.) Pich, is

TABLE I
BISINDOLE ALKALOIDS WITH ESTABLISHED STRUCTURES ISOLATED FROM *Catharanthus* SPECIES

Alkaloid	Source	Ref.
Anhydrovinblastine (**8**)	*C. roseus*	*27*
Catharanthamine (**9**)	*C. roseus*	*28*
Catharine (**10**)	*C. roseus*	*29–31*
	C. ovalis	*32*
	C. longifolius	*33*
Deacetoxyleurosine (**13**)	*C. roseus*	*34*
Deacetoxyvinblastine (**5**)	*C. roseus*	*34–36*
Deacetylvinblastine (**6**)	*C. roseus*	*37*
Demethylvinblastine (**7**)	*C. roseus*	*38,39–42*
4′-Deoxyvinblastine (**16**)	*C. roseus*	*43*
Leurocolombine (**20**)	*C. roseus*	*44*
Leurosidine (**18**)	*C. roseus*	*45*
Leurosidine N'_b-oxide (**19**)	*C. roseus*	*46*
Leurosine (**11**)	*C. roseus*	*47–49*
	C. ovalis	*32*
	C. longifolius	*33*
	C. lanceus	*50*
	C. pusillus	*51*
Leurosine N'^b-oxide (**12**)	*C. roseus*	*53,54*
5′-Oxoleurosine (**14**)	*C. roseus*	*52*
Pseudovinblastinediol (**21**)	*C. roseus*	*44*
Roseadine (**23**)	*C. roseus*	*53*
Vinamidine (catharinine) (**26**)	*C. roseus*	*44*
	C. ovalis	*55*
	C. longifolius	*55*
Vincadioline	*C. roseus*	*56*
Vinblastine (vincaleukoblastine) (**1**)	*C. roseus*	*47–49,57–60*
	C. ovalis	*32*
	C. longifolius	*33*
	C. trichophyllus	*61*
Vincathicine (**24**)	*C. roseus*	*37*
Vincovalicine (**30**)	*C. ovalis*	*62*
Vincovaline (**28**)	*C. ovalis*	*62*
Vincovalinine (**29**)	*C. ovalis*	*62*
Vincristine (leurocristine) (**2**)	*C. roseus*	*45,49*
Vindolicine (**32**)	*C. roseus*	*29,53*

TABLE II
ALKALOIDS WITH UNKNOWN STRUCTURES ISOLATED FROM *Catharanthus roseus*

Alkaloid	Composition[a]	Ref.
Carosine	$C_{46}H_{56}N_4O_{10}$	*64*
Carosidine	n.a.[a]	*64*
Catharicine	$C_{46}H_{54}N_4O_{10}$	*64*
Leurosinine	$C_{41}H_{54}N_4O_{11}$	*65,66*
Neoleurocristine	$C_{46}H_{54}N_4O_{11}$	*64*
Neoleurosidine	$C_{46}H_{58}N_4O_{10}$	*64*
Roseamine	n.a. (MW 810)	*53*
Rovidine	n.a.	*37*
Vinaphamine	n.a.	*37*
Vincamicine	n.a.	*29*
Vindolidine	$C_{48}H_{46}N_4O_{10}$	*64*
Vinosidine	$C_{44}H_{52}N_4O_{10}$	*65,66*
Vinsedicine	n.a. (MW 780)	*67*
Vinsedine	n.a. (MW 778)	*67*

[a] n.a., Not available.

listed in Table I. There are a number of reports on the isolation of additional bisindole alkaloids from *C. roseus* whose chemical structures are still unknown, and these are summarized in Table II.

II. Structure Elucidation and Spectroscopy

1. Vinblastine (Vincaleukoblastine) (1)

The almost simultaneous discovery of vinblastine (**1**) by Noble *et al.* (*57*) and by Svoboda and co-workers (*58*) is one of the the most publicized events in alkaloid chemistry. Since its initial discovery, vinblastine (**1**) has been reisolated from *C. roseus* several times (*47–49,57–60,*) and it has also obtained from *C. ovalis* (*32*), *C. longifolius* (*33*), and *C. trichophyllus* (*60*). The large-scale separation of vinblastine (**1**) and vincristine (**2**) from *C. roseus* received attention from pharmaceutical industries, and several procedures for the separation of these alkaloids have been reported in the patent literature (*61–71*).

The initial structure elucidation of vinblastine (**1**) has been reviewed (*1,3*); however, the complete assignment of its ^{1}H- and ^{13}C-NMR spectra was made possible only by recent developments in high-resolution NMR spectroscopy. An extensive ^{1}H-NMR study on vinblastine (**1**) has appeared (*72*) (Table III), reporting the ^{1}H-NMR data of **1** in $CDCl_3$,

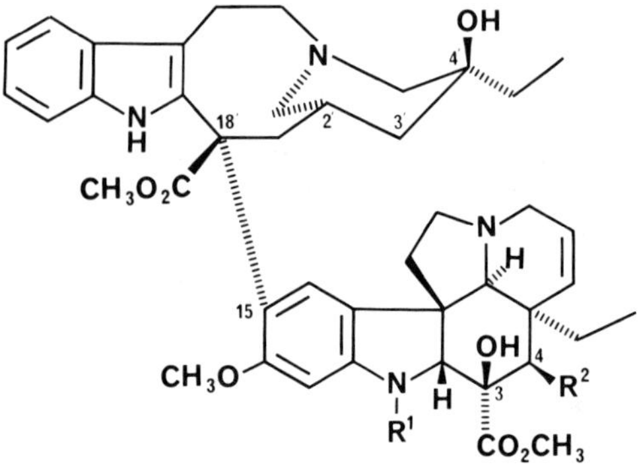

	R^1	R^2
1	CH_3	$OCOCH_3$
5	CH_3	H
6	CH_3	OH
7	H	$OCOCH_3$

acetone-d_6, benzene-d_6, as well as in a mixture of 75% benzene-d_6 and 25% $CDCl_3$. From the data gathered, it was possible to establish the chair conformation of the piperidine ring in the velbanamine moiety and explain the observed long-range interactions. For example, the axially placed 4′-hydroxyl group permits hydrogen bonding between the hydroxyl hydrogen atom and the lone pair of electrons on N'_b a phenomenon that could account for the reduced basicity observed for N'_b in vinblastine (**1**) (*73*). The ^{13}C-NMR spectra of **1** and related derivatives have been thoroughly examined (*74–77*.) The complete assignments for all carbon atoms of **1** have been made (Table IV). Data are consistent with the conformation shown in structure **1** based on chemical shift comparisons and accounting for the β- and α-substituent effects.

The absolute configuration of the stereo centers* of vinblastine (**1**) was determined from the X-ray crystal structure of vincristine (**2**) methiodide (*79,80*) in view of the known relationship between **1** and **2**. The absolute stereochemistry at C-18′ in vinblastine (**1**) and related derivatives can also be deduced by means of ORD and CD spectroscopy (*81,82*). The determination was made possible by the synthesis and structure elucidation of several compounds possessing the unnatural configuration at C-18′ (*82,84*). Because this stereo center controls the relative geometry of the

* Instead of "chirality center" we use the more recently introduced "stereo center" terminology (*78*).

TABLE III
^{1}H-NMR DATA OF VINBLASTINE (**1**) (*72*) AND 4-DEACETOXYVINBLASTINE (**5**) (*34*)

Dihydroindole unit			Indole unit		
Hydrogen	**1**	**5**	Hydrogen	**1**	**5**
2-H	3.74	3.64	1′-H_a	4.00	3.96
4-H	5.45	1.98 (2*H*)	1′-H_b	2.32	2.28
6-H	5.30	5.49	2′-H	0.85	0.87
7-H	5.85	5.69	3′-H_a	1.48	1.48
8-H_a	3.37	3.27	3′-H_b	1.44	1.40
8-H_b	2.86	2.71	5′-H_a	2.81	2.81
10-H_a	2.18	2.15	5′-H_b	2.81	2.81
10-H_b	1.85	1.81	7′-H_a	3.70	3.66
11-H_a	3.30	3.23	7′-H_b	3.12	3.14
11-H_b	2.47	2.40	8′-H_b	3.30	3.30
14-H	6.10	6.09	8′-H_b	3.12	3.10
17-H	6.65	6.62	11′-H		
19-H	2.68	2.81	12′-H	7.18–7.50[a]	7.15–7.54[a]
20-H_a	1.32	1.34	13′-H		
20-H_b	1.80	1.32	14′-H		
21-H	0.82 (3*H*)	0.92 (3*H*)	19′-H_a	3.39	3.42
			19′-H_b	2.41	2.40
CO_2CH_3	3.80 (3*H*)	3.80 (3*H*)	20′-H_a	1.32	1.32
$OCOCH_3$	2.12 (3*H*)	—	20′-H_b	1.32	1.32
OCH_3	3.62 (3*H*)	3.62 (3*H*)	21′-H	0.88 (3*H*)	0.89 (3*H*)
N-CH_3	2.81 (3*H*)	2.81 (3*H*)	CO_2CH_3	3.80 (3*H*)	3.80 (3*H*)

[a] No assignment of the aromatic protons was made.

indole and dihydroindole moieties in the molecule, the optical rotation, determined by the dipolar coupling between electronic transitions in these two chromophores, can be correlated with the absolute stereochemistry at C-18′. As a result, the two C-18′ epimeric series give opposite Cotton effects at lower wavelengths (one sign can be found around 210 nm, the other close to 225 nm). Configurational changes elsewhere in the molecule, e.g., at C-4′, do not significantly affect the ORD or CD spectra of the bisindole alkaloids.

2. Deacetoxyvinblastine (5)

Deacetoxyvinblastine (**5**) (formerly named desacetoxyvincaleukoblastine), $C_{46}H_{56}N_4O_7$, mp 183–190°C, was first isolated by Neuss *et al.* (*35*) in the course of purifying quantities of vinblastine (**1**) obtained from *C. roseus*. Recently, **5** was reisolated, and a thorough ^{1}H-NMR study was performed in order to establish its structure. The lack of an acetate

TABLE IV

^{13}C-NMR Data of Vinblastine (**1**), Deacetoxyvinblastine (**5**), Deacetylvinblastine (**6**), and Demethylvinblastine (**7**) (*76*)

Dihydroindole unit					Indole unit				
Carbon	**1**	**5**	**6**	**7**	Carbon	**1**	**5**	**6**	**7**
2	83.3	81.8	82.8	74.2	1′	34.4	34.5[a]	34.3	34.5
3	79.7	77.1	80.7	79.7	2′	30.1	30.3	30.2	30.4
4	76.4	38.1	74.1	76.5	3′	41.4	41.6	41.4	41.6
5	42.7	37.2	42.4	43.4	4′	69.4	69.7	69.5	69.8
6	129.9	135.9	130.0	129.9	5′	64.2	64.5	64.3	64.5
7	124.4	124.0[a]	124.2[a]	124.5	7′	55.8	55.7	55.8	55.7
8	50.2	50.1	50.4[a]	50.5	8′	28.2	28.8	28.7	28.8
10	44.6	44.7	44.7	42.9	9′	117.0	116.8	117.0	116.9
11	50.2	50.1	49.8[a]	51.1	10′	129.5	129.5	129.4	129.5
12	53.2	53.8	53.2	53.2	11′	118.4	118.4	118.4	118.5
13	122.6	121.8	122.8	122.7	12′	122.1	122.1	122.2	122.2
14	123.5	123.7[a]	123.9[a]	124.2[a]	13′	118.7	118.7	118.7	188.8
15	121.1	120.4	120.9	120.7	14′	110.4	110.3	110.4	110.3
16	158.0	157.8	158.0	157.9	15′	135.0	134.9	134.9	135.0
17	94.2	93.5	93.9	93.9	17′	131.4	131.6	131.3	131.7
18	152.5	152.8	152.5	149.1	18′	55.8	55.7	44.8	55.9
19	65.5	65.5	66.4	66.7	19′	48.0	48.4	48.1	48.4
20	30.8	34.3	32.9	31.2	20′	34.4	34.7[a]	34.3	34.3
21	8.3	8.6	8.6	8.3	21′	6.9	6.9	6.9	6.9
$\underline{C}O_2CH_3$	170.8	173.6	173.1	170.8	$\underline{C}O_2CH_3$	174.9	175.2	175.1	175.0
$CO_2\underline{C}H_3$	52.1	52.4	52.3	52.7	$CO_2\underline{C}H_3$	52.3	52.4	52.3	52.4
Ar-O$\underline{C}H_3$	55.8	55.7	55.8	55.7					
$O\underline{C}OCH_3$	171.6	—	—	172.7					
$OCO\underline{C}H_3$	21.1	—	—	21.0					
N-CH_3	38.3	38.1	—	38.1					

[a] Tentative assignment.

methoxy group compared to the ^{1}H-NMR spectrum of vinblastine (**1**) and the presence of a two-proton intense singlet at δ 1.98 suggested the 4′-deacetoxyvinblastine (**5**) structure (see Table III), which was also supported by ^{13}C-NMR measurements (Table IV) (*76*).

The absence of a function at C-4 was also clearly established by mass spectrometry. Major mass spectral fragmentations of **5** are illustrated in Schemes 1 and 2. Deacetoxyvinblastine (**5**) itself gives a molecular ion at *m/z* 752; however, an *m/z* 766 peak could also be observed in the mass spectrum since this type of alkaloid can undergo a facile intermolecular methylation by the C-18′ methoxycarbonyl group resulting in an increase in the molecular ion by 14 mass units. The high-resolution mass spectral data of the 18′-demethoxycarbonyl-3-hydrazide derivative, in which intermolecular transmethylation cannot occur, confirmed the absence of the acetoxy group at C-4, exhibiting an *m/z* 493 fragment instead of the *m/z* 509 one characteristic of the ion containing a hydroxyl at C-4 in the hydrazide derivative of vinblastine.

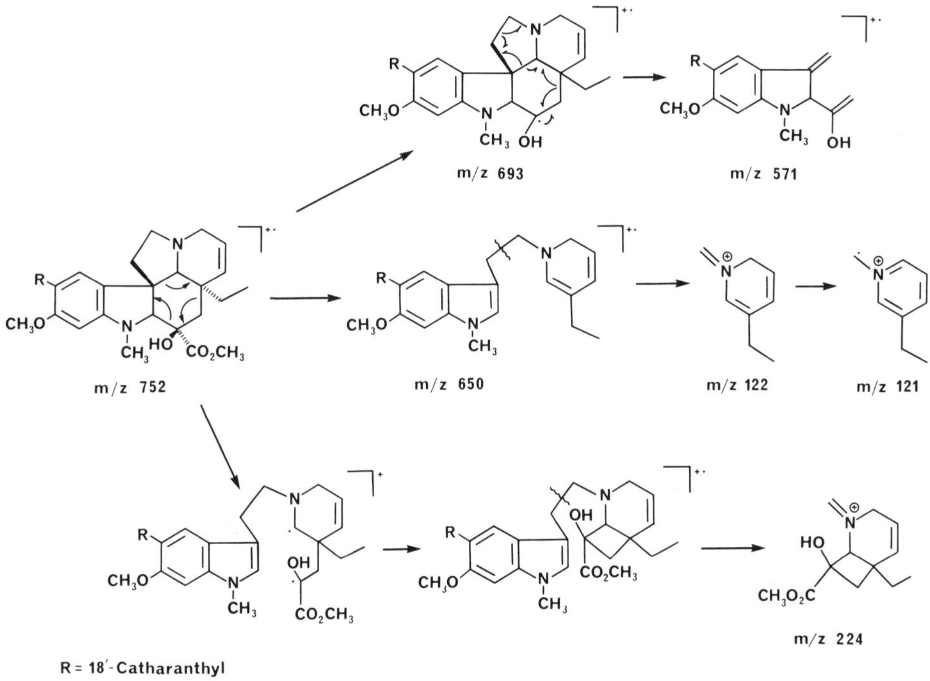

SCHEME 1. Mass spectral fragmentation of the indole part of deacetoxyvinblastine (**5**).

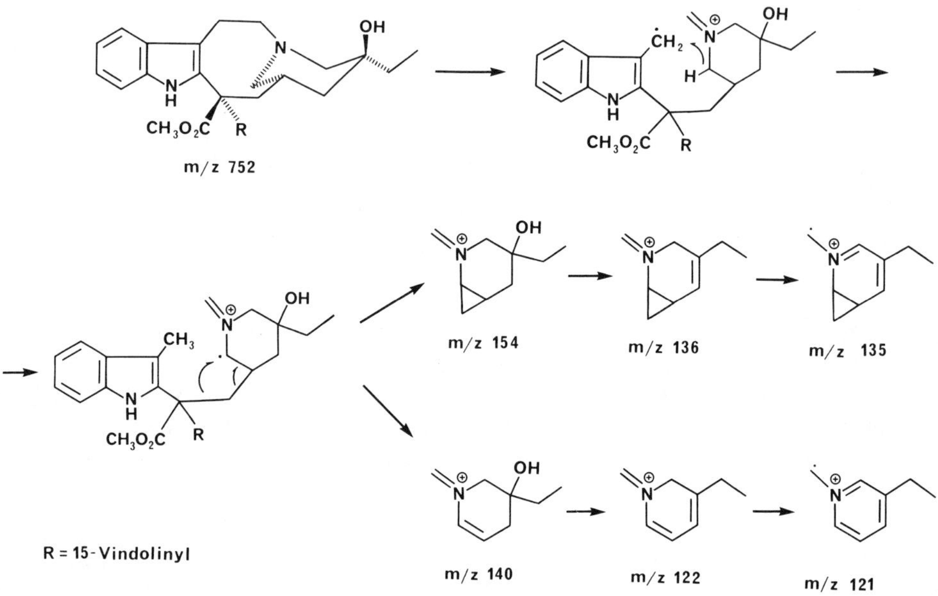

SCHEME 2. Mass spectral fragmentation of the dihydroindole part of deacetylvinblastine (5).

3. Deacetylvinblastine (6)

Deacetylvinblastine (**6**) (formerly named desacetylvincaleukoblastine), $C_{44}H_{56}N_4O_8$, mp 205–210°C, has been isolated from *C. roseus* as a minor alkaloid (*37*). The structure elucidation of **6** was based on comparison of its UV, IR, and low-resolution ^{1}H-NMR spectra with those of semi-synthetic deacetylvinblastine (**6**) obtained by mild hydrolysis of **1**. The ^{13}C-NMR data of **6** (see Table IV) are very similar to those of vinblastine (**1**); the only significant variation in the spectrum of **6** is that the C-20 resonance is deshielded relative to that of **1**. This indicates that C-20 of **1** is significantly shielded by the acetyl moiety, most probably by the carbonyl oxygen. Because such shielding is expected only when the acetyl group is close to C-20, these results indicate that ring C of these compounds is in the same boat conformation as observed in the crystal structure of vincristine (**2**) (*79*).

4. Demethylvinblastine (7)

Demethylvinblastine (**7**) (formerly named *N*-desmethylvincaleukoblastine and *N*-deformyl-vincristine), $C_{45}H_{56}N_4O_9$, mp 250°C (sulfate salt), has been isolated from *C. roseus* as a minor alkaloid (*38–42*). Nowadays, demethylvinblastine (**7**) is not isolated from *C. roseus,* but, in order to increase the yield of vincristine (**2**) obtainable from the plant, a procedure has been developed to formylate the crude alkaloid mixture of *C. roseus* prior to separation, resulting in the transformation of all **7** present in the mixture to the more valuable vincristine (**2**) (*40,41*). Demethylvinblastine (**7**) has also been obtained via microbiological demethylation of vinblastine (**1**) with *Streptomyces albogriseolus* (A 17178) (*85*).

The structure elucidation of **7** was completed by comparison of its ^{1}H- and ^{13}C-NMR spectra with those of vinblastine (**1**). In the ^{1}H-NMR spectrum two diagnostic differences were observed: the *N*-methyl signal of **1** at δ 2.78 is absent in the ^{1}H-NMR spectrum of **7**, and the signal at δ 3.68 assigned to 2-H in the spectrum of **1** is replaced by the X part of an AX system at δ 4.12 in the ^{1}H-NMR spectrum of **7**, indicating the presence of an NH proton. The structure of demethylvinblastine (**7**) was further confirmed by ^{13}C-NMR measurements. As shown in Table IV an upfield shift for C-2 (−6.2 ppm) and C-18 (−3.2 ppm) of **7**, both carbons in an α position to N-1, could be observed compared to the corresponding chemical shift values of **1**, with downfield shifts observed for C-17 (+0.2 ppm), C-13 (+0.4 ppm), C-12 (+3.1 ppm), and C-2 (+0.5 ppm), the carbons in β positions to N-1. Further minor changes could also seen at carbons 5, 10, 19, and 20, possibly reflecting a minor readjustment in the conformation of the alicyclic portion of the vindoline moiety (*76*).

5. Anhydrovinblastine (**8**)

Anhydrovinblastine (**8**), which was first obtained through the coupling of vindoline (**3**) and the N'^{b}-oxide of catharanthine (**4**) using the modified Polonovski reaction (*82,86*), is a very unstable substance. The presence of **8** in crude extracts of *C. roseus* as a natural product was established by feeding *C. roseus* plants with both [*acetyl*-^{14}C]vindoline and [*methoxy*-^{3}H]catharanthine, resulting in the detection of either the corresponding acetyl-^{14}C- or the carbomethoxy-^{3}H- labeled anhydrovinblastine (*87*). The facile air oxidation of anhydrovinblastine (**8**) leading to vinblastine (**1**), leurosine (**11**), leurosidine (**18**), and catharine (**10**) suggests that **8** has the potential to be a precursor of most, if not all, of the vinblastine-type bisindole alkaloids (*88*).

Recently, an improved pretreatment of the plant material combined with preparative HPLC separation has been published for the isolation of anhydrovincaleukoblastine (**8**) from the leaves of *C. roseus* (*27*). It has also be shown that the yield of **8** could be enhanced by treatment of acidic aqueous extracts of *C. roseus* with sodium borohydride, suggesting that

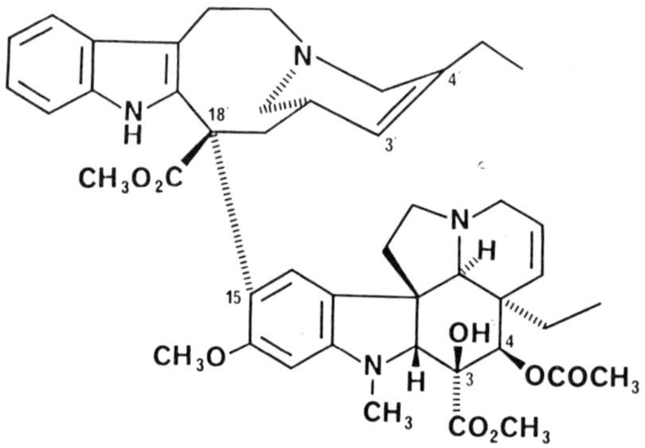

8

an iminium intermediate produced by enzymatic coupling of catharanthine (**3**) and vindoline (**4**) is also present in the crude extract (*27*).

6. Vincristine (Leurocristine) (**2**)

Vincristine (leurocristine) (**2**) is present in *C. roseus* in approximately 0.0003% yield, the lowest level of any medicinally useful alkaloid produced on commercial basis. Since the initial isolation of **2** (*45*), its structure elucidation has been reviewed (*1,3*). The final confirmation of structure **2** and the determination of the absolute configurations of the stereo centers of vincristine (**2**) were achieved by X-ray crystallography of its methiodide derivative (*79,80*).

7. Catharanthamine (**9**)

Chromatographic separation of the post-leurocristine alkaloid fraction of *C. roseus* yielded amorphous catharanthamine (**9**) with a molecular

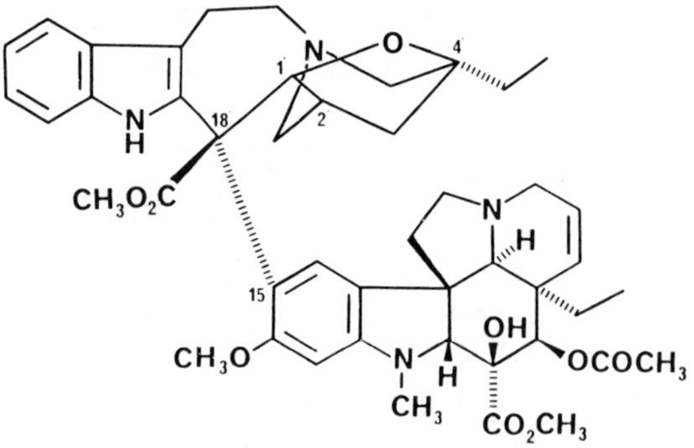

9

composition of $C_{46}H_{56}N_4O_9$ (*28*). The UV spectrum of **9**, λ_{max} (log ϵ) 213 (3.76), 223 (3.71), 265 (3.34), and 295 (3.04) nm, indicated the presence of both indole and dihydroindole moieties, and the IR spectrum showed only one carbonyl absorption at 1740 cm^{-1} due to a saturated ester group. Typical ^{1}H-NMR resonances were evident for a substituted vindoline building block possessing an acetyl methyl (δ 3.81) and a carbomethoxyl group (δ 3.70). Substitution of the vindoline moiety was established to be at the C-15 position owing to the singlet character of the 17-H resonance at δ 6.14. From the ^{1}H-NMR spectrum it was apparent that the ethyl side chains in the indole and dihydroindole units had remained intact (triplets at δ 0.68 and 1.02). In the indole half of the molecule, however, a multiplet was observed at δ 4.36 that could be attributed to a methine proton at the 1′ position, suggesting an unusual oxygenation at C-1′. The presence of

TABLE V

^{13}C-NMR DATA OF CATHARANTHAMINE (**9**) (*28*) AND CATHARINE (**10**) (*30*)

Dihydroindole unit			Indole unit		
Carbon	**9**	**10**	Carbon	**9**	**10**
2	83.48	83.41	1′	65.64	31.13
3	79.50	79.65	2′	38.01	204.26
4	76.35	70.32	3′	43.68	51.40
5	42.74	42.95	4′	69.40	95.18
6	130.0	130.19	5′	65.80	126.56
7	124.10	124.64	7′	55.80	56.32
8	50.50	50.85	8′	26.84	27.95
10	44.71	50.32	9′	117.55	117.81
11	51.20	50.32	10′	129.07	127.89
12	53.20	44.46	11′	117.76	118.41
13	123.12	124.12	12′	123.34	121.27
14	123.67	122.83	13′	119.10	118.67
15	121.43	123.11	14′	110.33	111.35
16	157.80	159.32	15′	134.52	134.05
17	93.19	94.92	17′	131.20	134.46
18	153.55	153.23	18′	58.06	52.41
19	66.64	66.58	19′	48.60	162.08
20	30.76	30.71	20′	33.36	29.04
21	8.30	8.21	21′	6.80	11.17
$\underline{C}O_2CH_3$	170.69	170.74	$\underline{C}O_2CH_3$	171.00	173.55
$CO_2\underline{C}H_3$	52.14	52.21	$CO_2\underline{C}H_3$	52.17	52.06
Ar-O$\underline{C}H_3$	55.33	n.a.[a]			
O$\underline{C}$OCH$_3$	171.82	n.a.			
OCO$\underline{C}H_3$	21.05	n.a.			
N-CH_3	38.28	n.a.			

[a] n.a., Not reported in the original literature.

an ether linkage between C-1′ and C-4′ in catharanthamine (**9**) was unambiguously confirmed by ^{13}C-NMR measurements (Table V). Catharanthamine (**9**) was found to be cytotoxic in the KB test system *in vitro* and displayed significant (*T/C* 176% at 1 mg/kg) activity in the P388 lymphocytic leukemia test system.

8. Catharine (10)

The isolation of catharine (**10**), $C_{46}H_{54}N_4O_{10}$, mp 271–275°C, an oncolytically inactive alkaloid, has been reported from several *Catharanthus* species: *C. roseus* (*29–31*), *C. ovalis* (*32*), and *C. longifolius* (*33*). The structure of catharine (**10**) has been elucidated by X-ray crystallography (*89–91*) of its acetone solvate. Catharine (**10**) can be obtained by mild oxidation of either leurosine (**11**) (*30*) or anhydrovinblastine (**8**) (*92–93*). In view of the ease of this oxidation, catharine (**10**) may be considered as an artifact of the isolation process.

In the ^{1}H-NMR spectrum of **10** the doubling of several absorption signals was observed (*31*), i.e., δ 2.21/2.25, 5.02/5.10, 5.41/5.45, and 7.64/8.14. These signals coalesced at 60°C and appeared as sharp singlets at

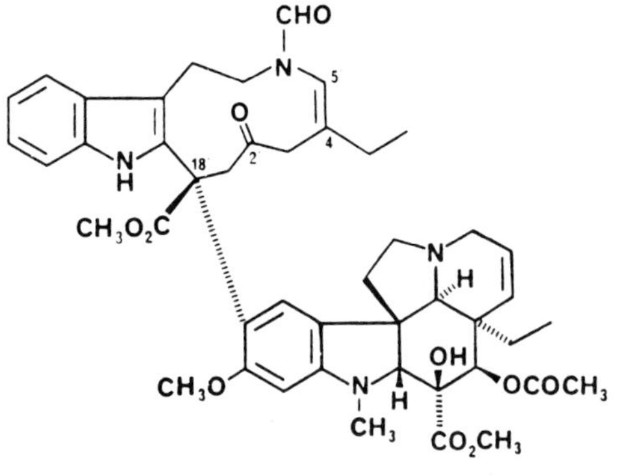

10

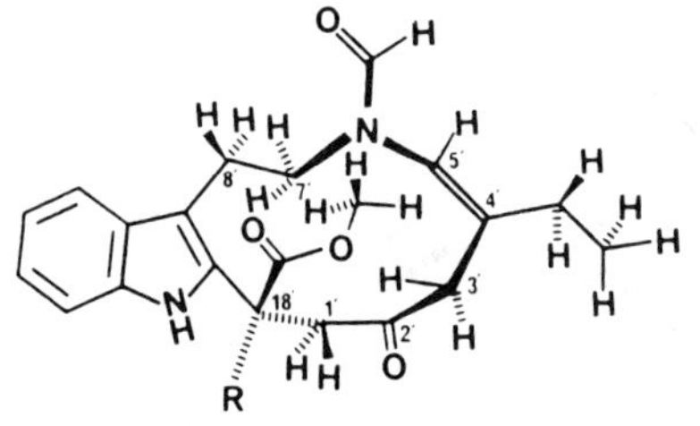

10A

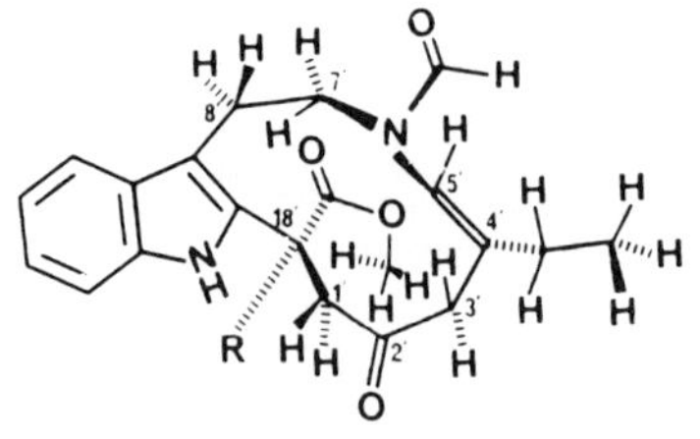

10B

R: 15-Vindolinyl

the midpoints of the previous doublets, indicating that catharine (**10**) could exist in two different conformations (**10A** and **10B**) at room temperature. Similar phenomena were observed in the ^{13}C-NMR spectrum of **10**, as shown in Table V together with the complete assignments. Study of the mass spectral fragmentation of catharine (**10**) (*30*) showed that the molecular ion (*m/z* 822) fragments directly to ions *m/z* 763 and 662 in accordance with the proposed structure (Scheme 3).

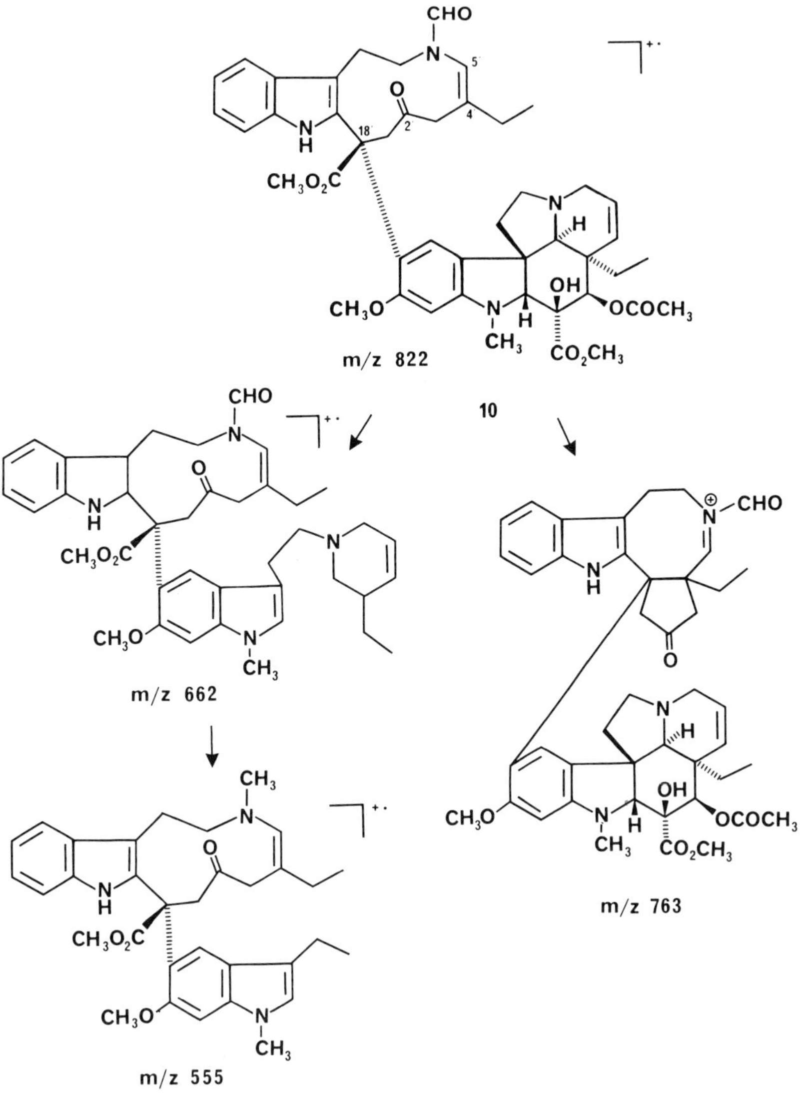

CHO N 5' O 4 N H 18' 2' CH_3O_2C N H OH CH_3O N H $OCOCH_3$ CH_3 CO_2CH_3 m/z 822 10

CHO N O N H CH_3O_2C N CH_3O N CH_3 m/z 662

$\oplus$ N CHO N H O N H OH CH_3O N H $OCOCH_3$ CH_3 CO_2CH_3 m/z 763

CH_3 N O N H CH_3O_2C CH_3O N CH_3 m/z 555

SCHEME 3. Major mass spectral fragmentation of catharine (**10**).

9. Leurosine (11)

Leurosine (**11**), $C_{46}H_{56}N_4O_9$, has been found in *C. roseus* (*47–49*), *C. ovalis* (*32*), *C. longifolius* (*32*), *C. lanceus* (*50*), and *C. pusillus* (*51*). Its structure was established by chemical means, and its relationship with vinblastine (**1**) has previously been summarized (**3**). The placement of the epoxide function at C-3′–C-4′ was verified by ^{1}H- and ^{13}C-NMR measurements (Tables VI and VII, respectively). From the coupling constants of $J_{2',19'a}$ = 1 Hz and $J_{2'19'b}$ = 2 Hz it could be determined that the projection of the C-2′–H-2′ bond is between the C-19′–H-19′a and C-19′–H-19′b bonds. In addition, the coupling constant of $J_{2',3'}$ is 2.6 Hz, which suggests that 2′-H and 3′-H are in a trans relationship (*89*). This means that, taking

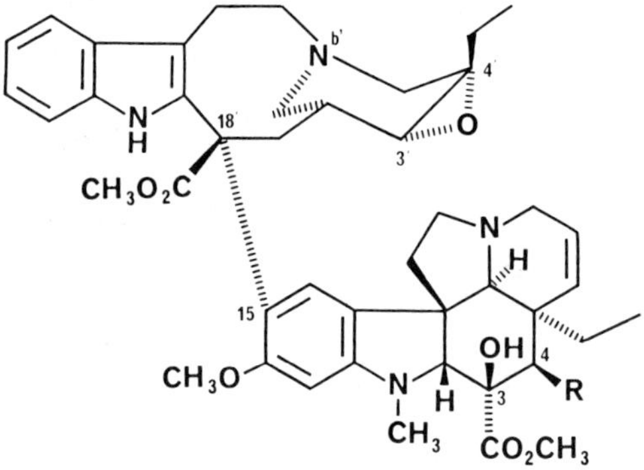

	R	
11	$OCOCH_3$	
12	$OCOCH_3$	N_b'-oxide
13	H	

into consideration the already known relative configurational data of 2′-H being in an α position and the half-chair conformation of the piperidine ring, the stereochemistry of the epoxide could be established as α. These findings agree completely with those suggested from the ^{13}C-NMR measurements. An initial ^{13}C-NMR study (*74*) revealed that C-4′ absorbs at anomalously high field and must be part of a small, strained ring. A refined study (*75*) showed that C-3′ also had an unusually high chemical shift which could only be explained by the presence of an epoxide function. This was further supported by the measurement of the C-3′–3′-H heteronuclear coupling constant, the large magnitude of which could only be explained by the presence of an epoxide at C-3′–C-4′ (*75*).

TABLE VI
^{1}H-NMR DATA OF LEUROSINE (**11**) (*89*) AND 4-DEACETOXYLEUROSINE (**13**) (*34*)

Dihydroindole unit			Indole unit		
Hydrogen	**11**	**13**	Hydrogen	**11**	**13**
2-H	3.74	3.64	1'-H_a	3.27	3.18
4-H	5.47	1.98 (2*H*)	1'-H_b	1.22	1.32
6-H	5.30	5.50	2'-H	1.22	1.32
7-H	5.86	5.63	3'-H	3.14	2.72
8-H_a	3.37	3.25	5'-H_a	3.32	3.32
8-H_b	2.81	2.69	5'-H_b	3.20	3.22
10-H_a	2.18	2.17	7'-H_a	3.37	3.22
10-H_b	1.76	1.73	7'-H_b	3.16	3.32
11-H_a	3.37	3.41	8'-H_a	3.32	3.32
11-H_b	2.43	2.39	8'-H_b	3.08	3.32
14-H	6.12	6.10	11'-H		
17-H	6.60	6.60	12'-H	7.15–7.51[a]	7.12–7.51[a]
19-H	2.65	2.50	13'-H		
20-H_a	1.76	1.32	14'-H		
20-H_b	1.37	1.20	19'-H_a	3.08	3.25
21-H	0.82 (3*H*)	0.93 (3*H*)	19'-H_b	3.78	2.77
CO_2CH_3	3.85	3.61 (3*H*)	20'-H	1.60 (2*H*)	1.61 (2*H*)
$OCOCH_3$	2.10	—	21'-H	0.96 (3*H*)	0.99 (3*H*)
OCH_3	3.63	3.82 (3*H*)	CO_2CH_3	3.85 (3*H*)	3.83 (3*H*)
N-CH_3	2.72	2.78 (3*H*)			

[a] No assignment of the aromatic protons was made.

10. Pleurosine (12)

Pleurosine (leurosine N'_b-oxide) (**12**), $C_{46}H_{56}N_4O_{10}$, mp 218–219°C, is one of the many bisindole alkaloids originally obtained by Svoboda and co-workers from *C. roseus* (*54*). The chemical relationship with leurosine (**11**) was established, but definitive evidence for the location of the *N*-oxide and the stereochemical integrity of the velbanamine unit was not presented at that time. High-resolution ^{1}H- and ^{13}C-NMR spectroscopic investigation permitted confirmation of structure **12** (*53*). The ^{1}H-NMR spectrum of **12** substantiated the presence of the vindoline moiety substituted at C-15, displaying characteristic resonances for the methyl group at C-21, the C-4 acetate and methine proton, the indoline *N*-methyl, the aromatic methoxy, the methoxycarbonyl, two cis-related olefinic protons, and two para aromatic protons. The ^{1}H-NMR resonances of **12** for the indole moiety, however, showed significant differences from those of leurosine (**11**). For example, the nonequivalent aminomethylene protons

TABLE VII
^{13}C-NMR DATA OF LEUROSINE (**11**) (*53*), LEUROSINE N'_b-OXIDE (**12**) (*53*), AND 5′-OXOLEUROSINE (**14**) (*52*)

Dihydroindole unit				Indole unit			
Carbon	**11**	**12**	**14**	Carbon	**11**	**12**	**14**
2	83.1	83.0	83.2	1′	31.7	32.1	29.4
3	79.4	79.6	79.4	2′	33.5	33.6	31.9
4	76.2	76.4	76.1	3′	60.3	59.6	61.4
5	42.5	42.6	42.5	4′	60.0	59.0	59.5
6	129.8	129.7	129.8	5′	53.9	71.9	167.9
7	124.3	124.5	124.4	7′	49.8	65.9	46.7
8	50.1	50.3	50.3	8′	24.8	28.0	23.5
10	44.4	44.2	44.7	9′	116.8	117.6	115.3
11	50.1	50.2	50.3	10′	129.2	130.3	128.9
12	53.0	53.2	53.1	11′	118.1	118.9	118.1
13	122.9	123.4	123.4	12′	122.2	119.4	123.2
14	123.4	123.7	122.7	13′	118.8	117.6	119.3
15	120.5	119.5	120.0	14′	110.3	111.8	110.5
16	157.7	157.6	157.8	15′	134.6	134.6	134.6
17	94.1	93.8	94.3	17′	130.7	130.7	131.1
18	152.8	153.1	153.2	18′	55.2	54.7	55.1
19	65.5	65.4	65.8	19′	42.3	59.0	44.0
20	30.6	30.7	30.7	20′	28.0	27.6	29.4
21	8.2	8.5	8.4	21′	8.4	8.4	8.7
$\underline{C}O_2CH_3$	170.7	170.9	170.8	$\underline{C}O_2CH_3$	174.2	173.8	173.4
$CO_2\underline{C}H_3$	52.0	52.6	52.1	$CO_2\underline{C}H_3$	52.2	52.8	52.5
Ar-O$\underline{C}H_3$	55.6	55.1	55.8				
O$\underline{C}O_2CH_3$	171.4	171.5	171.5				
O$CO_2\underline{C}H_3$	20.9	20.9	21.0				
N-CH_3	38.1	37.9	38.0				

of carbons 7′ and 19′ appeared more downfield (at δ 4.11 and 4.24, respectively) than those of the corresponding hydrogens in leurosine (**11**). The ^{13}C-NMR data of **12** (Table VII) were in good agreement with the proposed leurosine N'_b-oxide structure, since the methylene carbon resonances of C-5′, C-7′, and C-19′ had undergone substantial deshielding on comparison with leurosine (**11**) owing to the *N*-oxide function.

Leurosine N'_b-oxide (**12**) was found to be highly cytotoxic in the KB test system (ED_{50} 0.019 μg/ml) *in vitro*, but, more importantly, activity was also observed in two *in vivo* systems. In the P388 lymphocytic leukemia system, at doses in the range 4.0–32.0 mg/kg, either on day 1 or on days 1–9, typical test/control values were in the range of 160–180%. Exceptional activity was observed in the B16 melanoma test system.

Although toxicity was observed at 20 mg/kg, good dose responses were observed in the dose range 2.5–10.0 mg/kg. At the highest dose, test/control values in excess of 300% were consistently observed with five cures in one instance and four cures in another. In a parallel test, vinblastine (**1**) showed *T/C* 309% at 0.5 mg/kg with five survivors. According to these data leurosine N'_b-oxide is one of the most active compounds in the B16 test system isolated thus far.

11. Deacetoxyleurosine (13)

Deacetoxyleurosine (**13**) (formerly named desacetoxyleurosine), $C_{44}H_{54}N_4O_7$, mp 202°C, has been isolated as an impurity of crude leurosine (**11**) obtained from *C. roseus* (*34*). The final separation of **13** and **11** was performed by preparative HPLC on an octadecyl reversed-phase silica gel column using gradient elution from 50 to 85% methanol in water with 0.1% ethanolamine. The ^{1}H-NMR spectrum of **13** (Table VI) exhibited two characteristic ABCX spin systems typical for the 10-H_2 and 11-H_2 protons of the vindoline part and for the aromatic protons in the catharanthine moiety. The chemical shift and multiplicity of 3′-H at δ 2.72 established the presence of the epoxide function at C-3′–C-4′. The absence of an acetate methoxy group, compared to the ^{1}H-NMR spectrum of leurosine (**11**), was clearly indicated, and furthermore, instead of the easily recognizable 4-H methine signal a two-proton intense singlet was found at δ 1.98, which could only be assigned as 4-H_2. Mass spectral data of **13** were in accord with a structure derived from leurosine (**11**) by loss of acetic acid.

12. 5′-Oxoleurosine (14)

5′-Oxoleurosine (**14**), mp 212–215°C, has been isolated from *C. roseus* (*52*) and its structure elucidated by spectroscopic measurements. A molecular ion at *m/z* 822 analyzed as $C_{46}H_{54}N_4O_{10}$ indicated that the isolate contained an additional oxygen atom and two less hydrogens than leurosine (**11**). In addition, the mass spectrum displayed a fragmentation pattern typical of the presence of a vindoline moiety, particularly the ion at *m/z* 282. The assignment of vindoline as the dihydroindole portion of the isolate was supported by chemical shift assignments of all the carbon resonances of this unit in comparison with leurosine (**11**) and vinblastine (**1**); in this way, attachment of the indole unit to C-15 of a vindoline building block was confirmed. For the indole component, all of the aromatic and most of the aliphatic carbon resonances corresponded very closely with those of leurosine (**11**). In particular, an epoxide function was considered

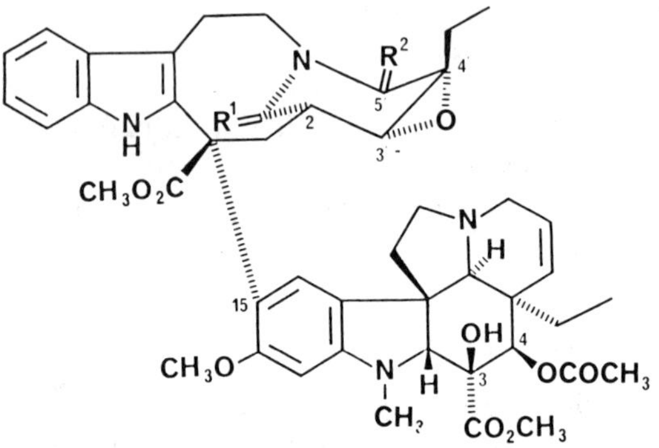

	R^1	R^2
14	H_2	O
15	O	H_2

to be present on the basis of the chemical shift values of C-3′ and C-4′ at δ 61.37 and 59.53, respectively (Table VII). The most substantial difference on comparison with the ^{13}C-NMR spectrum of leurosine (**11**) was the presence of an additional carbonyl carbon at δ 167.85 with a concomitant loss of an aminomethylene signal, suggesting the presence of a cyclic amide. Distinction between the placement of the carbonyl group at C-5′ or C-19′ was made by comparison of the ^{1}H-NMR spectrum of **14** with that of 19′-oxoleurosine (**15**) previously synthesized from leurosine (**11**) by oxidation with iodine under basic conditions (*90*). The latter compound exhibited a signal for the 2′-H at δ 4.76 quite different in multiplicity from that observed in the isolate at δ 4.66, permitting the placement of the carbonyl group of **14** at C-5′.

5′-Oxoleurosine (**14**) was evaluated for antitumor activity according to established protocols. No activity was observed in the dose range 0.25–6.25 mg/kg in the P388 lymphocytic leukemia system in mice, but the alkaloid was cytotoxic in the KB test system *in vitro* (ED_{50} 0.31 μg/ml) (*52*).

13. 4′-Deoxyvinblastine (**16**)

4′-Deoxyvinblastine (**16**) was first detected as an accompanying impurity of leurosine (**11**) (*43*). Initially, **16** was believed to be isomeric with leurosine (**11**) and was therefore named isoleurosine. However, high-resolution mass spectrometry proved that isoleurosine (**16**) corresponded to a molecular composition of $C_{46}H_{58}N_4O_8$ and, in fact, contains two more hydrogen atoms and one less oxygen atom than **11**. The structure of 4′-

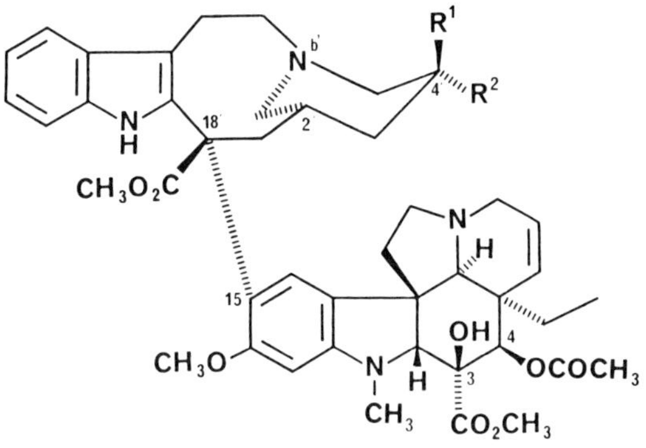

	R^1	R^2	
16	C_2H_5	H	
17	H	C_2H_5	
18	C_2H_5	OH	
19	C_2H_5	OH	N_b'-oxide

deoxyvinblastine (**16**) became evident from the reductive cleavage of the corresponding hydrazide leading to 20α-dihydrocleavamine. It was also found that treatment of leurosine (**11**) with Raney nickel in refluxing ethanol yielded two products: deoxyvincaleukoblastine B (**17**) and deoxyvincaleukoblastine A (**16**), the latter of which proved to be identical with the natural product (**16**) on the basis of spectral evidence (*43*). 4′-Deoxyvinblastine (**16**) was also prepared by catalytic hydrogenation of anhydrovinblastine (**8**) (*91*). The structure of the product obtained was fully established by ^{1}H-NMR spectral data (Table VIII), and the absolute configuration at the C-18′ stereo center was determined as *S* by means of CD measurement.

14. Leurosidine (**18**)

Leurosidine (vinrosidine) (**18**) has been isolated from *C. roseus* (*45*), and on the basis of its elemental composition, $C_{46}H_{58}N_4O_9$, spectroscopic properties, and functional group analysis, it was determined to be an isomer of vinblastine (**1**). From the ease of acetylation of leurosidine (**18**), and because its second pK_a value was found to be 1.4 log units higher than the corresponding pK_a of vinblastine (**1**), it was deduced that the hydroxyl group of the indole component is in a different environment, one in which hydrogen bonding with the lone pair electrons on N'_b is not permitted. Thorough examination of the ^{13}C-NMR spectrum of leurosi-

TABLE VIII
^{1}H-NMR DATA OF 4′-DEOXYVINBLASTINE (**16**) (*91*)

2-H	3.73	1′-H_a	3.23
4-H	5.47	1′-H_b	2.30
6-H	5.26	2′-H	1.12
7-H	5.89	3′-H_a	2.06
8-H_a	3.37	3′-H	1.71
10-H_a	2.18	5′-H_a	2.81
10-H_b	1.80	5′-H_b	2.81
11-H_a	3.30	7′-H_a	3.56
11-H_b	2.44	7′-H_b	3.15
14-H	6.12	8′-H_a	3.30
17-H	6.58	8′-H_b	2.90
19-H	2.74	11′-H	
20-H_a	1.78	12′-H	7.15–7.43[a]
20-H_b	1.23	13′-H	
21-H	0.09 (3*H*)	14′-H	
CO_2CH_3	3.80 (3*H*)	19′-H_a	3.18
$OCOCH_3$	2.12 (3*H*)	19′-H_b	2.80
OCH_3	3.62 (3*H*)	20′-H_a	1.34
N-CH_3	2.75 (3*H*)	20′-H_b	1.27
		21′-H	0.83
		CO_2CH_3	3.80 (3*H*)

[a] No assignment of the aromatic protons was made.

dine (**18**) showed a close similarity to that of vinblastine (**1**), including a downfield signal due to C-4′ (Table IX). The data supported the notion that leurosidine (**18**) is the C-4′ epimer of vinblastine (**1**) (*75*), a conclusion which was later substantiated by the reported epimerization of **1** to give leurosidine (**18**) (*92*).

15. Leurosidine N'_b-oxide (**19**)

Leurosidine N'_b-oxide (**19**), mp 215–218°C, has also been isolated from *C. roseus* (*46*). A molecular ion at *m/z* 826 corresponding to $C_{46}H_{58}N_4O_{10}$ indicated the isolate to contain one more oxygen atom than vinblastine (**1**) or leurosidine (**18**), and it could be reduced by aqueous $FeSO_4$ to a product identical with leurosidine (**18**) but not vinblastine (**1**). The structure elucidation of **19** was further supported by ^{13}C-NMR measurements (Table IX). The most significant features of the ^{13}C-NMR spectrum of **19** were the presence of three deshielded methylene carbons at δ 77.4 (C-5′), 69.6 (C-7′), and 66.5 (C-19′). Such downfield shifts have been reported previously for the aminomethylene carbons in vincaleukoblastine N'_b-ox-

TABLE IX
^{13}C-NMR DATA OF LEUROSIDINE (**18**) AND LEUROSIDINE N'^{b}-OXIDE (**19**) (*46*)

Dihydroindole unit			Indole unit		
Carbon	**18**	**19**	Carbon	**18**	**19**
2	83.1	83.3	1′	35.1	37.8
3	79.5	79.9	2′	29.8	29.8
4	76.2	76.6	3′	40.4	40.6
5	42.6	76.6	4′	71.8	72.3
6	129.7	131.4	5′	55.1	77.4
7	124.3	124.7	7′	53.9	69.9
8	50.2	50.7	8′	21.4	21.2
10	44.5	44.8	9′	116.6	117.9
11	50.2	50.9	10′	128.9	128.6
12	53.1	53.3	11′	117.9	117.9
13	123.0	123.5	12′	122.0	123.5
14	123.4	124.1	13′	118.6	119.8
15	120.4	120.2	14′	110.2	110.8
16	157.6	158.1	15′	134.5	135.2
17	94.0	94.4	17′	130.2	130.1
18	152.8	153.6	18′	55.4	55.6
19	65.5	66.0	19′	43.9	66.5
20	30.7	31.0	20′	38.5	38.0
21	8.3	8.4	21′	7.1	6.9
$\underline{C}O_2CH_3$	170.7	171.0	$\underline{C}O_2CH_3$	173.9	173.8
$CO_2\underline{C}H_3$	51.9	52.2	$CO_2\underline{C}H_3$	52.2	52.6
Ar-O$\underline{C}H_3$	55.7	55.9			
O$\underline{C}O_2CH_3$	171.4	171.8			
$OCO_2\underline{C}H_3$	20.9	20.0			
N-CH_3	38.2	38.0			

ide (*75*) and leurosine N'_b-oxide (*53*). When evaluated for antitumor activity, leurosidine N'_b-oxide (**19**) was found to be cytotoxic in both the P388 (ED_{50} 2.7 μg/ml) and KB (ED_{50} 0.26 μg/ml) test systems *in vitro* (*46*).

16. Leurocolombine (**20**)

Leurocolombine (**20**) was isolated from *C. roseus* by gradient pH purification followed by preparative HPLC separation (**44**). The mass spectrum of **20** indicated a molecular weight of 826, and high-resolution mass measurement verified the molecular formula $C_{46}H_{58}N_4O_{10}$. Deuterium exchange in combination with mass spectrometry suggested that **20** contained one more exchangeable proton, most likely attached to an oxygen atom, than vinblastine (**1**), indicating that an additional hydroxyl group

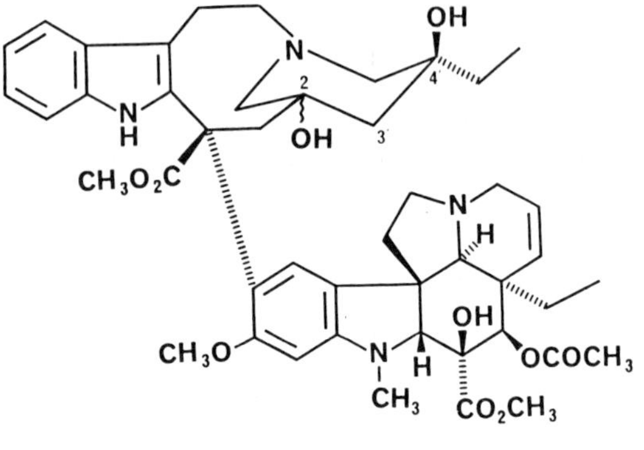

20

was present in the molecule. The vindoline half of **20** appeared to be intact as evidenced by peaks at *m/z* 469, 282, 135, 122, and 121. Since typical indole fragments at *m/z* 143, 144 were present, the extra hydroxyl was assumed to be in the alicyclic portion of the indole half of the molecule. The placement of the additional hydroxyl group in **20** was established by the inversion-recovery method in conjunction with off-resonance decoupled ^{13}C-NMR measurements (Table X). The aromatic carbons of the indole half and the ester carbonyl and methoxy were readily assigned by comparison of chemical shifts with those of vinblastine (**1**). One singlet carbon signal corresponded well to C-4′ of **1**, and one was in the range where C-18′ could be expected. The third singlet at δ 71.2 was in the range of carbons singly bonded to oxygen. Since the methine doublet corresponding to C-2′ in vinblastine (**1**) was absent in **20**, the additional hydroxyl group must be placed at C-2′. The configuration at C-4′ in leurocolombine (**20**) is considered to be the same as in vinblastine (**1**) in view of the similarity of the chemical shifts of the neighboring carbon atoms in **1** and **20**; however, no configuration has yet been determined for C-2′. Leurocolombine (**20**) exhibited antimitotic activity and marginal antitumor activity against the Ridgeway osteogenic sarcoma (27% inhibition at 15 mg/kg) (*44*).

17. Pseudovinblastinediol (**21**)

Pseudovinblastinediol (**21**) (formerly named pseudovincaleukoblastinediol) has been isolated from *C. roseus* as a minor constituent (*44*). The high-resolution mass spectrum of **21** established a molecular formula of $C_{44}H_{56}N_4O_8$. The characteristic vindoline fragments of *m/z* 469 and 282 were shifted 58 mass units to *m/z* 411 and 224, respectively, indicating the

TABLE X
^{13}C-NMR DATA OF LEUROCOLOMBINE (**20**) AND VINCADIOLINE (**31**) (*76*)

Dihydroindole unit			Indole unit		
Carbon	**20**	**31**	Carbon	**20**	**21**
2	83.3	83.4	1′	40.7	32.8[a]
3	79.5	79.6	2′	71.3	39.2
4	76.3	76.5	3′	50.5	75.2
5	42.7	42.7	4′	69.9	71.3
6	129.9	130.0	5′	63.6	60.3
7	124.5[a]	124.5[a]	7′	56.0	55.6
8	50.5	50.4	8′	27.3	28.5
10	44.4	44.6	9′	117.2	116.9
11	50.5	50.4	10′	129.3	129.4
12	53.2	43.3	11′	118.5	118.5
13	124.0	123.0	12′	122.4	122.3
14	123.7[a]	123.6[a]	13′	119.0	118.9
15	120.0	120.6	14′	110.6	110.5
16	158.7	158.1	15′	135.1	134.9
17	94.6	94.2	17′	130.4	131.6
18	153.7	152.7	18′	55.8	55.4
19	65.8	65.7	19′	56.0	43.2
20	30.9	30.9	20′	34.0	29.2[a]
21	8.4	8.4	21′	6.9	6.2
$\underline{C}O_2CH_3$	173.6	170.9	$\underline{C}O_2CH_3$	174.4	174.8
$CO_2\underline{C}H_3$	52.4	52.2	$CO_2\underline{C}H_3$	52.6	52.4
Ar-OCH_3	55.4	55.8			
$O\underline{C}OCH_3$	171.6	171.7			
$OCO\underline{C}H_3$	21.1	21.1			
N-CH_3	38.2	38.3			

[a] Tentative assignment.

presence of a deacetoxyvindoline building block as the dihydroindole half of the molecule. Further mass spectral studies established an extra hydroxyl group in the alicyclic portion of the indole half of **21** compared to vinblastine (**1**). Pseudovinblastinediol (**21**) formed a diacetate with one acetate group in the vindoline half and the other acetate in the indole half, as evidenced by an ion at *m/z* 371 shifting to *m/z* 413. The two acetates were observed in the ^{1}H-NMR spectrum at δ 1.88 and 1.90. The resonance at δ 1.90 was assigned to the acetate at C-3 in the vindoline part, identical in chemical shift to the C-3 acetate in deacetoxyvinblastine acetate. The other acetate signal at δ 1.88 was quite similar to the one observed for C-3′ acetate in vincadioline acetate (*93*). The acetate derivatives of both pseudovinblastinediol (**21**) and vincadioline (**31**) exhibited a characteristic one-proton singlet at δ 4.47–4.48 assigned to 3′-H. From

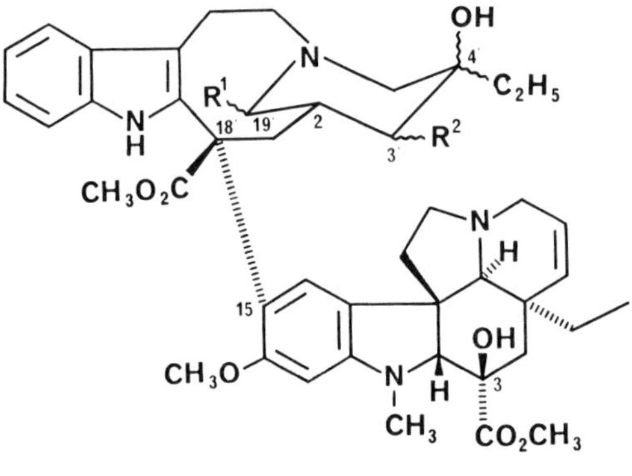

	R^1	R^2
21	H	OH
22	OH	H

these data, structure **21** was proposed for pseudovinblastinediol; however, based on the evidence available at present, the alternative structure **22** cannot be excluded from consideration.

18. Roseadine (**23**)

The bioassay-guided fractionation of *C. roseus* using the P388 lymphocytic leukemia and Eagle's carcinoma of the nasopharynx (KB) systems lead to the isolation of roseadine (**23**), $C_{46}H_{56}N_4O_9$, whose structure had previously been intimated in the course of defining the structure of an acid rearrangement product of leurosine (**11**) (*75*). Typical fragments of the vindoline moiety were observed at *m/z* 282, 174, 135, 122, and 121 in

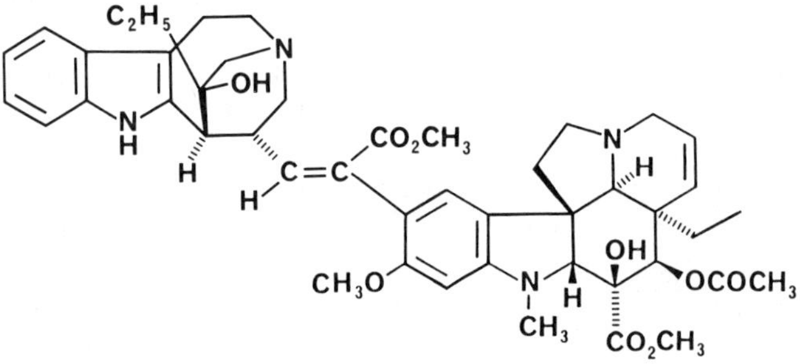

23

the mass spectrum of **23**, establishing the presence of this structural unit in the molecule.

Examination of the ^{13}C-NMR spectra of roseadine (**23**) (Table XI) through comparison with vindoline (**3**) and leurosine (**11**) permitted the assignment of all carbons of the dihydroindole unit. The carbons of the indole nucleus were assigned by comparison with vinblastine (**1**), and the presence of three deshielded carbons, a methine carbon at δ 142.9 and two quaternary carbons at δ 133.2 and 169.2, were observed. The latter was assigned to the methoxycarbonyl carbon, which is shielded somewhat from its characteristic chemical shift of δ 174 ± 1 ppm in the vinblastine series by attachment of an olefinic unit. The other two deshielded carbons at δ 133.2 and 142.9 could be assigned as C-18′ and C-17′, respec-

TABLE XI
^{13}C-NMR DATA OF ROSEADINE (**23**) (*53*) AND VINCATHICINE (**24**) (*94*)

Dihydroindole unit			Indole unit		
Carbon	**23**	**24**	Carbon	**23**	**24**
2	83.2	83.0	1′	142.9	38.7
3	79.2	79.6	2′	34.2	28.9
4	76.1	76.3	3′	48.2	49.2
5	42.3	42.8	4′	74.4	75.8
6	129.8	130.5	5′	58.9	63.1[a]
7	123.9	124.6	7′	57.8	55.0[a]
8	50.7	50.6	8′	33.2	32.2
10	43.4	43.6	9′	109.5	57.1[a]
11	50.7	50.9	10′	128.8	144.2
12	52.1	53.0	11′	117.2	124.6[a]
13	122.8	122.8	12′	121.0	127.7[a]
14	123.9	123.6	13′	118.7	124.1[a]
15	119.0	120.6	14′	110.1	121.3[a]
16	157.7	159.1	15′	134.3	154.0
17	92.8	94.3	17′	131.6	187.0
18	153.2	152.8	18′	133.2	63.1
19	66.1	65.7	19′	51.5	52.6[a]
20	20.4	30.6	20′	24.3	34.6
21	7.6	7.6	21′	6.7	7.0
<u>C</u>O$_2$CH$_3$	170.5	170.7	<u>C</u>O$_2$CH$_3$	169.2	174.8
CO$_2$<u>C</u>H$_3$	51.9	52.2	CO$_2$<u>C</u>H$_3$	51.6	52.6
Ar-OCH$_3$	55.4	55.4			
O-<u>C</u>OCH$_3$	171.6	171.7			
O-CO<u>C</u>H$_3$	20.7	21.1			
N-CH$_3$	38.1	38.4			

[a] Tentative assignment.

tively, in accordance with the proposed carbon framework of roseadine (**23**). The stereochemistry of the double bond has not been determined unambiguously thus far; however, model studies indicate that in the fragmentation reaction of leurosine (**11**) leading to roseadine (**23**) the proton lost from C-1′ should come from the α side of the molecule resulting in a *Z* configuration for the double bond (*75*) (see Scheme 4). Roseadine (**23**) was active in the P388 test system, showing *T/C* 176% at 2.0 mg/kg.

19. Vincathicine (**24**)

The amorphous alkaloid vincathicine (**24**), $C_{46}H_{56}N_4O_9$, initially isolated from *C. roseus* (*37*), was suspected to contain an oxindole chromophore. No further work was reported on the structure elucidation of **24** until it was observed that vincathicine (**24**) could be prepared from leurosine (**11**) by acid treatment (*94,95*). Efforts followed by the reisolation of vincathicine (**24**) from *C. roseus* led to the unambiguous structure elucidation of this alkaloid (*94*).

From the mass, ^{1}H-, and ^{13}C-NMR spectra of **24** it became evident that the molecule contained a vindoline building block. Twenty-five of the res-

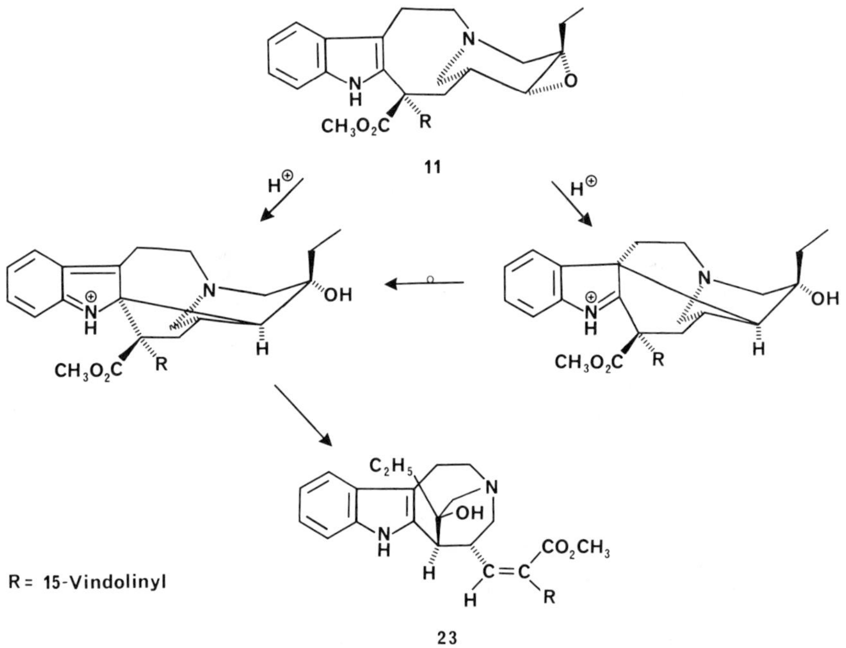

R = 15-Vindolinyl

SCHEME 4. Chemical fragmentation of leurosine (**11**) leading to roseadine (**23**).

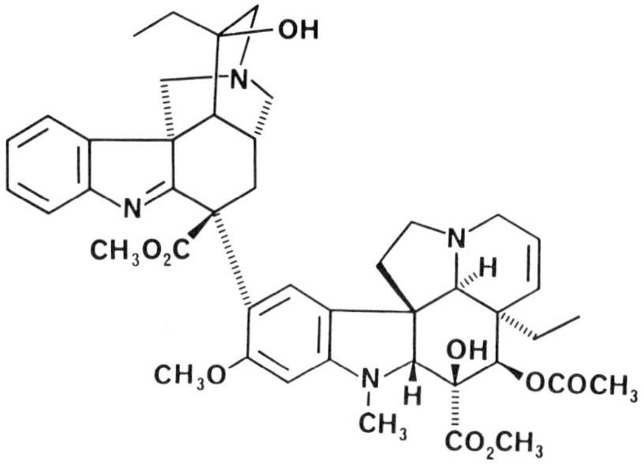

24

onances in the ^{13}C-NMR spectrum of **24** were consistent with the spectrum of the 15-vindolinyl portion of vinblastine (**1**), and the remaining 21 carbon resonances established the presence of seven sp^2-hybridized carbon atoms instead of the eight required for a substituted indole derivative. In view of the production of vincathicine (**24**) by acid treatment of leurosine (**11**), two structures are possible for the initial product, depending on whether a 9′,3′ or a 9′,4′ bond is formed. Both of the final products formed after subsequent deprotonation would explain the multiplicities of the carbon resonances of the indolenine component of vincathicine (**24**) (i.e., three quaternary carbons apart from the sp^2-hybridized carbons, two methines, six methylenes, and two methyl groups), but absorptions assigned to C-3′ and C-4′ are in favor of structure **24** in which initially a 9′,3′ bond has been formed (see Scheme 5).

20. Vinamidine (Catharinine) (**26**)

Vinamidine was first isolated from *C. roseus* by Svoboda and co-workers (*44*). The mass spectrum of the isolate indicated a molecular weight of 824, and the high-resolution mass measurement established the formula $C_{46}H_{56}N_4O_{10}$. The IR spectrum had an intense band at 1660 cm^{-1} characteristic of an amide. ^{13}C-NMR studies indicated the presence of a 15-vindolinyl moiety and a ketone carbonyl at δ 210.4 in the indole half of the molecule. The presence of another carbonyl at δ 163.4 was also observed, which appeared as a doublet in the inversion-recovery/off-resonance decoupling measurement, indicating that this carbon bears a proton. Based on chemical shift theories, the presence of a formamide group was suggested, and structure **25** was proposed for vinamidine (*44*).

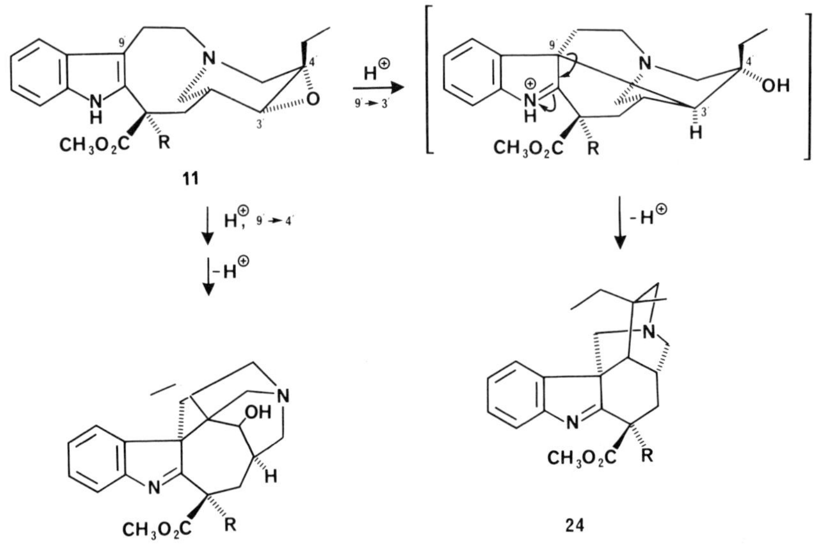

11

24

R= 15-Vindolinyl

SCHEME 5. Formation of vincathicine (**24**) by acid treatment of leurosine (**11**).

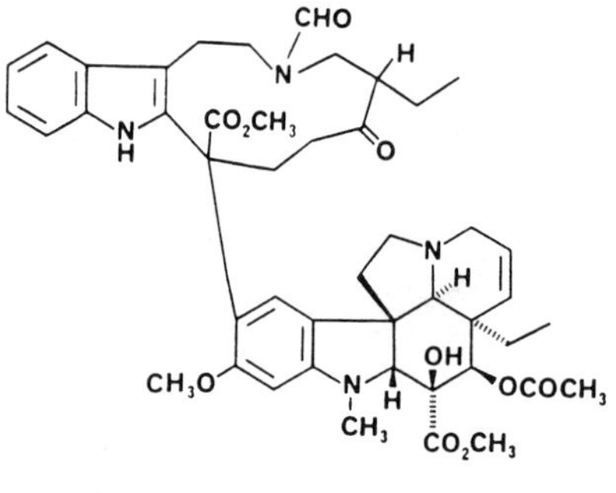

25

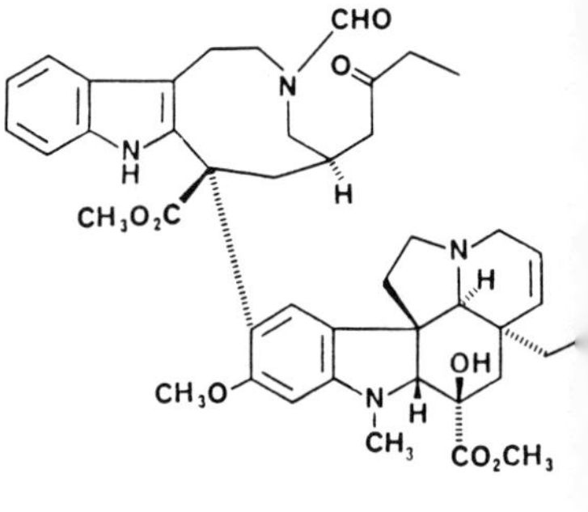

26

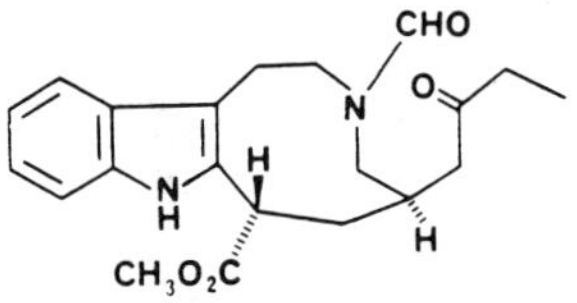

27

Subsequently, catharinine, an amorphous base, was obtained from *C. longifolius* and *C. ovalis* (*55*), and the isolate was so named because of a suspected structural resemblance with catharine (**10**). Later, it was shown that catharinine is identical with vinamidine, and the structure of the alkaloid was revised to **26**. Chemical evidence strongly suggested that structure **25** is equivalent of dihydrocatharine, but authentic dihydrocatharine on sodium borohydride reduction lactonized whereas compound **26** under similar conditions did not. The final structure confirmation was achieved by X-ray analysis of derivative **27** obtained by reductive cleavage (Sn–$SnCl_2$–HCl) of isolate **26**. The absolute configuration of the C-18′ stereo center of vinamidine (**26**) has been determined as (*S*) according to the ORD and CD data of **26** on comparison with data of **27**, considering the fact that this product exerts a (18*S*) configuration, which is clearly the thermodynamically preferred isomer (*55*). Vinamidine (**26**) may well be formed by a biogenetic cleavage of the piperidine ring of leurosine (**11**) as shown in Scheme 6, and an alternative fragmentation of **11** may led to catharine (**10**).

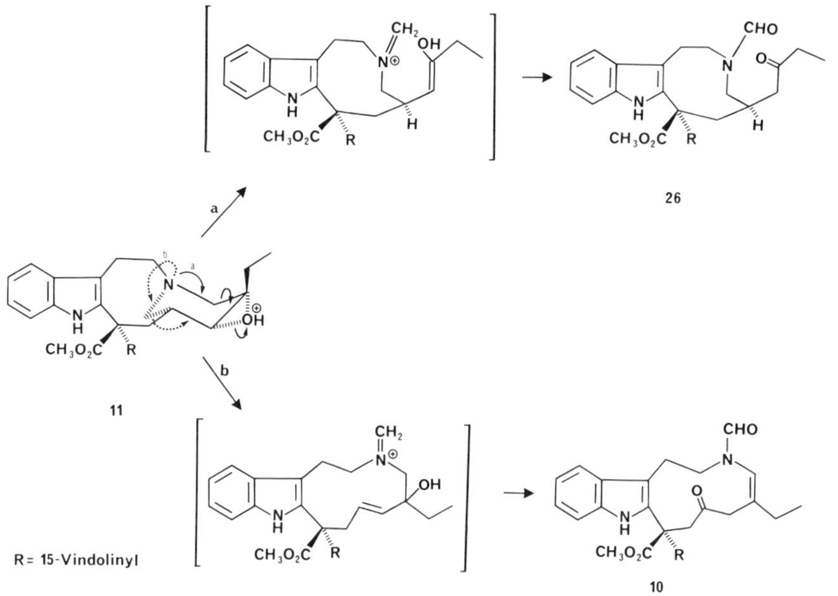

R= 15-Vindolinyl

SCHEME 6. Formation of vinamidine (catharinine) (**26**) and catharine (**10**) from leurosine (**11**).

21. Vincovaline (**28**)

Vincovaline (**28**), $C_{46}H_{58}N_4O_9$, isolated from *C. ovalis* (*62,63*), is an isomer of vinblastine (**1**) and appears to be the first bisindole alkaloid in which vindoline (**3**) is coupled to an indole component of the coronaridine series. The mass spectrum of vincovaline (**28**) is very similar to that of vinblastine (**1**), and there are also close similarities in the ^{1}H-NMR spectra. One significant difference from the ^{1}H-NMR spectrum of **1** is the much lower chemical shift for the C-21 methyl protons in the spectrum of vincovaline (**28**), which indicates the somewhat different environment for this group. The resistance of the C-4′ hydroxyl group to acetylation suggests that it may have the same configuration (β) as vinblastine (**1**). Comparison of the CD spectrum of **28** with those of various synthetic bisindole derivatives epimeric at stereo centers C-2′ and C-18′ established the (2′*S*,18′*R*) absolute configuration for the corresponding stereo centers of vincovaline (**28**).

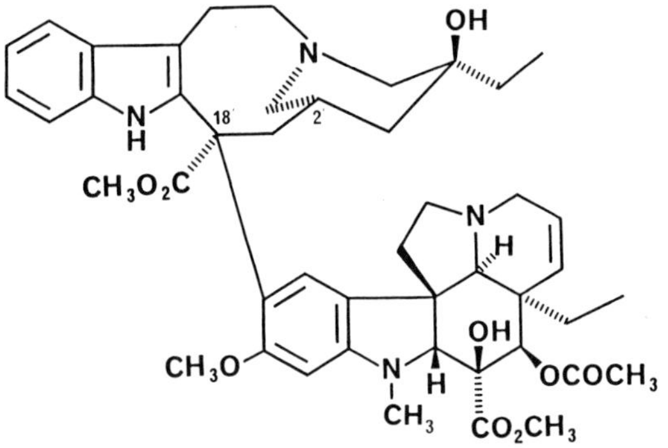

28

22. Vincovalinine (**29**)

Vincovalinine (**29**), $C_{44}H_{56}N_4O_7$, has been isolated from *C. ovalis* as an amorphous material (*62*). The mass spectrum of **29** suggested the presence of a vindoline moiety showing the characteristic fragments at *m/z* 282, 135, 122, and 121, which was supported by the ^{1}H-NMR spectrum. Comparison of the ^{1}H-NMR spectrum of **29** with that of leurosine (**11**) indicated the absence of a methoxycarbonyl resonance, and a new signal was observed at δ 6.10 that could be attributed to 18′-H, establishing the 18′-demethoxycarbonylleurosine structure for **29**.

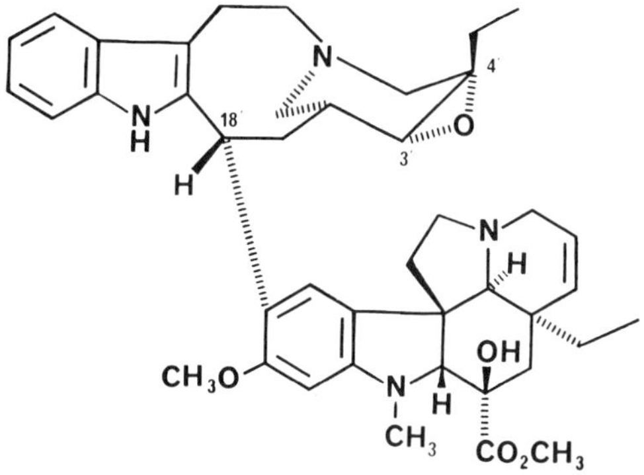

29

23. Vincovalicine (30)

Vincovalicine (**30**), $C_{46}H_{54}N_4O_{10}$, was isolated from *C. ovalis* (*62*) and exhibited a UV spectrum characteristic for the superposition of indole and indolenine chromophores, λ_{max} (log ϵ) 220 (4.52), 256 (4.11), and 300 (3.99) nm. ^{1}H-NMR data verified that a similar *N*-demethyl-*N*-formyl-vindoline building block is present in **30** as in vincristine (**2**). Based on available data from the mass, UV, and ^{1}H-NMR spectra structure **30** was suggested for vincovalicine, but this may need refinement.

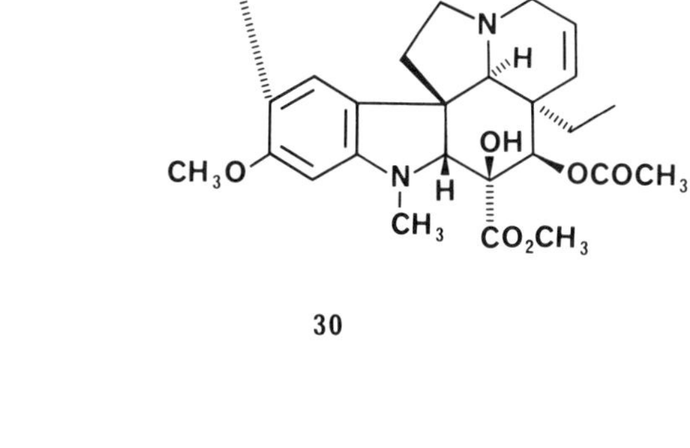

30

24. Vincadioline (**31**)

The isolation of vincadioline (**31**) from *C. roseus* has been reported (*96*) only in the patent literature. The ^{13}C-NMR data of **31** (*76*) (Table X) support the 3′-hydroxyvinblastine structure. Both C-2′ and C-4′ absorptions shifted downfield compared to the corresponding values of vinblastine (**1**). In addition, upfield shifts of C-1′ and C-19′ were observed as a result of the γ-effect of the new oxygen atom at C-3′. The resonance of C-19′ shifted 4.8 ppm, a magnitude consistent with an axial hydroxyl at C-3′. Such circumstances should lead to a nearly equal shielding effect at C-5′, as is indeed observed. These data allow the assignment of the α relative steric position of the C-3′ hydroxyl group.

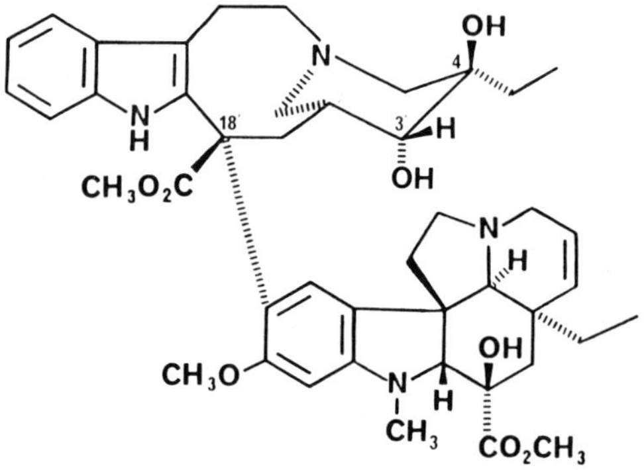

31

25. Vindolicine (**32**)

Vindolicine (**32**) is an unusual alkaloid, isolated from *C. roseus* (*29,53*), whose structure reflects the joining of two vindoline units through con-

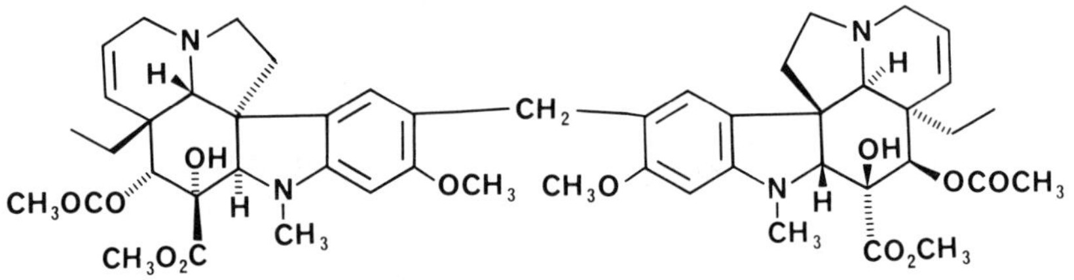

32

densation with a one-carbon fragment. Although there are other bisindole alkaloids known to be formed by the union of two *Aspidosperma* units (*3*), vindolicine (**32**) is the only example linked in the above manner. The structure of **32** was confirmed by UV, IR, ^{1}H-, and ^{13}C-NMR measurements. No molecular ion was observed in the mass spectrum, but the fragment ions found were typical of those in the mass spectrum of vindoline (**3**). According to the ^{13}C-NMR spectrum of **32** (Table XII) the molecule is not symmetrical in space, based on the observed conformationally induced interaction between the two units, which causes some carbons to be shifted downfield while others are shifted upfield.

TABLE XII
^{13}C-NMR DATA OF VINDOLICINE (**32**) (*53*) AND VINDOLINE (**3**) (*74*)

Carbon	**32**		**3**
2	83.0	83	83.2
3	79.3	79.5	79.5
4	76.0	75.9	76.2
5	42.5	42.4	42.8
6	130.2	130.0	130.2
7	123.9	124.4	123.9
8	50.9[a]	51.0	50.9
10	42.7	41.5	43.9
11	50.8[a]	50.6	51.9
12	52.7	52.8	52.6
13	123.4	123.4	124.9
14	123.7	124.1[a]	122.4
15	124.4[a]	124.4[a]	104.5
16	159.0	159.0	161.1
17	93.5	93.0	95.6
18	152.1	152.1	153.6
19	67.0	67.0	67.0
20	30.5	30.3	30.6
21	7.1	7.1	7.5
$\underline{C}O_2CH_3$	170.6	170.6	170.4
$CO_2\underline{C}O_3$	52.0	51.8	51.9
$O\underline{C}OCH_3$	171.6	171.4	171.7
$OCO\underline{C}H_3$	20.7	20.7	20.8
$N\text{-}CH_3$	38.0	37.7	38.0
$Ar\text{-}OCH_3$	55.1	55.1	55.1
$Ar\text{-}CH_2\text{-}Ar$	34.0		—

[a] Tentative assignment.

III. Biosynthesis

Since the last major review of the biosynthesis of the monoterpenoid indole alkaloids (*97*), there have been several full and partial (*98–104*) reviews of various aspects of the work that has been conducted since 1974. Two major developments have dominated the field in this period, namely, the demonstrations that (i) strictosidine (**33**) is the universal precursor of the monoterpenoid indole alkaloids and (ii) selected cell-free systems of *C. roseus* have the ability to produce the full range of alkaloid structure types, including the bisindoles. This section traces some aspects of these developments, paying particular attention to work been carried out with *C. roseus,* and omitting work, important though it may be, on other monoterpenoid indole alkaloid-producing plants, e.g., *Rauwolfia*, *Camptotheca*, and *Cinchona*.

In order to understand some of these developments in context it is necessary to briefly review for the reader the status of the field as it was in 1974 (*97*). During the years prior to that time, tremendous effort had gone into the demonstration, using intact plants for the most part, that the overall sequence for the biosynthesis of the indole alkaloids began with the condensation of a secoiridoid, specifically secologanin (**34**), with tryptamine to form a glucoside, whose C-3 stereochemistry was uncertain. Preliminary feeding experiments had shown this compound to be a precursor for a wide range of indole alkaloid skeleta. From this point, the pathway was suggested to proceed through a Corynanthe intermediate, such as geissoschizine (**35**), to a *Strychnos* compound, such as stemmadenine (**36**). *Iboga* and *Aspidosperma* alkaloids were regarded as being produced through an intermediate acrylic ester, dehydrosecodine (**37**), whose skeleton had recently been reported. The primordial *Aspidosperma* alkaloid tabersonine (**38**) was envisaged to undergo a series of oxidations to afford vindoline (**3**), one of the major alkaloids of *C. roseus* and one half of vinblastine (**1**). No enzymes in the pathway had been isolated, and relatively few of the chemical steps involved in the long and complicated process had been investigated either *in vivo* or *in vitro*.

This overall biosynthetic scheme is summarized in Scheme 7, and it is well to remember that *C. roseus* is almost alone as a plant in which this whole pathway can be viewed in its entirety, for most indole alkaloid-containing plants produce only one or two of the major alkaloid classes, and not all four. In addition, *C. roseus* is without doubt the most economically important of the indole alkaloid-containing plants, and thus studies were, and continue to be, driven by the goal of increasing the availability of the commercially significant alkaloids ajmalicine (**39**), vinblastine (**1**), and vincristine (**2**).

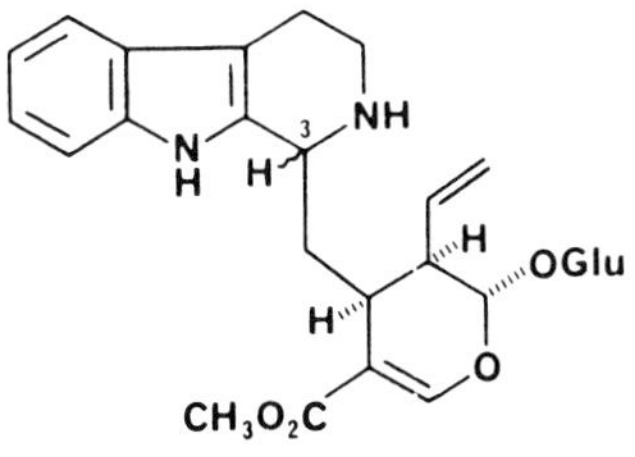

33 C-3Hα

85 C-3Hβ

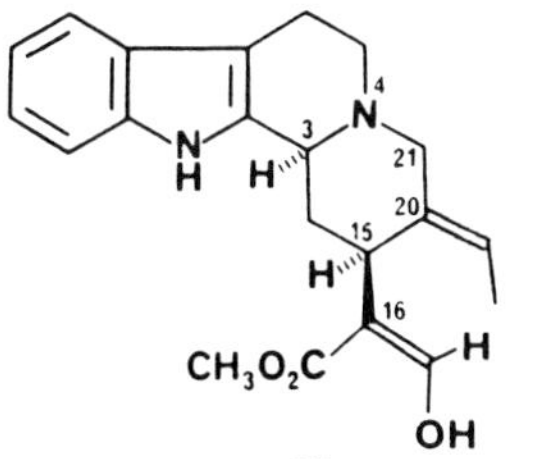

35

96 4,21-dehydro

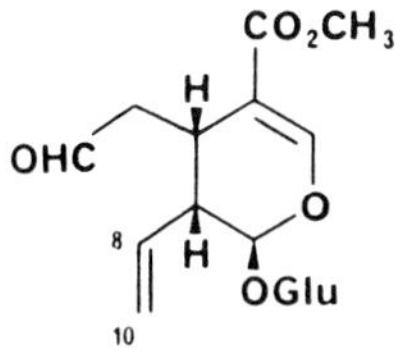

34

91 8,10-dihydro

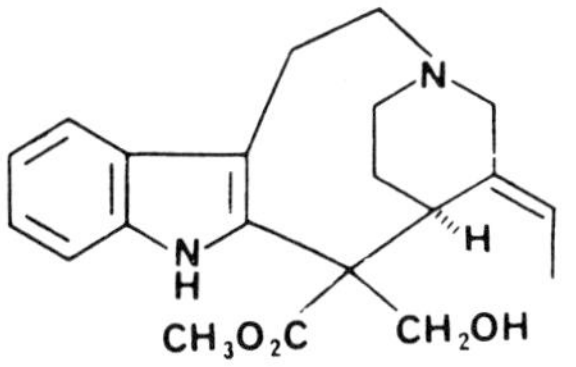

36

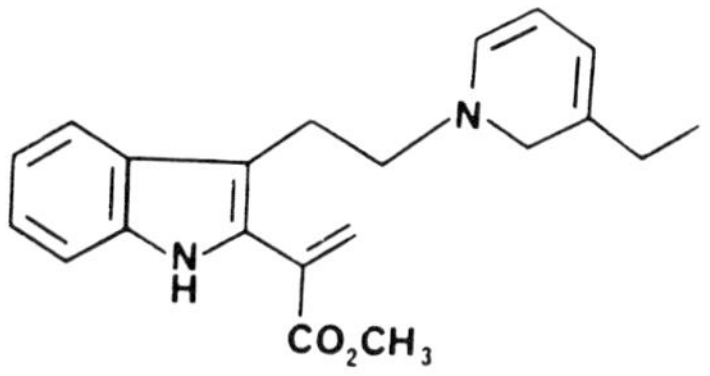

37

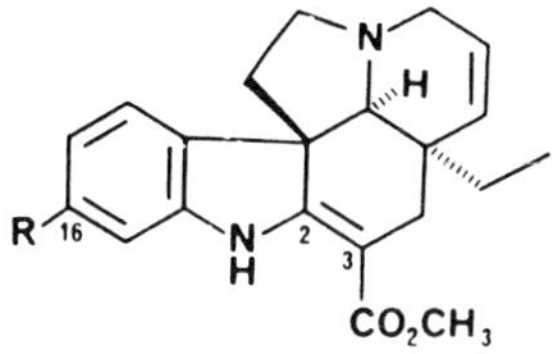

	R
38	H
44	OH
45	OMe
46	OMe 2,3-dihydro

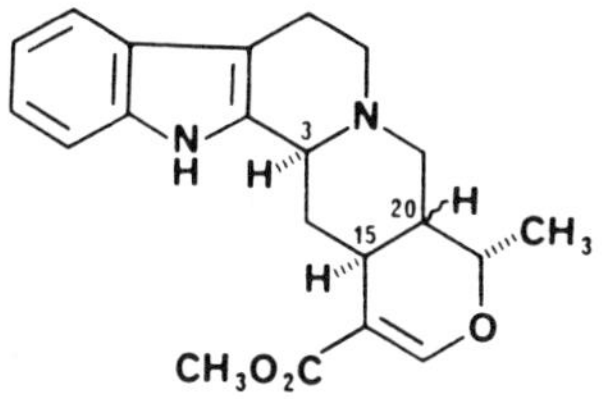

39 C-20Hβ

75 C-20Hα

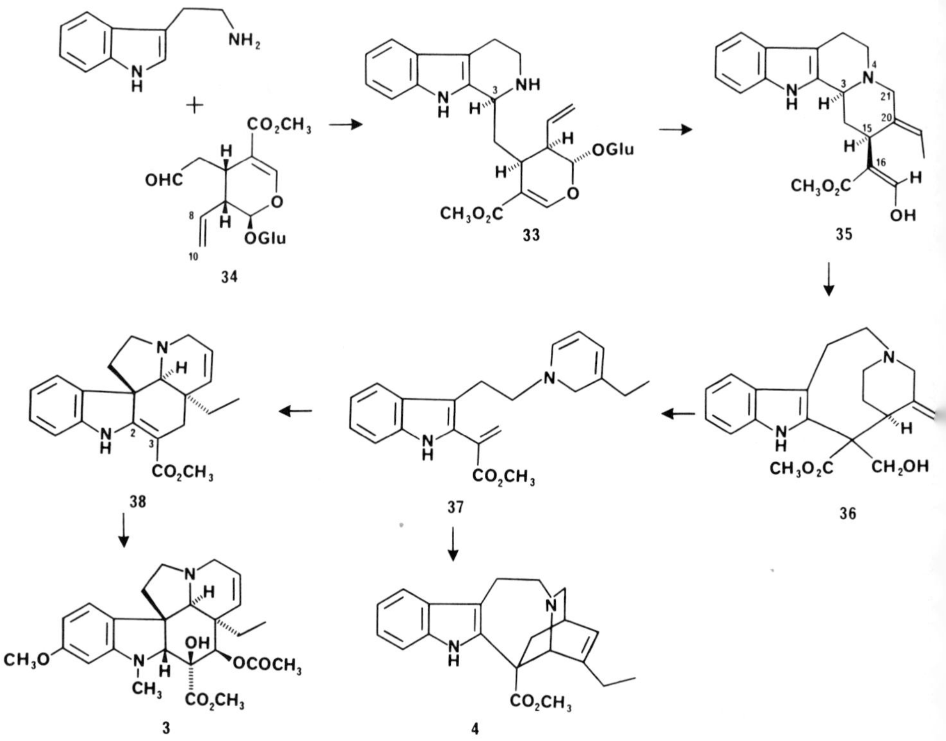

SCHEME 7. Biogenesis of the indole alkaloids in 1974.

Additionally, there had, for many years, been a vast synthetic effort underway aimed at the synthesis of the two monomeric units, where it was anticipated that the two units could be joined to form the vinblastine-type bisindole alkaloids. Coincidentally, as it transpired, 20 years of effort in the areas of synthesis and biosynthesis converged, at almost the same time, on the compound 3′,4′-anhydrovinblastine (**8**).

This review of the biosynthesis of the bisindole alkaloids of *C. roseus* is organized along a developing biosynthetic pathway, as far as possible, and relies on the notion that the most sophisticated studies are those utilizing the purified enzyme systems. Biosynthetic studies on the other monoterpene indole alkaloids are not reviewed here.

A. WHOLE PLANT STUDIES

Reda (*105*) investigated the distribution and accumulation of the total alkaloids in *C. roseus* during six different stages of flowering and fruiting. The highest concentration of alkaloids (using perivine as the standard) was found in the roots at the start of flowering, and the minimum in the stems during the full fruiting stage. The rate of alkaloid accumulation decreased during fruit maturation, and the most active stage of alkaloid biosynthesis was at the start of flowering in all vegetative organs. This is in accord with previous studies (*106*) where a higher level of vinblastine (**1**) was reported in extracts of leaves of young *C. roseus* plants than in those of older plants. Analysis of the fruit latex of *C. roseus* indicated an absence of vindoline (**3**) (*107*). Other workers have found no correlation between the dry weight of the roots and either total or specific alkaloids (*108*).

Mersey and Cutler (*109*) examined the differential distribution of specific indole alkaloids in the leaves of *C. roseus* following the observation by Constabel (*110*) suggesting that certain leaf protoplasts exhibited unusual properties, indicating that they might be idioblasts, cells having a propensity for the accumulation of secondary metabolites (*111*). Idioblast protoplasts were readily distinguishable from other leaf cells by the unusually high refractivity of their vacuolar contents and a characterisitc yellow fluorescence (serpentine). Following enrichment of the idioblast protoplasts in the lower regions of a dextran gradient, four alkaloids, vindoline (**3**), catharanthine (**4**), serpentine (**40**), and strictosidine lactam (**41**), were always detected in high amounts (20–45% of total alkaloids). Given the previous failures to obtain vindoline (**3**) from cell suspension cultures, it was thought that some degree of morphological differentiation may be required.

The alkaloid pattern of light- and dark-grown seedlings of *C. roseus* cv. Little Delictata has been examined by Balsevich *et al.* (*112*). The dark-grown seedlings had low levels of vindoline (**3**), deacetylvindoline (**42**), deacetoxyvindoline (**43**), and 11-hydroxytabersonine (**44**), whereas the levels of tabersonine (**38**), 11-methoxytabersonine (**45**), and catharanthine (**4**) were relatively high. When the dark-grown seedlings were subjected to light for 4 days, vindoline (**3**) levels increased rapidly (from 5 to 54 mg/kg fresh weight) with a concomitant loss of tabersonine (**38**) (from 40 to 13 mg/kg). Levels of catharanthine (**4**) stayed essentially constant at about 30–35 mg/kg fresh weight. Based on these experiments the pathway between tabersonine (**38**) and vindoline (**3**) was suggested to involve 11-hydroxytabersonine (**44**), 11-methoxytabersonine (**45**), 11-methoxy-2,16-dihydrotabersonine (**46**), deacetoxyvindoline (**43**), deacetylvindoline (**42**), and vindoline (**3**). Experiments designed to examine the enzymes involved in the latter stages of the formation of vindoline (**3**) are described subsequently.

Hutchinson and co-workers (*113*) in the course of studies aimed at the investigation of the formation of the bisindole alkaloids in whole plants of

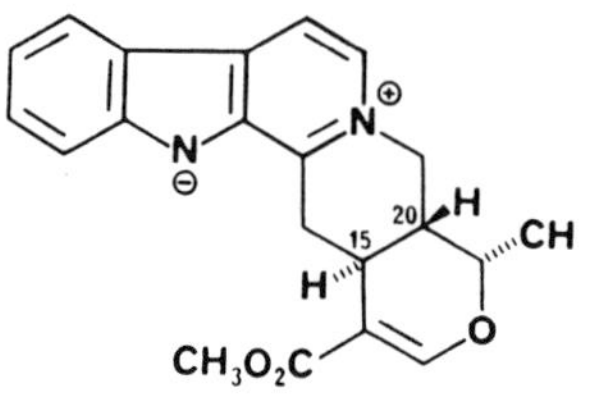

40

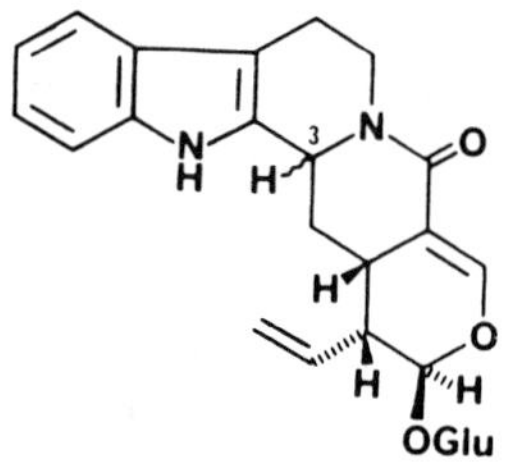

41 C-3Hα
92 C-3Hβ

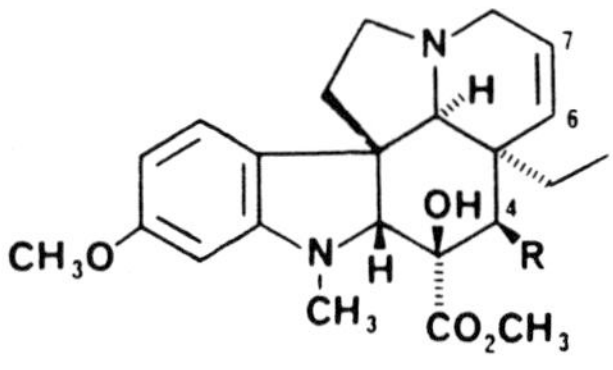

	R	
3	$OCOCH_3$	
42	OH	
43	H	
51	$OCOCH_3$	6,7-dihydro

C. roseus also examined the catabolism of some of the principal alkaloids, including vindoline (**3**) and catharanthine (**4**), using $^{14}CO_2$ and monitoring the level of incorporation with time. Turnover occurred much more rapidly in apical cuttings than in intact plants, possibly because of the energy required for the general anabolism of the plant and root generation. In any instance, it is clear that these biosynthetically advanced secondary metabolites are turned over in a quite dynamic fashion, in agreement with studies on other classes of alkaloids.

B. Callus Tissue Studies

Studies of conditions for the growth of callus tissue of *C. roseus* were first reported in 1962 (*114*) by Babcock and Carew, and analytical work commenced when Carew and co-workers (*115*) demonstrated that in the presence of 0.5 mg/liter kinetin callus tissue grew rapidly and probably produced vindoline (**3**), as well as a number of other alkaloids. At the same time, a group at Eli Lilly (*116*) reported an analysis of the alkaloids of crown gall cultures which had been maintained by Hildebrandt's group since 1943. Alkaloids were found in the tissue and in the culture medium, and vindoline (**3**) was positively identified. Similar results were reported by Mothes' group (*117*) using leaf and stem callus material, for they were able to determine quantitatively the presence of both vindoline (**3**) and vindolinine (**47**).

Leaf organ cultures of *C. roseus* have also been described (*118*). A typical 2.5-g fresh weight inoculum produced 29 g fresh weight of leaf material after 35 days; dedifferentiated tissue was absent. The alkaloids found included ajmalicine (**39**), sitsirikine (**48**), tetrahydroalstonine (**39**), serpentine (**40**), and vindoline (**3**).

Goodbody and co-workers (*119*) have examined the production of alkaloids in root and shoot cultures induced from seedlings of *C. roseus*. The pattern of alkaloids in the root cultures was similar to that of the roots from intact plants. Thus ajmalicine (**39**) and catharanthine (**4**) were produced, but no vindoline (**3**), a major leaf alkaloid, and no bisindole alkaloids. Similarly, the pattern of the alkaloid content of the shoot cultures was like that of the leaves of the intact plant, showing the presence of vindoline (**3**), catharanthine (**4**), and ajmalicine (**39**), with **3** predominating. A search for the bisindole alkaloids in the cultures indicated the presence of anhydrovinblastine (**8**) and leurosine (**11**) in the shoot cultures (2.6 and 0.3 μg/g fresh weight, respectively), but no vinblastine (**1**) or vincristine (**2**).

The effect of light on alkaloid production in these cultures was also evaluated (*119*). More catharanthine was produced in the light than in the dark, and the same observation was made for ajmalicine (**39**). Knobloch *et al.* (*120*) examined the production of anthocyanins, ajmalicine (**39**), and serpentine (**40**) in cell suspension cultures and found that although serpentine levels increased 18-fold in the light, ajmalicine levels decreased 50%. In the work of Goodbody *et al.* (*119*) the biosynthesis of vindoline (**3**)

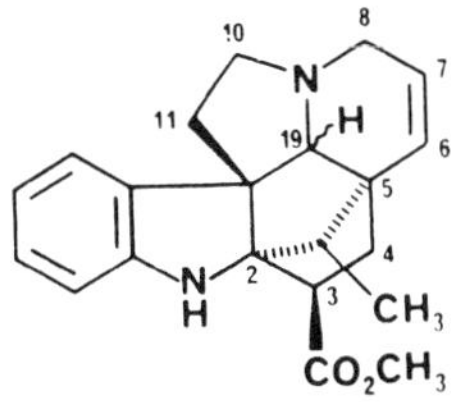

47	C-19Hα
52	C-19Hα N-oxide
53	C-19Hβ N-oxide

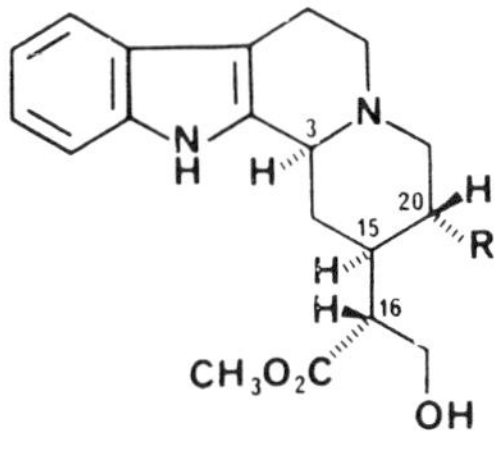

	R
48	$-CH=CH_2$
59	$-C_2H_5$

seemed to be dependent on both organogenesis and light; root cell suspension cultures failed to produce vindoline (**3**) in either the light or the dark. Although the shoot cultures produced vindoline (**3**) in the light, levels dropped dramatically in the dark. The inferences of these results for the formation of the bisindoles where both monomeric structural elements are *de facto* required are clear.

Miura and co-workers (*121*) have successfully induced multiple shoot cultures of *C. roseus* from seedlings in the presence of 1.0 mg/liter of the cytokinin benzyladenine. Vindoline (**3**) and catharanthine (**4**) were predominating alkaloids in the MSC-B-1 line, showing levels of 1.8 and 0.37 mg/g dry weight, respectively in the leaf tissue. In the case of catharanthine (**4**) this represented a 10-fold increase over the parent plant tissue, and such levels were sustained in the regenerated plants. When the benzyladenine was eliminated, overall growth was reduced, but vindoline (**3**) and catharanthine (**4**) concentrations increased to 3.2 and 1.1 mg/g dry weight, respectively.

It has taken many years, however, for a callus culture to be described which produces vinblastine (**1**). Success was achieved by Miura and co-workers (*122*) as a result of the screening of callus tissues with the HeLa cell line. Vinblastine (**1**) was detected at a level of 1 μg/g dry weight by HPLC and characterized by retention time and mass spectrometry. This level is below that in the whole plant, and thus its presence was probably overlooked by previous workers.

Continued work by the same group (*123*) has led to the first isolation of vinblastine (**1**) from a multiple shoot culture of *C. roseus*. The most productive line, MSC-B-1, consisted of two distinctly different tissues, multiple shoots and unorganized tissue, and was maintained growing and productive for 30 months. Vinblastine (**1**) was isolated by HPLC, and the content was estimated to be 15 μg/g dry weight. Production of this alkaloid was greater than that in the callus culture but less than that observed for the parent plant, even though the levels of catharanthine (**4**) and vindoline (**3**) were about the same.

C. Cell Suspension Studies

Studies using suspension cultures of *C. roseus* were first reported by Carew and Patterson in 1969 (*124*). Substantial variability was observed in the ability of the tissue to grow on repeated subculturing, but, in spite of this, quite large fermentations (3 liters) were carried out. The alkaloids produced included sitsirikine (**48**), akuammicine (**49**), and lochneridine (**50**), representing three of the four major alkaloid classes. Further studies on the Wc13S culture line, following subculture on a defined agar medium

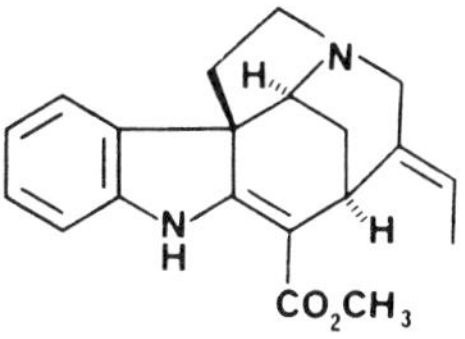

49

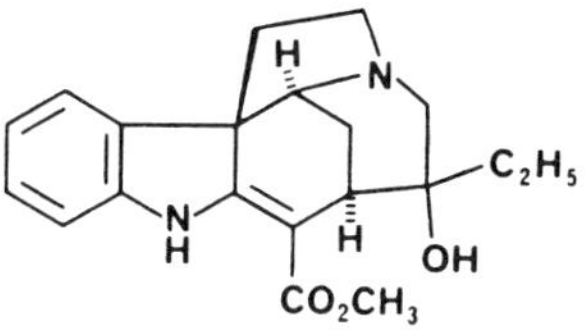

50

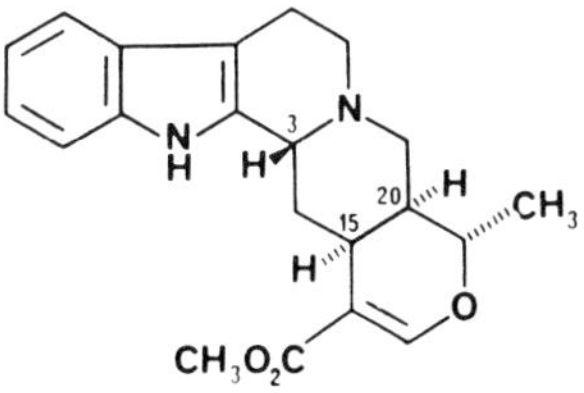

54

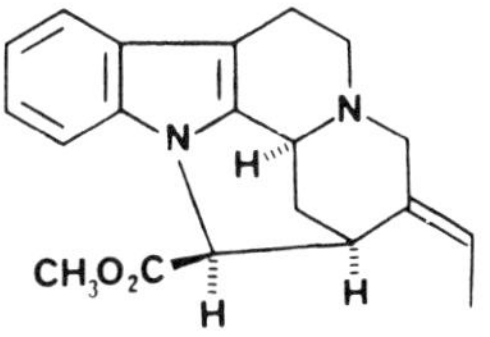

55

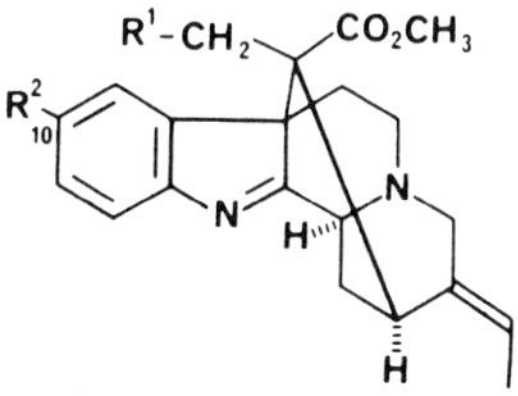

57

	R^1	R^2
56	$OCOCH_3$	H
64	OH	H
68	OH	OH

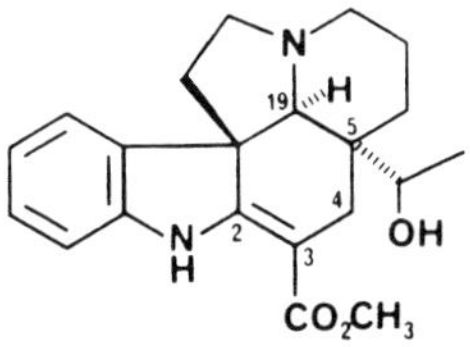

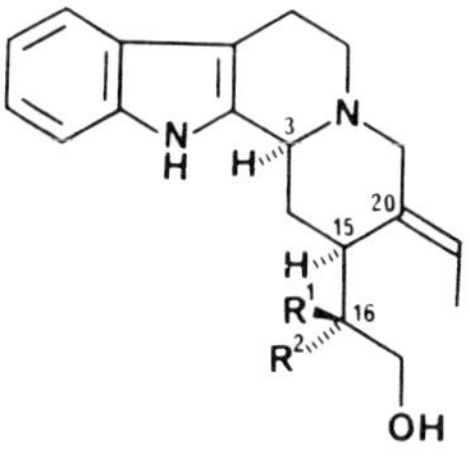

	R^1	R^2
58	H	CO_2CH_3
95	CO_2CH_3	H

and suspension culturing, yielded cell cultures which were used to study the metabolism of vindoline (**3**), catharanthine (**4**), and vinblastine (**1**) (*125*). Dihydrovindoline (**51**) and deacetylvindoline (**42**) were produced from vindoline (**3**), but no metabolites were produced from catharanthine (**4**). Three unidentified metabolites were formed from vinblastine (**1**).

Scott and co-workers have also reported on the isolation of alkaloids from *C. roseus* cell suspension cultures (*126*). The cell line used, identified as CRW, afforded akuammicine (**49**), catharanthine (**4**), and strictosidine (**33**), and feeding experiments with labeled tryptophan led to incorporation into ajmalicine (**39**), akuammicine (**49**), catharanthine (**4**), and vindoline (**3**). The ability to produce alkaloids was carried through 8 successive generations.

Stöckigt and Soll were the first to analyze cell suspension cultures of *C. roseus* in detail (*127*). Twelve alkaloids representing each of the main structure types were identified (Table XIII). Potier and co-workers (*128*) found a cell line which produced tabersonine (**38**) as the predominating alkaloid. On the other hand Höfle and co-workers (*129*) obtained 16 different alkaloids (Table XIII) from hexane extracts, including several alkaloids not isolated previously, e.g., vindolinine *N*-oxide (**52**), 19-epivindolinine *N*-oxide (**53**), akuammigine (**54**), pleiocarpamine (**55**), akuammiline (**56**), and minovincinine (**57**). The methanol extract of the same culture afforded sitsirikine (**48**), isositsirikine (**58**), dihydrositsirikine (**59**), yohimbine (**60**), 3-iso-19-epiajmalicine (**61**), pleiocarpamine (**55**), and antirhine (**62**) (*130*). Pétiard *et al.* (*131,132*) found a strain which produced over 50 alkaloids as detected by TLC, and 13 of these were identified (Table XIII), including 3 new to *C. roseus,* 3-epiajmalicine (**63**), deacetylakuammiline (**64**), and dihydrocondylocarpine (**65**), and 3 new natural products, 7-hydroxyindolenine ajmalicine (**66**), pseudoindoxyl ajmalicine (**67**), and 10-hydroxy-deacetylakuammiline (**68**). Another cell suspension system (*133*) produced (16*R*)-19,20-(*E*)-isositsirikine (**58**), (16*R*)-19,20-(*Z*)-isositsirikine (**69**), and a new sarpagine-type alkaloid, 21-hydroxycyclolochnerine (**70**).

The concurrent development of analytical techniques for the isolation, detection, and quantitation of indole alkaloids in small amounts of cellular material has been addressed by Höfle and co-workers (*134*), who developed a reversed-phase system using either methanol–water–triethylamine or acetonitrile–triethylammonium formate buffer as the eluent. Two prepacked cartridge systems were found to be very effective for the rapid preparation of the alkaloid mixture from the cell contents.

The detection of very low levels of alkaloids is a fundamental requirement for the analysis of cell contents and enzymatic products. One of the important compounds is of course vindoline (**3**), and a combined French

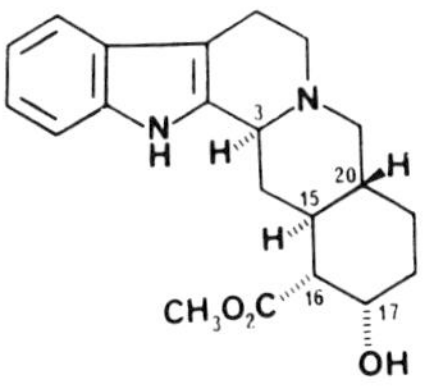

60

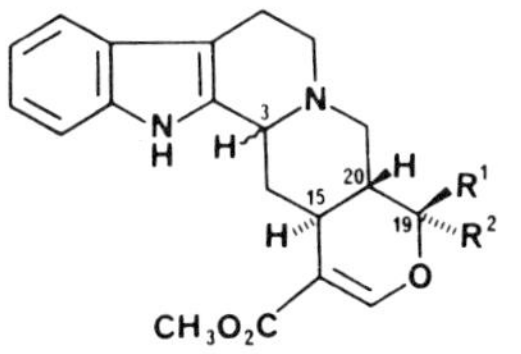

		R^1	R^2
61	C-3Hβ	CH_3	H
63	C-3Hβ	H	CH_3
74	C-3Hα	CH_3	H

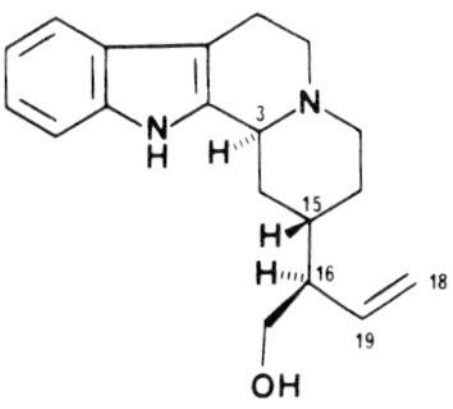

62

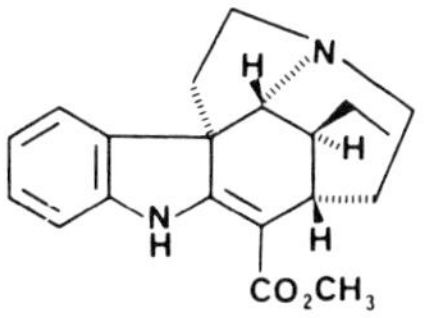

65

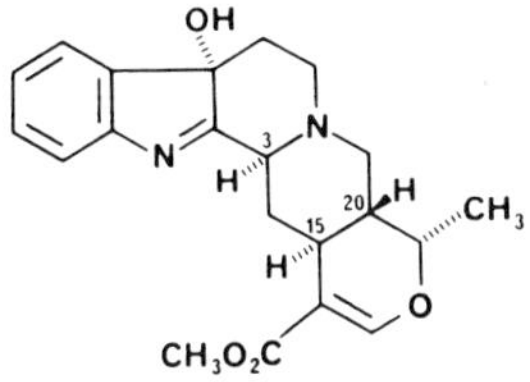

66

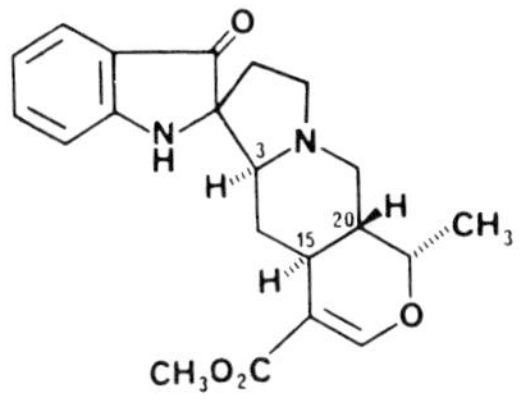

67

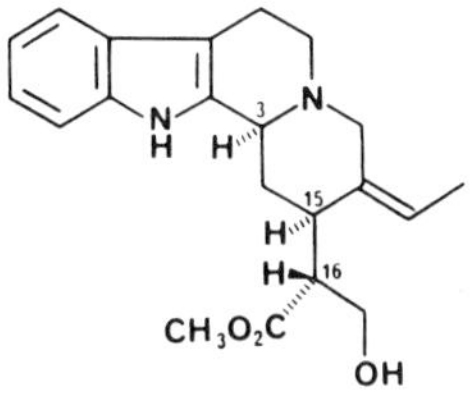

69

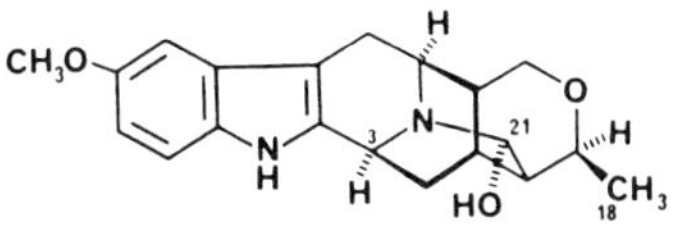

70

TABLE XIII

ISOLATION AND/OR DETECTION OF INDOLE ALKALOIDS IN SUSPENSION CULTURES OF *Catharanthus roseu*

Alkaloid	943 (*154*)	953 (*155,156*)	200GW (*155,159,160*)
Strictosidine (**33**)			+ +
Strictosidine lactam (**41**)		+	
Vallesiachotamine	+	+	+
Isovallesiachotamine			+
Ajmalicine (**39**)	+	+	+
3-Epiajmalicine (**63**)			
Tetrahydroalstonine (**75**)			
3-Epi-19-epiajmalicine (**61**)			
Akuammigine (**54**)			
Pseudoindoxylajmalicine (**67**)			
Mitraphylline			
7-Hydroxyindolenine ajmalicine (**66**)			
Serpentine (**40**)			
Sitsirikine (**48**)			
Isositsirikine (**58**)	+	+	
Dihydrositsirikine (**59**)			
Antirhine (**62**)			
Yohimbine (**60**)	+	+	
Akuammiline (**56**)			
Deacetylakuammiline (**64**)			
10-Hydroxy-deacetylakuammiline (**68**)			
Pleiocarpamine (**55**)			
Akuammicine (**49**)			
Xylosyloxyakuammicine	+		
Dihydrocondylcarpine			
Vinervine			
Tabersonine (**38**)			
19-Hydroxytabersonine			
19-Hydroxy-11-methoxytabersonine	+	+	
19-Acetoxy-11-hydroxytabersonine	+		
19-Acetoxy-11-methoxytabersonine	+	+	
Lochnericine (**83**)		+	
Lochnerinine			
Hörhammericine	+	+	+
Hörhammerinine	+	+	+
Minovincinine			
Vindolinine (**47**)	+	+	+
19-Epivindolinine	+	+	+
Vindolinine *N*-oxide (**52**)			
19-Epivindolinine *N*-oxide (**53**)			
Catharanthine (**4**)			+
N,*N*-Dimethyltryptamine		+	

Table XIII (Continued)

Cell line (Ref.)							
176G (161)	299Y (161)	340Y (161)	951G (161)	SR28 (127)	SR28 (129)	SR28 (130)	SR28 (131)
	+ +						
	+		+			+	
	+						+
+	+	+	+	+	+		+
					+		+
				+	+		+
						+	
					+		
							+
							+
							+
				+			
						+	
+						+	
						+	
						+	
+						+	
						+	
							+
							+
					+	+	
				+			
							+
				+			
					+		+
					+		
+	+			+	+		
	+						
+	+	+		+	+		
+	+	+		+			
					+		
+	+	+				+	+
+		+		+	+		
					+		
					+		
				+	+		
		+					

and German team has described a radioimmunoassay method for the detection of **3**. The hapten for coupling to the protein was prepared from deacetylvindoline (**42**) with glutaric anhydride, which was then coupled with bovine serum γ-globulin. Eleven molecules of the vindoline glutarate were bound to one molecule of protein. Antiserum specificity was determined through cross-reactivity studies, and it was shown that none of the natural alkaloids nor any of the dimers had any appreciable cross-reactivity. The average vindoline content of the total plant was 0.4%, but vindoline levels of up to 7.0% were found at the intercostal regions of the middle of the leaf. Root, flower, and shoot tissue contained almost no vindoline (**3**) (*135*).

The standard carbon source in the Schenk and Hildebrandt medium is sucrose (*136*). Berlin and co-workers (*120*) investigated the effect of sucrose concentrations on alkaloid production. Raising the level of sucrose to 11% increased the serpentine (**40**) levels by more than 300% but reduced ajmalicine (**39**) levels. Raab and Lee (*137*) have recently investigated three alternative carbon sources, glucose, lactose, and galactose. Total alkaloid concentration increased when glucose or galactose (10 and 19%, respectively) were used. With glucose, the levels of catharanthine (**4**) and ajmalicine (**39**) were increased compared to sucrose, but this was not the case with galactose. Cell growth was delayed when lactose was used as the carbon source, and alkaloid levels dropped substantially (17% of control for ajmalicine). It appears that monosaccharides are better precursors than the disaccharides.

The effect of phosphate on alkaloid production has also been evaluated (*138*). Using a modified induction medium devoid of phosphate and other essential growth factors, production of secondary compounds was more rapid than when phosphate was present. A broader study of the phenomenon has been reported by a French group (*139*) where, using three alkaloids as markers, the disappearance of the major nutrients from the medium and the evolution of phosphates, nitrates, ammonium ions, glucose, and starch in the cells were observed over time. It was not possible to relate alkaloid accumulation to the appearance or disappearance of any one metabolite in particular. However, other workers have found that the rate of biomass accumulation was directly related to the rate of formation of cellular serpentine (**40**) (*140*).

Kreuger and Carew (*141*) examined the effects of a number of alkaloid precursors on alkaloid production in suspension cultures and found that at 100 mg/liter tryptamine hydrochloride enhanced alkaloid production. The "alkaloids" produced, however, were *N*-acetyltryptamine and *N,N*-dimethyltryptamine, rather than the monoterpenoid indole alkaloids. Added geraniol and mevalonic acid had no effect on alkaloid production, but secologanin and tryptamine added in combination gave sitsirikine (**48**) and a related, unidentified metabolite. Using the C_{30} cell line of *C. roseus*, Guern and co-workers (*142*) studied the accumulation of tryptamine and concluded that this was due at least in part to the pH difference between the vacuole and the culture medium. *Catharanthus roseus* cells cultivated in liquid medium contained a large endogenous pool of tryptamine such that high isotopic dilution occurred when [^{14}C]tryptamine was used as a precursor. This followed work with intact plants which had led to poor incorporation into the alkaloids (*143,144*), and work of Zenk *et al.* (*145*) demonstrated that tryptamine had no effect on alkaloid production in a cell suspension system.

Both Kutney's (*146*) and Scott's (*147*) groups have investigated the effects of bioregulators on the biosynthesis of indole alkaloids in cell culture. Scott tested five known carotenoid inducers in promoting indole alkaloid formation, the idea being that these compounds were likely to enhance monoterpenoid production. Three of these, 1,1-dimethylpiperidine, 2-diethylaminoethyl-3,4-dichlorophenylether (**71**), and 2-diethylaminoethyl-β-naphthylether, increased the total alkaloid content, with concomitant increases in the levels of ajmalicine (**39**) and catharanthine (**4**), and **71** at doses in the range of 2–5 ppm gave increases in catharanthine (**4**) content from 82 to 146%. In Kutney's studies (*146*), several of the same bioregulators were used with the *C. roseus* cell line PRL 200. 1,1-Dimethylpiperidine and 2-dimethylaminoethyl-3,4-dimethylphenyl ether (**72**) were found to enhance alkaloid levels, including those of ajmalicine (**39**) and catharanthine (**4**). In the latter case, with the bioregulator **72**, the yield of catharanthine (**4**) was raised from 0.001% after 6 days to 0.009%.

The long-term production of indole alkaloids by high-yielding strains of *C. roseus* has been examined by Deus-Neumann and Zenk (*148*). Serpentine (**40**), which exhibits a bright blue fluoresence, was chosen as the alkaloid to monitor, and the level of this compound was followed in six cell

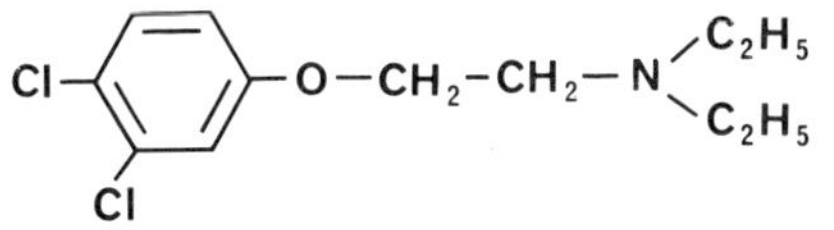

Cl, Cl–C6H3–O–CH2–CH2–N(C2H5)2

71

CH3, CH3–C6H3–O–CH2–CH2–N(CH3)2

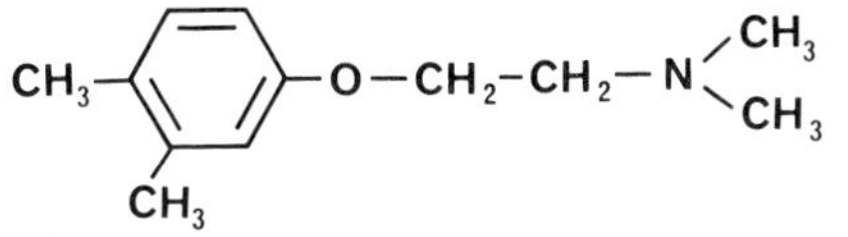

72

lines transferred every 14 days for 8 years. One of the strains produced up to 565 mg/liter of heteroyohimbine alkaloids initially, but the typical range was 311–372 mg/liter for serpentine (**40**) and ajmalicine (**39**). For one of the strains examined, alkaloid yields after only 3 years had dropped to 36 and 5 mg/liter. Clearly, this is a major obstacle to the biotechnological production of alkaloids unless a process for stabilization is developed or there is repeated clonal selection for high-yielding strains.

All of the previous studies with cell suspension cultures of *C. roseus* have led to the conclusion that not all of the cells in suspension produce alkaloids, i.e., that some differentiation occurs. Neumann and co-workers at Halle (*149*) used fluorescence and electron microscopy to show that, like the intact plants, indole alkaloid accumulation occurs in the vacuoles of particular cells. Yet there appears to be no ultrastructural difference between these cells and those which do not produce alkaloids. It had been suggested earlier that basic alkaloids were accumulated by some kind of ion trap mechanism in acidic vacuoles (*150*). Indeed, a substantial pH difference was observed between those vacuoles which do accumulate alkaloids (pH 3) and those which do not (pH 5). It was concluded that the tonoplast of the alkaloid cells seemed to be highly permeable to the neutral form of the alkaloids, but only slightly permeable to the protonated forms. Cell lines which did not exhibit a difference in their vacuolar pH did not accumulate alkaloids.

Work by Zenk (*151,152*) has focused on the localization of the alkaloids in the ajmalicine (**39**) pathway and factors affecting their transport. Serpentine (**40**), a major alkaloid of *C. roseus* cells, was found by radioimmunoassay to be located inside the vacuole whereas two key enzymes in the pathway were located in the cytosol. As a result, while protoplasts took up established indole alkaloid precursors such as tryptamine and strictosidine (**33**) and incorporated them into indole alkaloids, isolated vacuoles were incapable of producing alkaloids. The transport of the alkaloids across the tonoplast from the cytoplasm to the vacuolar space for storage was found to be sensitive to the pH of the surrounding medium, and transport was stimulated by potassium and magnesium ions. It was also an energy-requiring process which may involve ATP. Vindoline (**3**) was not bound inside the vacuoles as a nondiffusible salt but was readily exchanged with excess vindoline in the medium. This would lead to the conclusion that an ion trap mechanism is perhaps not operating in the accumulation of alkaloids in *C. roseus*.

D. Selection of Alkaloid-Producing Cell Lines

With the success achieved in demonstrating that both serially cultured callus and suspension cultures of *C. roseus* produced variable levels of alkaloids, even from highly uniform explants, it became apparent that more genotype selection was required for alkaloid production to be optimized. Kutney's and Constabel's collaborative efforts, which have been summarized (*102*), began (*153*) by describing how from 458 serially cultured cell suspensions, 312 accumulated alkaloids at a detectable level. Of these, 76.6% accumulated *Aspidosperma* alkaloids, while only 13.1% contained Corynanthe, *Strychnos*, and *Aspidosperma* alkaloids. The particular alkaloid spectrum produced by a cell line seemed to be almost independent of the source (cultivar) of the cell line, treatment with a mutagen, or the ability of the callus to root, although the *Aspidosperma* alkaloids originated only from the pink-flowering cultivar. Apparently, none of the cell lines produced any of the *Iboga* alkaloids. Studies were then initiated on the production of alkaloids from selected cell lines.

The 943 cell line, after 3 weeks of growth in a bioreactor, or in shake flasks, gave alkaloid yields of 0.44 and 0.39% dry cell weight, respectively. A broad range of alkaloids was produced, representing three of the main structure types, but, again, no *Iboga* alkaloids were detected (*154*). A slightly different spectrum of alkaloids was detected in the cell line 953 (*155,156*). After 21 days the shake flask method afforded higher (0.37%) yields of alkaloids than the bioreactor and was optimized at about 0.63% after 6 weeks. Paralleling the earlier (*157*) work of Scott *et al.* on the appearance of alkaloids in *C. roseus* seedlings, it was found that the more primitive alkaloids such as ajmalicine (**39**) and yohimbine (**60**) peaked at 10–15 days, whereas the *Aspidosperma* alkaloid vindolinine (**47**) peaked at 18 days. A summary of the alkaloids produced by this cell line is given in Table XIII. Alkaloid production was found to be pH dependent (0.17% at pH 5.5 and 0.33% at pH 7.0), and incubation of the cell line with added catharanthine (**4**) and vindoline (**3**) did not yield bisindole alkaloids but rather led to intact recovery (up to 3 days) or extensive catabolism (up to 88%) after 2 weeks (*158*).

The 200GW line proved to be quite different, and of particular interest was the discovery that this line produced catharanthine (**4**) at levels about three times that of the intact plant (0.005%) (*155,159,160*). Curiously, the predominant alkaloid (60.48%) was strictosidine lactam (**41**), which is not normally seen in extracts of intact plants. Variation of the pH and added phytohormones did not significantly alter the pattern of alkaloids produced by this cell line (*160*). Further cell line studies (*161*) afforded one line (176G) which produced mainly ajmalicine (**39**) and lochnericine (**73**) and one (299Y) which apparently contained relatively inactive β-glucosidases, since the major alkaloids produced were strictosidine (**33**) (83%) and strictosidine lactam (**41**) (Table XIII).

Two groups (*162–164*) have reported on the isolation of cell lines with enhanced tryptophan levels, the notion being that such cell lines might also produce higher levels of alkaloids. Schallenberg and Berlin (*163*) looked for cell lines resistant to 5-methyltryptophan, and, although higher tryptophan and tryptamine levels were observed, alkaloid levels were not increased. Scott *et al.* (*162*) found that tryptophan synthase and anthranilate synthase from the resistant cells differed from the normal enzymes by being more resistant to feedback inhibition by tryptophan. Accumulation of alkaloids in these resistant cell lines was not observed. Cell lines resistant to 4-methyltryptophan have also been obtained (*164*) that had enhanced tryptophan decarboxylase activity and also accumulated alkaloids in the medium.

E. Cell-Free Enzyme Systems

The work by Scott and Lee (*165*) on the isolation of a crude enzyme system from a callus tissue culture of *C. roseus* was followed by studies of Zenk *et al.* on an enzyme preparation from a cell suspension system which produced indole alkaloids (*166*). The cell-free preparation was incubated with tryptamine and secologanin (**34**) in the presence of NADPH to afford ajmalicine (**39**), 19-epiajmalicine (**92**), and tetrahydroalstonine (**55**) in the ratio 1 : 2 : 0.5. No geissoschizine (**35**) was detected. In the absence of NADPH, an intermediate accumulated which could be reduced with a crude homogenate of *C. roseus* cells in the presence of NADPH to ajmalicine (**39**). Thus, the reaction for the formation of ajmalicine is critically dependent on the availability of a reduced pyridine nucleotide.

The intermediate which accumulated in the absence of NADPH was identified as 20,21-dehydroajmalicine (**76**) and given the trivial name cathenamine (*167*). It was labeled to the extent of 52% after feeding [2-^{14}C]tryptamine in the absence of NADPH, and reduction with $NaBH_4$ afforded tetrahydroalstonine (**75**). Intermediacy in the enzymatic reaction was also established through conversion to the ajmalicine isomers with

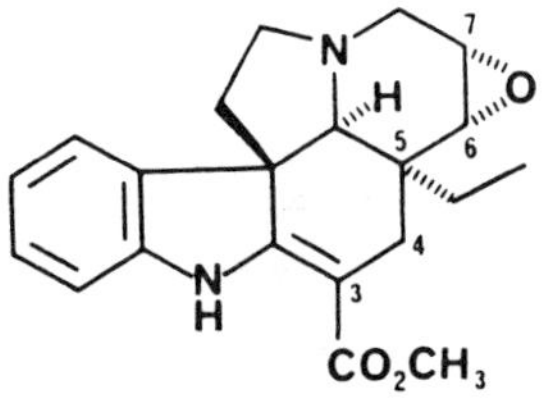

73

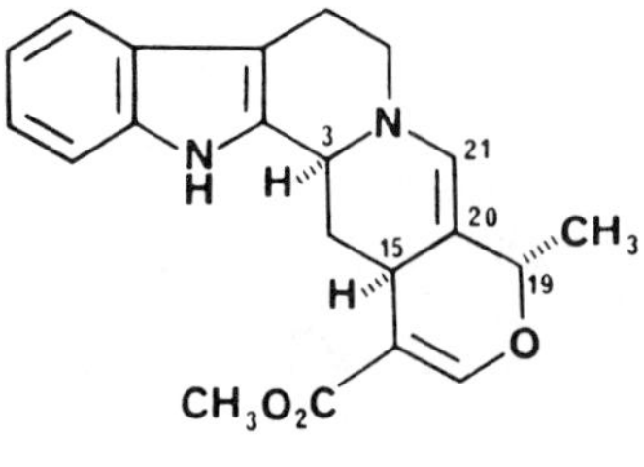

76

an enzyme preparation. Coincidentally, the same alkaloid was isolated from the leaves of *Guettarda eximia* (Rubiaceae) (*168*).

In 1978, Kutney and co-workers reported on the preparation of cell-free extracts from mature *C. roseus* plants which produced vindoline (**3**) (*169*). Leaves from the plants were homogenized in buffer in the presence of Polyclar AT, and the supernatant produced by centrifugation at 30,000 *g* was incubated with [2-^{14}C]tryptamine and secologanin for 2 hr at 34°C in the presence of FAD and NADPH. Vindoline (**3**) was isolated by TLC, diluted with cold material, and recrystallized to constant activity. Incorporations were in the range 1.10–1.36%, and no incorporation was observed when boiled enzyme preparation was used. Stemmadenine (**36**) was not incorporated into vindoline (**3**) using this preparation. These dramatic results were subsequently published in full (*170*).

Two groups have attempted to repeat these experiments (*135,171*). Zenk and co-workers (*135*) had modified vindoline in order to develop a radioimmunoassay for vindoline (**3**). The assay was sensitive in the range 0.2–45 ng, and several hundred assays could be performed in a day. The distribution of vindoline (**3**) in a single plant could be readily determined. But, more importantly, this sensitive assay was used to reexamine the cell-free formation of vindoline (**3**). Using the protocol described by Kutney *et al.* (*170*), no vindoline (**3**) was observed, although the system was capable of forming cathenamine (**76**) in high yield. This was followed by work by Stöckigt *et al.* (*171*) which employed more precisely the method described in detail in the later publication of Kutney *et al.* Interestingly, the initial purification step gave an incorporation level of 1.16%, comparable to that reported by Kutney *et al.*, but hydrolysis and esterification followed by repurification gave an incorporation level of 0.0019%. The conclusion was that a crude enzyme extract *does not* catalyze vindoline (**3**) formation from tryptamine and secologanin (**34**).

F. The Monoterpene Unit

It is well established that the iridoids are derived from two units of mevalonic acid (*97*), which itself is derived from acetyl-CoA. Mevalonate is also known to be a metabolic product of leucine (*172*), and the latter is a precursor of the monoterpene linalool (*173*). Wigfield and Wen (*174*) pursued the incorporation of leucine into the monoterpene unit in both vindoline (**3**) and catharanthine (**4**), where levels of 0.07 and 0.02%, respectively, were found, irrespective of the amount of precursor fed. This was important because, although initial results were obtained with [2-^{14}C] leucine, the specificity of incorporation was determined with 2-^{13}C-labeled precursor. Two carbons in vindoline (**3**), C-8 and C-24, were enriched to the extent of 16 and 25%, respectively, together with five other carbons, C-12, C-14, C-16, C-5, and C-23. This pattern of labeling did not support the notion of an aberrant rather than a direct precursor relationship for leucine into mevalonate in *C. roseus*. Judging from some of the previous work in which acetate also showed random incorporation into the alkaloids (*175*), the precursors of mevalonate remain elusive.

The incorporation of acetate into the monoterpene unit of the indole alkaloids has recently been reexamined (*176*). Using [1,2-$^{13}C_2$]acetate it was established that no intact incorporation occurred, and a similar labeling pattern to that from [2-$^{13}C_2$]acetate was observed, i.e., C-3, C-4, C-20, C-22, and C-23. Extensive scrambling of the acetate occurred via the Krebs cycle to label the 1 and 2 positions of acetate prior to incorporation. [2-^{13}C]Mevalonate was incorporated equally into C-17 and C-22 of ajmalicine (**39**), indicating that an equilibration occurs at some point in the pathway, as had been established previously with radiolabeled precursors (*176*).

Secologanin (**34**) is well established as the unit which condenses with tryptamine to form the first indole alkaloid, but details of the formation of **34** remain incomplete, particularly in two areas, the transformation of 10-hydroxygeraniol/nerol (**77/78**) to loganin (**79**) and the conversion of loganin (**79**) to secologanin (**34**). A group at the NRC in Canada has addressed the first of these areas in cell suspension cultures of *C. roseus*. 10-[1,1,7-$^2H_3$]Hydroxygeraniol (**77**) and 10-[1,1,7-$^2H_3$]hydroxynerol (**78**) were used as precursors of strictosidine lactam (**41**) and ajmalicine (**39**) (*177*). Mass spectrometric determination of the two products showed that the labels were at C-15 and C-21, indicating intact incorporation. Since the levels of incorporation were high, it was felt that the intermediacy of the 10-hydroxyl derivatives was established. Further work by Balsevich and Kurz (*178*), using a wide range of oxo- and hydroxygeraniol derivatives in a cell suspension culture and monitoring incorporation into ajmalicine (**39**), showed that the direct pathway involves the hydroxylation of geraniol (**80**) to 10-hydroxygeraniol (**77**) which is then further hydroxylated at C-9 to afford 9,10-dihydroxygeraniol (**81**). Subsequent steps appear to involve oxidation of the hydroxyl groups to give 9,10-dioxogeranial (**82**), in which the sequence of oxidation is thought to be C-9, C-1, C-10 based on competitive feeding experiments. One of the observations made at this time concerned the ability of the cell system to transform the precursors to other monoterpenoids, and this point has been further pursued by Balsevich (*179*) with the cell line 615. Details of the steps between 9,10-dioxogeranial (**82**) and deoxyloganic acid (**83**) are not known, but the trialdehyde **84** is regarded as a probable intermediate.

Coscia and Guarnaccia first reported the isolation of an enzyme system

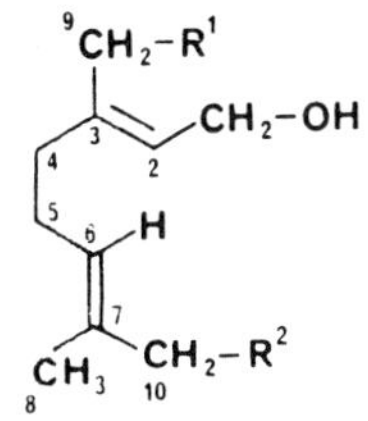

	R^1	R^2
77	H	OH
80	H	H
81	OH	OH

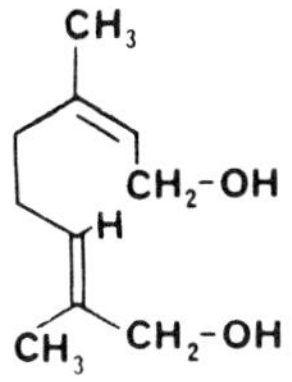

78

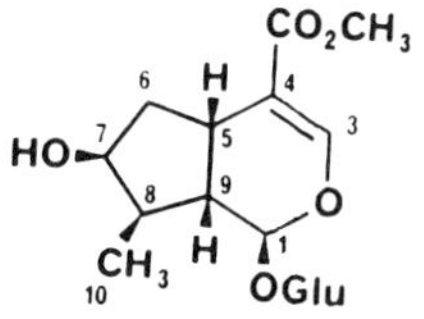

79

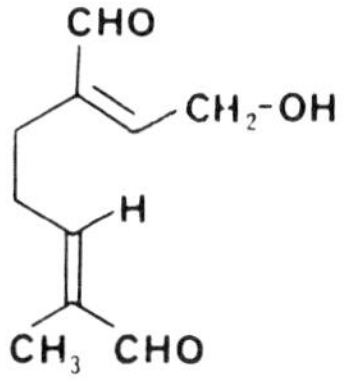

82

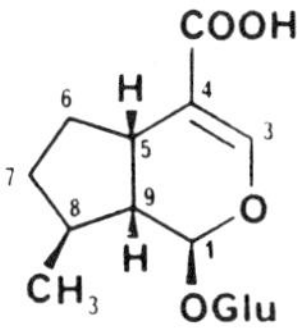

83

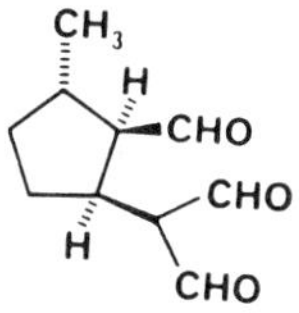

84

from *C. roseus* (*180*), an *O*-methyltransferase which catalyzed the transformation of loganic and secologanic acids to their corresponding methyl esters.

G. Strictosidine Synthase

As indicated previously, Battersby and co-workers (*181,182*) had provided evidence which suggested that a single precursor was involved in the elaboration of the alkaloids. Following the revision of the stereochemistry at C-3 in vincoside (*183–186*), however, the reality of this compound as a universal precursor was regarded as improbable (*97*), because it required inversion of the C-3 configuration with retention of the proton during the process of incorporation into the alkaloids (*187*). The matter was resolved by the work of Zenk and Stöckigt (*188,189*). They reported on a crude enzyme preparation from suspension cultures of *C. roseus* which was capable of transforming tryptamine and secologanin (**34**) to strictosidine (**33**). It was demonstrated that up to 50% of the label administered as tryptamine could be trapped in strictosidine (**33**) when δ-D-glucuronolactone was used to block ajmalicine synthesis (*188,189*). Dilution studies followed by isolation and CD analysis established that the isolated β-carboline glucoside was the compound with the 3α stereochemistry, uncontaminated with the corresponding compound, vincoside (**85**), having the 3β stereochemistry (*240,243*). Cell suspension systems of several other apocynaceous plants confirmed these findings regarding the formation of strictosidine (**33**) as the sole enzymatic condensation product of tryptamine and secologanin (**34**) (*188*).

In vivo feeding experiments with singly and doubly labeled strictosidine (**33**) in *C. roseus* shoots afforded labeled ajmalicine (**39**), serpentine (**40**), vindoline (**3**), and catharanthine (**4**). Vincoside (**85**, page 37) was not incorporated into the alkaloids, suggesting that it was biologically inert (*188*). Brown and co-workers (*190*) conducted somewhat parallel studies examining the precursor relationship of strictosidine in *C. roseus*. Incorporation into tetrahydroalstonine (**75**), ajmalicine (**39**), catharanthine (**4**), and vindoline (**3**) was observed.

The next step was to conduct precursor studies of the incorporation of strictosidine (**33**) and vincoside (**85**) into alkaloids having the 3α and 3β stereochemistries from different plants in different families. Once again Zenk's group found that only strictosidine (**33**), and not vincoside (**85**), was a precursor of α-yohimbine (**86**) and reserpiline (**87**) in *Rauwolfia canescens* and of mitragynine (**88**) and speciociliatine (**89**) in *Mitragyna speciosa* (*191,192*). A critical test involved the use of [3-^{3}H,6-^{14}C]strictosidine where the alkaloids with the 3α configuration retained the ^{3}H, whereas those with the 3β configuration [e.g., reserpiline (**87**), isoreserpiline (**90**), and speciociliatine (**89**)] substantially lost the ^{3}H label at C-3 (*192*).

Following Zenk's demonstration that strictosidine (**33**) is the key precursor of the indole alkaloids in a variety of plants in the family Apocynaceae, as well as several plants in other families, Battersby *et al.* (*193*) repeated experiments with both the 3α and 3β isomers of the primordial glucoside. The earlier results were not confirmed, for it was the 3α isomer which was specifically incorporated. Thus, the cycle of experiments is now complete for the involvement of strictosidine as the key intermediate in both enzyme systems and intact plants.

The enzyme responsible for the stereospecific condensation of tryptamine and secologanin (*34*) was called strictosidine synthase, and its presence was demonstrated by Treimer and Zenk (*194*) in a number of indole alkaloid-producing plants, including *Amsonia salicifolia, Catharanthus roseus, Ochrosia elliptica, Rauwolfia vomitoria, Rhazya orientalis, Stemmadenia tomentosa, Vinca minor,* and *Voacanga africana*. Enzyme activity as high as 1698 pkat/mg protein was observed for *O. elliptica*. No

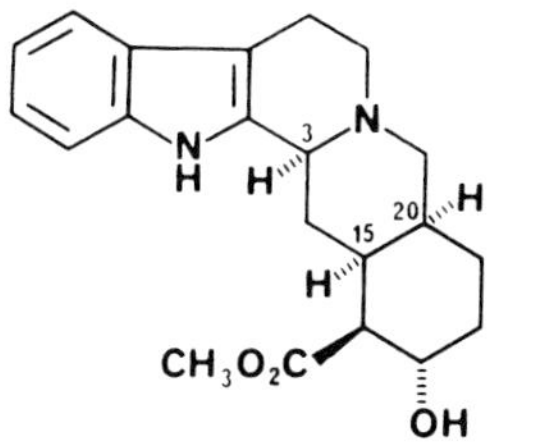

86

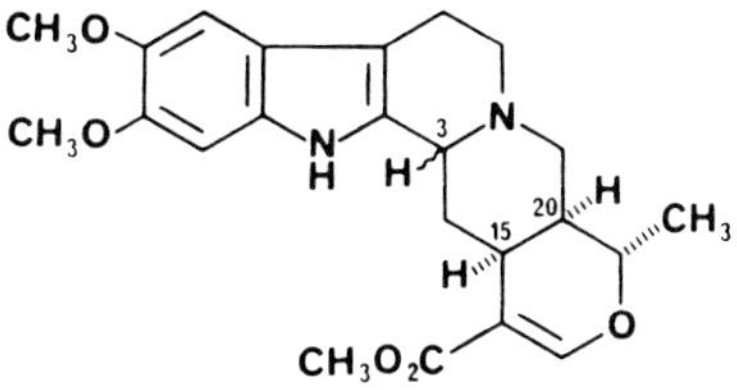

87 C-3Hβ
90 C-3Hα

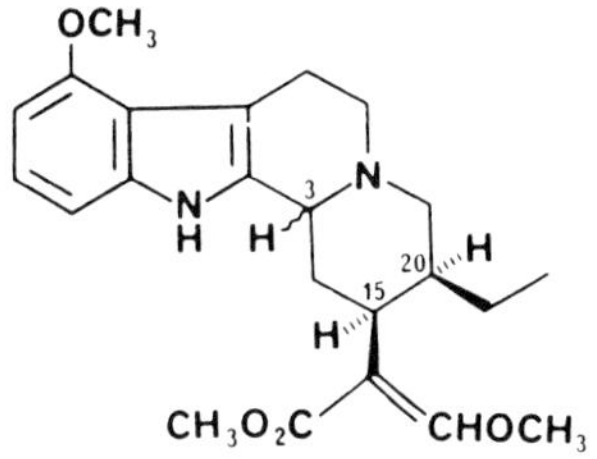

88 C-3Hα
89 C-3Hβ

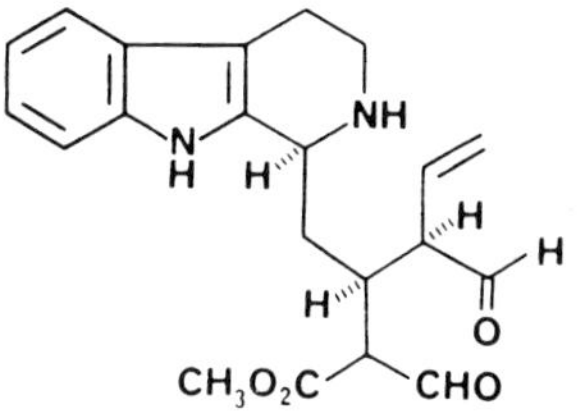

93

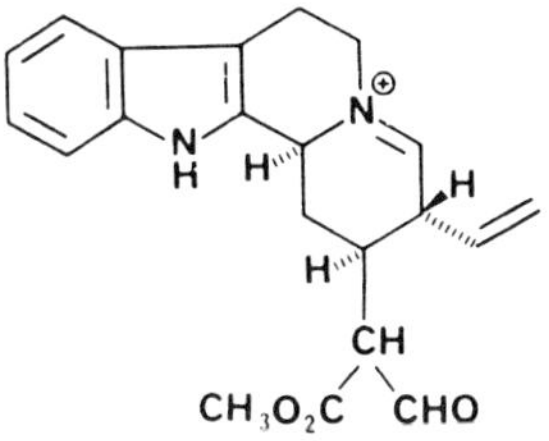

94

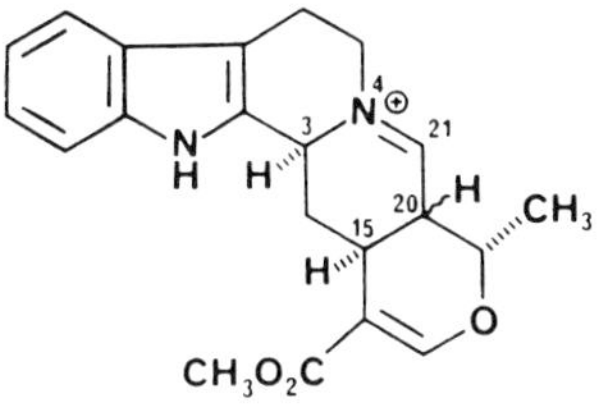

97

enzyme activity was observed for *Nicotiana tabacum* and *Trifolium pratense*, which do not produce indole alkaloids. A number of optimum conditions were established for the enzyme including temperature (45°C), pH (6.5 in phosphate buffer), and substrate concentration (2.6 m*M* for secologanin and 5.8 m*M* for tryptamine). Strictosidine synthase from seven different plant species ranged in molecular weight from 26,000 to 33,000.

Scott's group has also reported on the isolation of strictosidine synthase from *C. roseus* cell cultures (*195*). The enzyme, which they purified through $(NH_4)_2SO_4$ precipitation, DEAE-cellulose, hydroxylapatite, and Sephadex G-75 chromatography, and finally isoelectric focusing, showed an activity of 5.9 nkat/mg of protein. The molecular weight was found to be 38,000 with a pH optimum in the range 5.0–7.5.

Continuing work by Treimer and Zenk (*196*) yielded an enzyme purified about 50-fold, with recovery being about 9%. K_m values of 2.3 and 3.4 m*M* were found for tryptamine and secologanin, respectively, whereas Scott *et al.* (*195*) had found values of 0.46 and 0.83 m*M*, respectively. Substrate specificity using a number of different aromatic and iridoid compounds was evaluated, and only 7-methyltryptamine (15% of tryptamine activity) and 2′- and 3′-*O*-methylsecologanins (**34** and dihydrosecologanin (**91**) were effective substrates; D- and L-tryptophans and secologanic acid were not. A molecular weight of about 34,000 was estimated, with some variation (26,000–34,000) being noted between species. The Stokes' radius also varied between species (2.10–2.75 nm), as did the optimal pH for catalysis. More tropical species, such as *R. vomitoria, R. orientalis,* and *V. africana,* showed a higher temperature optimum (50°C). No evidence was found for the formation of any vincoside (**85**) during the course of these studies when its potential formation was monitored through the analysis of its decomposition product, vincoside lactam (**92**).

Pfitzner and Zenk (*197*) were successful in immobilizing strictosidine synthase on Sepharose following a modified enzyme purification process. Remarkably, the enzyme demonstrated superior stability (325-fold) compared with the soluble enzyme, having a half-life of about 68 days at 37°C, and 45% activity was observed after 2 years of storage at 4°C. Almost quantitative conversion of tryptamine and secologanin (**34**) to give strictosidine (**33**) was observed over prolonged periods (12 days) to yield multigram quantities of **33**. The potential therefore exists for the large-scale production of this important compound.

H. Formation of Ajmalicine

Although ajmalicine (**39**) is not on the pathway to the bisindole alkaloids, it is a compound of substantial commercial interest, and several of the intermediates in its formation are probable intermediates in the extended biosynthetic pathway. This work is therefore reviewed for the purpose of completeness of studies on *C. roseus*. Considerable progress has been made on the biosynthesis of ajmalicine (**39**), and the studies on the formation of strictosidine (**33**) and cathenamine (**76**) have already been described. One of the preparations described by Scott and Lee was a supernatant from a suspension of young seedlings of *C. roseus* which afforded ajmalicine (**39**) (*165*). The second preparation was a supernatant obtained from leaf and stem callus tissue. This latter preparation showed an 18% incorporation of [2-^{14}C]tryptamine into ajmalicine (**39**) and also allowed the demonstration that secologanin (**34**) was well incorporated into ajmalicine (**39**) and geissoschizine (**35**). The formation of biosynthetically more advanced metabolites was not observed in this preparation. Following the initial work (*165*) with a crude enzyme system, further fractionation work using a *p*-nitrophenyl-β-glucoside assay led to the isolation of four β-glucosidase enzymes, one of which was capable of the overall transformation of tryptamine and secologanin (**34**) to ajmalicine (**39**) in the presence of NADPH. Attempted additional purification led to loss of the alkaloid-synthesizing activity (*198,199*).

The next area for study was the pathway between strictosidine (**33**) and cathenamine (**76**), where the initial step is viewed as hydrolysis by β-glucosidase and opening of the hemiacetal to a dialdehyde **93**. Attempts to trap this intermediate (*200*) have thus far failed, and it appears that recyclization and dehydration occur too rapidly, giving rise to 4,21-dehydrocorynantheine aldehyde (**94**). Thus, an enzyme preparation was incubated at pH 7.0 in the presence of KBH_4 to afford the C-16 epimers **69** and **95**, thereby impuning the existence of **94**.

Using a radioimmunoassay (RIA) for ajmalicine (**39**) (*201*), Zenk and co-workers selected plants and cells in culture capable of producing high yields of ajmalicine (**39**) and 19-epiajmalicine (**74**) from tryptamine and secologanin (**34**) (*202*). The assay again demonstrated the importance of a β-glucosidase in the overall process and the involvement of a NADPH-dependent enzyme which reduces cathenamine (**76**). Thus, it was concluded that the RIA had the same ability to examine biosynthetic processes as radiotracer analysis. Somewhat surprisingly, the crude enzyme system was capable of carrying out the condensation of a variety of substituted tryptamines with secologanin (**34**). The principal disadvantage to using RIA in biosynthesis was concluded to be the time-consuming process of establishing the RIA.

The glucosidases involved in the biosynthetic pathway have been studied in detail by Hemscheidt and Zenk (*203*), and two of the isolated enzymes were specific for the hydrolysis of strictosidine (**33**) and were purified 120-fold. The enzyme was isolated from a number of indole alkaloid producing plants, including *C. roseus*, *C. pusilus*, and *C. trichophyllus*. The pH optimum was 6.5, and the K_m values were 0.2 m*M* for enzyme I and 0.1 m*M* for enzyme II. Molecular weights were estimated at 230,000 for enzyme I and 450,000 for enzyme II. Unlike the case of the enzyme system of Scott *et al.* (*198,199*), tryptamine did not activate either of the two enzymes. The enzymes were highly substrate specific; vincoside (**85**) was hydrolyzed at about 10% of the rate of strictosidine (**33**), and secologanin (**34**), strictosidine lactam (**41**), and vincoside lactam (**92**) were not hydrolyzed. Zenk and Hemscheidt regard the isolation of an enzyme complex by Scott which carries out the formation of ajmalicine (**39**) as "fortuitous" and the naming of the enzyme complex as ajmalicine synthase as without foundation (*203*).

The enzyme system of Scott and Lee from *C. roseus* was reported (*165*) to synthesize geissoschizine (**35**) and to convert this alkaloid to ajmalicine (**39**), supporting the earlier work (*204*) on the intermediacy of geissoschizine (**35**) using intact plants. However, Zenk *et al.* found that a cell-free system of *C. roseus,* which produced substantially more of the ajmalicine isomers than the intact plant, did not produce geissoschizine (**35**) (*166*). On reinvestigating this precursor situation (*205*), Stöckigt found a 0.12% incorporation of geissoschizine (**35**) into ajmalicine (**39**) in intact plants. But using a cell-free enzyme system the rate of incorporation of tryptamine into ajmalicine (**39**) was much higher than that of geissoschizine (**35**). This led Stöckigt to suggest that geissoschizine was not a true intermediate in the pathway.

The optimum concentration of geissoschizine (**35**) for conversion to ajmalicine (**39**) and 19-epiajmalicine (**74**) was determined using an RIA, and this allowed an experiment to be conducted in which tryptamine and secologanin (**34**) were incubated in the presence of geissoschizine (**35**). Since the formation of the ajmalicine isomers from tryptamine and secologanin (**34**) and from geissoschizine (**35**) was additive, the conversion of geissoschizine (**35**) to the isomers must be at least partly independent of their synthesis from tryptamine and secologanin (**34**). Additional experiments demonstrated that this set-up conversion requires both $NADP^+$ and NADPH, but the conversion of tryptamine and secologanin (**34**) to the ajmalicine isomers is independent of $NADP^+$, further evidence that geissoschizine (**35**) is not a true intermediate in the biosynthesis of ajmalicine (**39**). How, then, is geissoschizine (**35**) converted to 19-epiajmalicine (**74**)? Stöckigt *et al.* examined this question (*206*) and concluded that geissoschizine (**35**) is oxidized by $NADP^+$ to a 4,21-dehydro derivative **96** (*207*), followed by cyclization to the cathenamine isomers (at C-19) and reduction to the ajmalicine isomers.

Scott and co-workers have reached exactly the opposite conclusion (*208*). Reduction of cathenamine (**76**) with $NaBH_4$ gave 19β-H products ajmalicine (**39**) and tetrahydroalstonine (**75**) (*167*). Enzymatic (*189*) incubation, however, gave 19-epiajmalicine (**39**), which is a new natural product. Was the enzyme system disturbed during the isolation? Using their previously reported cell-free system, the incorporation of [2-^{14}C]tryptamine and [*aryl*-^{3}H]-geissoschizine (**35**) into ajmalicine (**39**) was studied. No significant amount of 19-epi-ajmalicine (**74**) was formed in these incubations, and it was therefore concluded that cyclization of the E ring occurs most rapidly by proton attack at C-20 on the *si* face. The question of the intermediacy of cathenamine (**76**) was not resolved by these experiments.

For the biosynthetic conversion of cathenamine (**76**) to the 19- and 20-epi derivatives, an equilibrium should exist between the enamine and iminium forms of cathenamine (i.e., **76** and **97**) (*167*). This was examined (*209*) with deuterium labeling studies of incorporation into tetrahydroalstonine (**75**), whereupon C-21 was labeled from both the enamine and iminium forms. When the enamine form was present, a second deuterium was incorporated (presumably at C-20) on reduction with $NaBD_4/D_2O$. Sulfate was effective in pushing the equilibrium toward the iminium species (*209*).

Incubation of geissoschizine (**35**) with a cell-free extract from *C. roseus* (*210*) in the presence of NADPH caused the accumulation of an isomer of isositsrikine whose structure was established chemically to be the (16*R*) isomer **58**. None of the 16-epi isomer **95** was detected in the cell-free incubations or in feeding experiments with intact plants. Additionally, Stöckigt has reviewed enzymatic studies on the formation of strictosidine (**33**) and cathenamine (**76**) (*211*), and Zenk has provided a very elegant summary of the enzymatic synthesis of ajmalicine (**39**) (*212*).

The work of Zenk and co-workers on the immobilization of strictosidine synthase was mentioned previously. The same group also reported the immobilization of a crude cell preparation of *C. roseus* capable of transforming tryptamine and secologanin (**34**) into a mixture of the ajmalicine isomers (*213*). Cells were immobilized on alginate beads, and a solution of the precursors, including radiolabeled tryptamine, was circulated through the column. After about 90 hr, 10% of the product was a mixture of ajmalicine (**39**), 19-epiajmalicine (**74**), and tetrahydroalstonine (**75**), indicating that the endogenous NADPH is recycled within the entrapped cells. In addition, unlike the case of cell suspension systems where secondary metabolites are held within the vacuoles, in this instance of immobilization, a majority of the radioactivity (85%) was recovered. It was postulated that the release of the metabolites could be due to traces of chloroform used for the continuous extraction of the product.

I. Formation of Vindoline

Almost nothing is known about the biosynthetic pathway between a Corynanthe intermediate and the first *Aspidosperma* alkaloid tabersonine (**38**) and, indeed, the first *Iboga* alkaloid catharanthine (**4**). Thus, the focus of further work has been the intermediates involved in the pathway between **38** and vindoline (**3**). Two groups have examined these enzymes. Stöckigt *et al.* have isolated two of the enzymes involved in this pathway, acetyl CoA:4-*O*-deacetylvindoline 4-*O*-acetyltransferase (*214*) and *S*-adenosyl-L-methionine:16-*O*-demethyl-4-*O*-deacetylvindoline 16-*O*-methyltransferase (*215*). Given the previous discussion on the formation of vindoline (**3**) in cell suspension cultures, any biochemical information regarding the biosynthesis of vindoline (**3**) is important if the goal of vinblastine (**1**) formation is to be realized. Success in this area was the result of a different strategy by Stöckigt. If it is accepted, based on the prior whole plant feeding studies, that tabersonine (**38**) is the *Aspidosperma* alkaloid precursor of vindoline (**3**), the steps involved may be summarized as an aromatic hydroxylation and O-methylation, N_a-methylation, 3,4-dihydroxylation, and O-acetylation.

The last enzyme in the pathway, acetyl-CoA:4-*O*-deacetylvindoline 4-*O*-acetyltransferase, was obtained (*214*) in a preparation from *C. roseus* leaves following ammonium sulfate precipitation and gel chromatography. The acetylation reaction was critically dependent on the presence of acetyl-CoA as a cofactor, and high substrate specificity was demonstrated with a variety of closely related *Aspidosperma* alkaloids. Only a small number of related compounds were acetylated with this enzyme, namely, 4-*O*-deacetyl-N_a-demethylvindoline (**98**), 4-*O*-deacetylvindorosine (**99**), N_a-demethyldeacetylvindorosine (**100**), and 16-*O*-demethyl-4-*O*-deacetylvindoline (**101**). Neither deacetylvindoline (**42**) nor 4-hydroxytabersonine (**102**) were substrates. The probable next to last enzyme in the pathway is an *O*-methyltransferase capable of methylation of the 16-hydroxyl group. At pH 8 16-*O*-demethyl-4-*O*-deacetylvindoline (**101**) as the substrate in the presence of radiolabeled *S*-adenosyl-L-methionine afforded 4-*O*-deacetylvindoline (**42**). 16-*O*-Demethylvindoline (**103**) was not a substrate and neither was 4-*O*-deacetyl-N_a-demethylvindoline (**98**), establishing that O-methylation precedes O-acetylation, and that the *N*-methyl- and *O*-methyltransferases are different.

DeLuca and co-workers have reached somewhat different conclusions. In support of their pathway they have isolated an *N*-methyltransferase which catalyzes the N_a-methylation of 16-methoxy-2,3-dihydro-3-hydroxytabersonine (**104**) to 4-deacetoxyvindoline (**43**) (*216*). The *N*-methyltransferase activity was found to co-chromatograph with chlorophyll on Sephadex 100, but subsequent centrifugation gave a supernatant which possessed the *N*-methyltransferase activity and a pellet which contained the chlorophyll. Highest substrate specificity was shown by **105** (0.07 pkat/100 μl enzyme), followed by 2,3-dihydrotabersonine (**106**). 4-*O*-Deacetyl-N_a-demethylvindoline (**98**) was not a substrate, but, unfortunately, 16-*O*,N_a-bisdemethyl-4-*O*-deacetylvindoline (**107**), which is a key intermediate in Stöckigt's proposed pathway, was not tested (*216*).

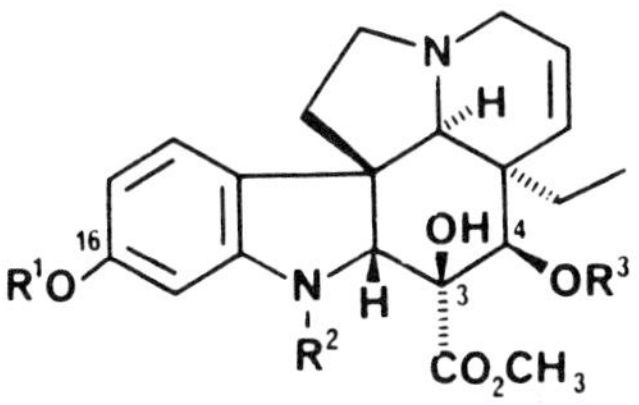

	R^1	R^2	R^3
98	CH_3	H	H
101	H	CH_3	H
103	H	CH_3	$COCH_3$
107	H	H	H

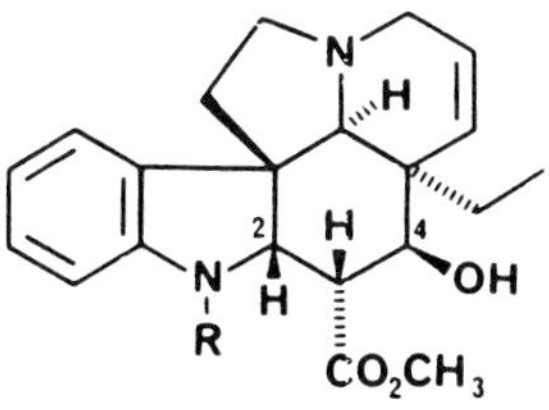

	R
99	CH_3
100	H

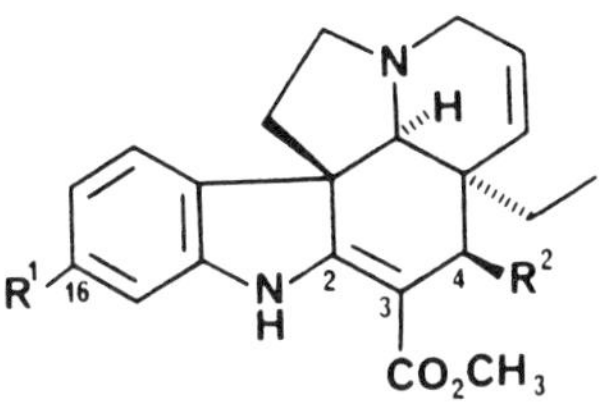

	R^1	R^2	
102	H	OH	
104	OCH_3	H	2,3-dihydro
106	H	H	2,3-dihydro

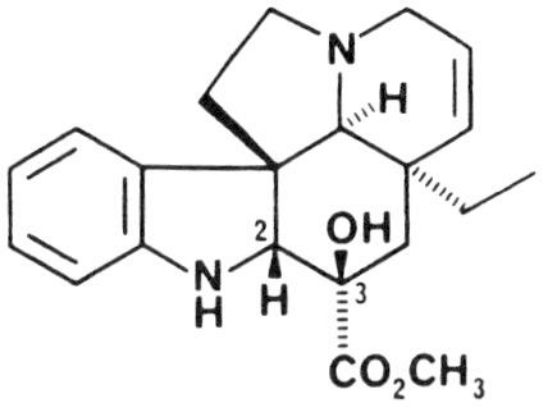

105

J. Formation of Bisindole Alkaloids

Dadonna and Hutchinson (*217*) investigated the incorporation of [3,6-$^3H_2$,*methoxy*-^{14}C]loganin into vinblastine (**1**), catharanthine (**4**), and vindoline (**3**) in whole plants of *C. roseus*. The specific incorporation into vinblastine (**1**) was in accord with intact incorporation into both units. Interpretation of the experiment is so complex, however, that further conclusions regarding intermediates or the stereospecificity of the processes are extremely speculative.

The only study of the biosynthesis of bisindole alkaloids in intact plants using advanced precursors was reported by Scott and co-workers in 1978

(*87*). Blank incubations of vindoline (**3**) and catharanthine (**4**) at pH 6–8 indicated the trace production of unidentified bisindole alkaloids, possibly through catharanthine *N*-oxide (**108**) as an intermediate. Separate feedings of labeled vindoline (**3**) and catharanthine (**4**) gave labeled anhydrovinblastine (**8**), and when the labeled precursors were fed concurrently and with rapid work-up, anhydrovinblastine was labeled to the extent of 2.63%. Levels of incorporation into vinblastine (**1**) were not affected by changing the work-up procedure, whereas those of anhydrovinblastine (**8**) were increased with a more rapid isolation. This suggests that anhydrovinblastine (**8**) decomposes to something other than vinblastine (**1**), and, given the known propensity of **8** to yield catharine (**10**) and leurosine (**11**) (*218*), this is perhaps not too surprising. Feeding labeled anhydrovinblastine (**8**) to *C. roseus* plants gave leurosine (**11**) above the level of the blank experiment. No incorporation was observed into vinblastine (**1**) itself, which could be due to either compartmentalization of the vinblastine-synthesizing enzymes or the instability of anhydrovinblastine (**8**).

The enzyme-catalyzed formation of anhydrovinblastine (**8**) from catharanthine (**4**) and vindoline (**3**) was first examined by Kutney and co-workers (*170,219*) using a cell-free preparation. [*aryl*-^{3}H]Catharanthine (**4**) and [*acetyl*-^{14}C]vindoline (**3**) were incubated for 3–8 hr, both separately and jointly with a preparation from *C. roseus,* which led to the isolation of labeled anhydrovinblastine (**8**) and leurosine (**11**); incorporations were of the order of 0.54 and 0.36%, respectively. On this basis, anhydrovinblastine (**8**) was proposed as the key biosynthetic intermediate en route to vinblastine (**1**) and vincristine (**2**).

More recently, Kutney and co-workers (*220*) have investigated whether the same dihydropyridinium intermediate **109** is involved in the enzymatic conversion of catharanthine (**4**) and vindoline (**3**) to anhydrovinblastine (**8**) as is involved in the chemical conversion. Use of a cell-free preparation from a 5-day culture of the AC3 cell line gave 18% of the bisindole alkaloids leurosine (**11**), catharine (**10**), vinamidine (**25**), and hydroxyvinamidine (**110**), with **10** predominating. When the incubations were carried out for only 5–10 min, the dihydropyridinium intermediate was detected followed by conversion to the other bisindole alkaloids, with FAD and $MnCl_2$ required as cofactors. Clearly a multienzyme complex is present in the supernatant, but further purification led to substantial loss of enzymatic activity. The chemical formation of anhydrovinblastine (**3**) is carried out with catharanthine *N*-oxide (**107**), but when this compound was used in the enzyme preparation described, no condensation with vindoline (**3**) occurred to give bisindole alkaloids. This has led Kutney and co-workers to suggest (*220*) that the *N*-oxide **108** is not an intermediate in the biosynthetic pathway, but rather that a 7-hydroperoxyindolenine

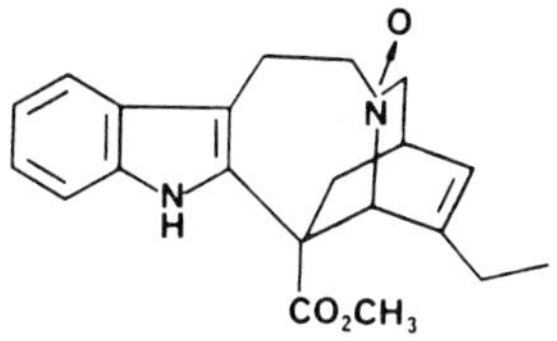

108

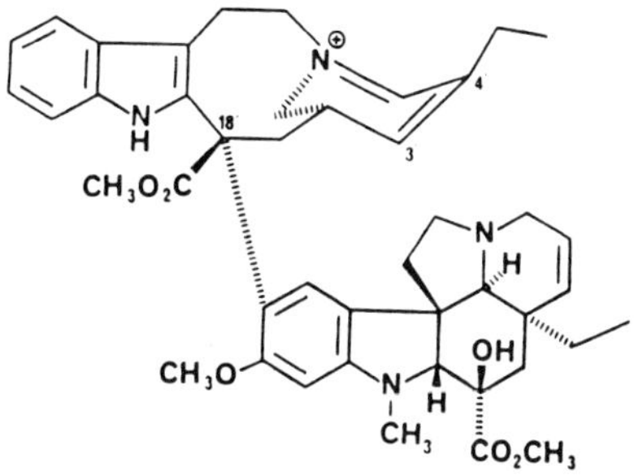

109

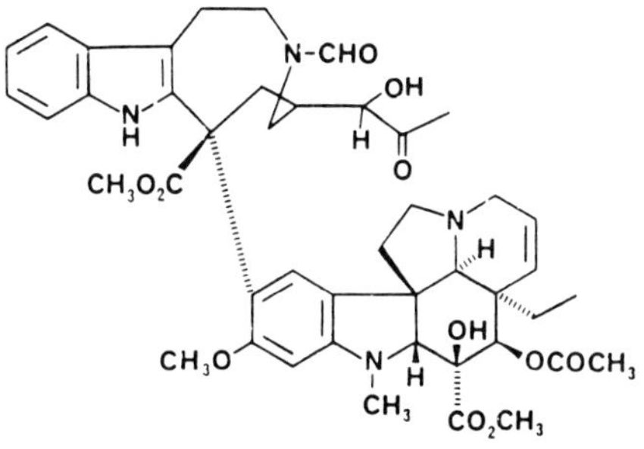

110

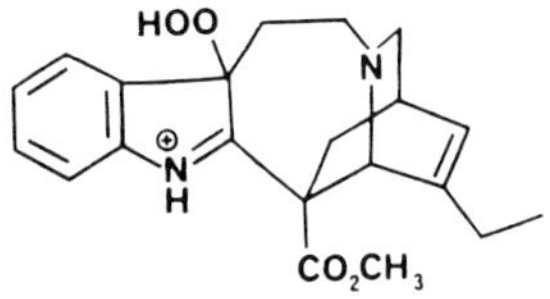

111

intermediate (**111**) is produced which then undergoes cleavage to give the same intermediate as from the chemical route for coupling with vindoline (**3**).

Detailed studies of the conditions required for the cell-free suspension system to produce the bisindole alkaloids have recently been described (*221*), which led to the conclusion that FAD or flavin mononucleotide (FMN) and $MnCl_2$ are effective in enhancing the yields of the bisindole alkaloids vinamidine (**25**) and (3*R*)-hydroxy vinamidine (**110**). Optimum conditions involved 1 m*M* $MnCl_2$, 1.0 m*M* FAD, pH 6.8, and MES buffer at 30°C. After 1.5 hr the yield of anhydrovinblastine (**8**) maximized at 25%, and longer incubations led to more highly oxidized products. The partial purification of the enzyme system which leads to anhydrovinblastine (**8**) has also been reported (*27*). After ammonium sulfate precipitation, DEAE Fractogel chromatography, Sephacryl 200 gel filtration, and isoelectric focusing, a series of four nonspecific isozymes coupling vindoline (**3**) and catharanthine (**4**) was obtained. The activity of at least one of these was enhanced in the presence of hydrogen peroxide rather than

FMN. Similar conversion yields for each of the isozymes were in the range 34–50%. Paralleling these data was the observation that horseradish peroxidase was also capable of converting vindoline (**3**) and catharanthine (**4**) to anhydrovinblastine (**8**) with the correct C-18′ stereochemistry (*222*).

Following the failure of Scott's (*223*) and Hutchinson's groups to observe incorporation of catharanthine (**4**) and vindoline (**3**) into vinblastine (**1**) in whole plants, Hassam and Hutchinson (*224*) attempted a series of experiments with apical cuttings. Some success was achieved although the levels were extremely low, being only 0.056% for catharanthine (**4**) and 0.049% for vindoline (**3**). Prophetically, it was suggested that anhydrovinblastine (**8**) might be the initial intermediate and that its presence in *C. roseus* should be determined.

The conversion of anhydrovinblastine (**8**) to vinblastine (**1**) has been examined by several different groups, using intact plants, cell suspension systems, and cell-free preparations. From the studies discussed above it was clear that 3′,4′-anhydrovinblastine (**8**) was probably the initially formed intermediate in the condensation of vindoline (**3**) and catharanthine (**4**) prior to vinblastine (**1**). Kutney and co-workers have reported (*225,226*) on the biotransformation of 3′,4′-anhydrovinblastine (**8**) using cell suspension cultures of the 916 cell line from *C. roseus;* a line which did not produce the normal spectrum of indole alkaloids. After 24 hr the major alkaloid products were leurosine (**11**) and catharine (**10**) in 31 and 9% yields, respectively, with about 40% of the starting alkaloid consumed.

When 3′,4′-[*aryl*-^{3}H]anhydrovinblastine (**8**) was incubated with a cell-free preparation at pH 6.3 for 50 hr, leurosine (**11**) and catharine (**10**) were labeled to the extent of 8.15 and 15.15%, respectively, and vinblastine (**1**) was labeled to 1.84% (*170,227*). Approximately the same level of incorporation was obtained by Scott's group, using 3′,4′-[21′-^{3}H]anhydrovinblastine (**8**) and isolating vinblastine (**1**) from cell-free extracts of *C. roseus* (*228*). Scott's failure (*87*) to observe incorporation of the same precursor into vinblastine (**1**) in whole plants was explained by the established (*82*) instability of anhydrovinblastine (**8**).

The question of the possible artifactual nature of leurosine (**11**) was examined by Kutney and co-workers (*229*) using a cell-free preparation at pH 6.3. After a 3-hr incubation, a 22% yield of leurosine (**11**) was formed from anhydrovinblastine (**8**), and when anhydrovinblastine (**8**) was incubated with horseradish peroxidase in the presence of H_2O_2, leurosine (**11**) was formed in 65% yield.

Guéritte and co-workers (*230*), in examining the role of a potential late precursor of vinblastine (**1**), showed that 4′-[*aryl*-^{3}H]- and 4′-[*acetyl*-^{14}C]deoxyleurosidines (**17**) were incorporated into vinblastine (**1**) in intact

C. roseus plants to the extent of 0.3–0.6%. Although the experiments do not indicate that **17** is in the pathway, since the conversion of **17** to vinblastine (**1**) did not occur *in vitro,* it is established that vinblastine (**1**) is a true alkaloid.

To examine whether a dihydropyridinium intermediate is involved in the conversion of anhydrovinblastine (**8**) to vinblastine (**1**), Scott *et al.* fed [5′α-^{3}H,*methyl*- ^{14}C]anhydrovinblastine (**8**) to a cell-free suspension system of *C. roseus* (*231*). The idea was that if a dihydropyridinium intermediate was formed at some stage as a result of a biosynthetic grid being established, then the 5′α-H should be lost. In fact, vinblastine was produced in the cell-free system without loss of the label, indicating that a trans elimination from an intermediate *N*-oxide does not occur. Three alternative pathways for the conversion of anhydrovinblastine (**8**) to vinblastine (**1**) are possible: (i) direct hydration of the 3′,4′-double bond, (ii) reduction of anhydrovinblastine (**8**) to 4′-deoxyvinblastine (**16**) and hydroxylation with retention of configuration, or (iii) reduction of anhydrovinblastine (**8**) to 4′-deoxyleurosidine (**17**) and hydroxylation with inversion of configuration. In a further study (*232*), [5′α-^{3}H]anhydrovinblastine (**8**) was incorporated into vinblastine in a cell-free system (EU4A) of *C. roseus* to the extent of 0.49% even though the cell system was not normally capable of producing the bisindole alkaloids.

The extremely low yield of vincristine (**2**) from intact plants has made pursuit of its biosynthesis a very challenging problem, which at this point in time remains unsolved. Kutney *et al.* have used both anhydrovinblastine (**8**) (*227*) and catharanthine *N*-oxide (**107**) (*233*) as precursors to vincristine (**2**) in a cell-free preparation, but incorporation levels were extremely low. Therefore, the question of whether vinblastine (**1**) is an *in vivo,* as well as an *in vitro,* precursor remains to be answered. Several possibilities exist for the overall oxidation of vinblastine (**1**) to vincristine (**2**), including a direct oxidation of the *N*-methyl group or oxidative loss of the *N*-methyl group followed by N-formylation.

IV. Summary

Higher plants are a very important source of medicinally useful compounds. *Catharanthus roseus* is one of the most important of these plants, and this chapter focuses on the further isolation work directed at the identification of new potentially more active and/or less toxic bisindole alkaloids. In addition, the biosynthesis of the indole alkaloids of *C. roseus* is reviewed. While the former area of research has been dominated by sophisticated high-field NMR and high-resolution mass spectral analyses,

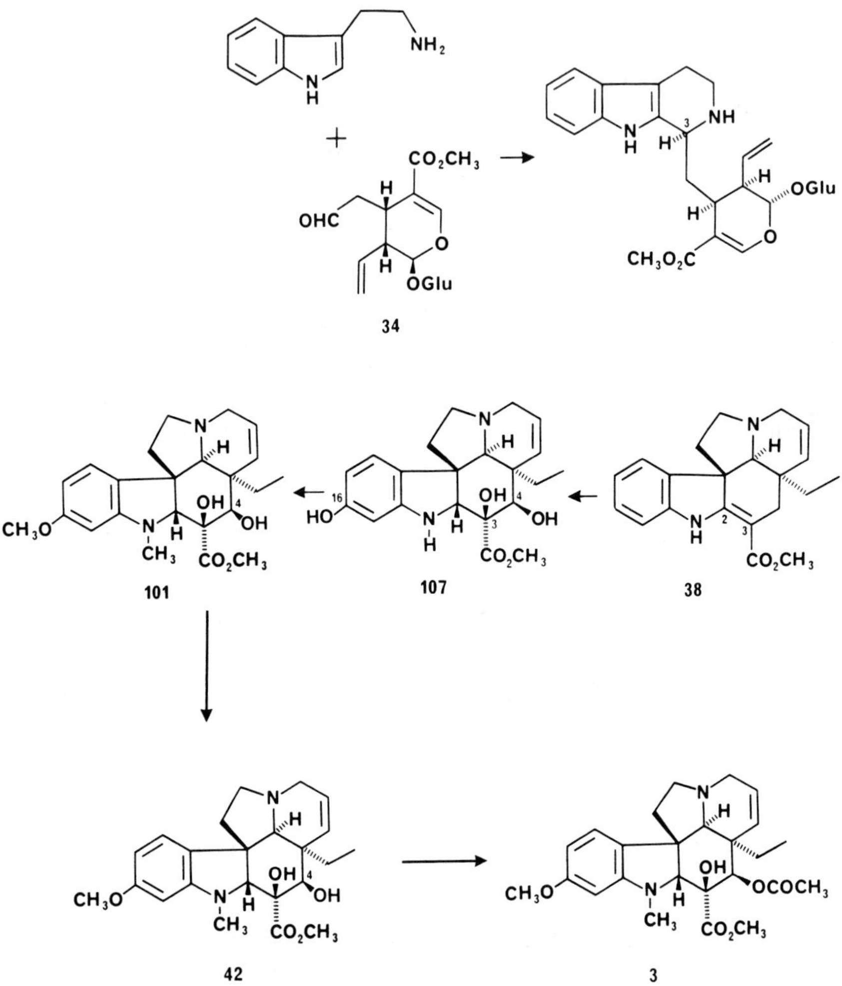

SCHEME 8. Current biogenetic ideas for the bisindole alkaloids of *Catharanthus roseus*.

the latter area has moved to the realm of biotechnology where the exploitation of plant cells in culture and the enzymes derived from them are a crucial strategy for the successful production of secondary metabolites (*234*). As a result of these extensive studies it is now possible to propose an improved biosynthetic pasthway for the formation of vindoline (**3**), catharanthine (**4**), and the bisindole alkaloids, as shown in Scheme 8.

SCHEME 8. (*continued*)

Acknowledgments

Work in the authors' laboratory was supported, in part, by a grant from the Division of Cancer Treatment, National Cancer Institute, Bethesda, Maryland. One of us (G.B.) was on leave from the Central Research Institute for Chemistry, Hungarian Academy of Sciences, Budapest, Hungary, during the preparation of the mansuscript.

REFERENCES

1. W.I. Taylor and N. R. Farnsworth, eds., "*Catharanthus* Alkaloids: Botany, Chemistry, Pharmacology and Clinical Uses." Dekker, New York, 1975.
2. N. R. Farnsworth, *Lloydia* **24**, 105 (1961).

3. G. H. Svoboda, I. S. Johnson, M. Gorman, and N. Neuss, *J. Pharm. Sci.* **51**, 707 (1962).
4. N. Neuss, *Bull. Chim. Fr.*, 1509 (1963).
5. I. S. Johnson, J. G. Armstrong, M. Gorman, and J. P. Burnett, *Cancer Res.* **23**, 1390 (1963).
6. N. Neuss, I. S. Johnson, J. G. Armstrong, and C. J. Jensen, *Adv. Chemother.* **1**, 133 (1964).
7. P. Obrecht, *Chemotherapy Cancer, Proc. Int. Symp., Lugano, Switzerland,* 165 (1964).
8. G. H. Svoboda, *Proc. 1st G. E. C. A. Symp., Paris,* 9 (1965).
9. W. I. Taylor, *in* "The Alkaloids" (R. H. F. Manske and H. L. Holmes, eds.), Vol. 8, p. 269. Academic Press, New York, 1965.
10. G. H. Svoboda, *Abh. Dtsch. Akad. Wiss. Berlin,* 465 (1966).
11. K. Stolle and D. Gröger, *Pharm. Zentral. Dtsch.* **106**, 185 (1967).
12. M. Manzoor-I-Khuda, *Sci. Ind.* **5**, 137 (1967); *Chem. Abstr.* **68**, 93467z (1968).
13. W. I. Taylor, *in* "The Alkaloids" (R. H. F. Manske and H. L. Holmes, eds.), Vol. 11, p. 99. Academic Press, New York, 1968.
14. O. F. Offelie, *Pharm. Weekbld.* **104**, 321 (1969).
15. G. H. Svoboda, *in* "Current Topics in Plant Science" (J. E. Gunckel, ed.), p. 303. Academic Press, New York, 1969.
16. G. H. Svoboda, *in* "Pharmacognosy and Phytochemistry" (H. Wagner and L. Hörhammer, eds.). Springer-Verlag, New York, 1971.
17. K. Wilms, *Planta Med.* **22**, 324 (1972).
18. I. Hansen and N. I. Nissen, *Ugeskr. Laegr.* **134**, 1315 (1972); *Chem. Abstr.* **77**, 134986 (1972).
19. J. P. Kutney, *Lloydia* **40**, 107 (1977).
20. J. M. Cassady and J. D. Douros eds., "Anticancer Agents Based on Natural Product Models." Academic Press, New York, 1980.
21. G. A. Cordell, *Korean J. Pharmacog.* **12**, 65 (1981).
22. G. A. Cordell and J. E. Saxton, *in* "The Alkaloids" (R. H. F. Manske and R. G. A. Rodrigo, eds.), Vol. 20, p. 1. Academic Press, New York, 1981.
23. G. A. Cordell, *in* "Indoles: Monoterpenoid Indole Alkaloids" (J. E. Saxton, ed.), p. 539. Wiley, New York, 1983.
24. M. Suffness and G. A. Cordell, *in* "The Alkaloids" (A. Brossi, ed.), Vol. 25, p. 1. Academic Press, New York, 1985.
25. G. Blaskó and G. A. Cordell, *in* "Economic and Medicinal Plant Research" (H. Wagner, H. Hikino, and N. R. Farnsworth, eds.), Vol. 2, p. 119. Academic Press, New York, 1988.
26. W. T. Stearn, *Lloydia* **29**, 169 (1966).
27. A. E. Goodbody, C. D. Watson, C. C. S. Chapple, J. Vukovic, and M. Misawa, *Phytochemistry* (in press).
28. A. El-Sayed and G. A. Cordell, *J. Nat. Prod.* **44**, 289 (1981).
29. G. H. Svoboda, M. Gorman, N. Neuss, and A. J. Barnes, Jr., *J. Pharm. Sci.* **50**, 409 (1961).
30. J. P. Kutney, J. Balsevich, and B. R. Worth, *Heterocycles* **9**, 493 (1978).
31. Atta-ur-Rahman, I. Ali, and M. I. Chaudhary, *Z. Naturforsch.* **40B**, 543 (1985).
32. N. Langlois and P. Potier, *Phytochemistry* **11**, 2617 (1972).
33. P. Rasoanaivo, N. Langlois, and P. Potier, *Phytochemistry* **11**, 2616 (1972).
34. A. De Bruyn, L. De Taeye, R. Simonds, M. Verzele, and C. De Pauw, *Bull. Soc. Chim. Belg.* **91**, 75 (1982).

35. N. Neuss, A. J. Barnes, Jr., and L. L. Huckstep, *Experientia* **31**, 18 (1975).
36. N. Neuss and A. J. Barnes, Jr., U.S. Patent 3,954,773 (1975); *Chem. Abstr.* **85**, 108887 (1976).
37. G. H. Svoboda and A. J. Barnes, Jr., *J. Pharm. Sci.* **53**, 1227 (1964).
38. R. Simonds, A. De Bruyn, L. De Taeye, M. Verzele, and C. De Pauw, *Planta Med.* **50**, 274 (1984).
39. G. Richter Co., Fr. Patent 2,210,392 (1974); *Chem. Abstr.* **82**, 116076 (1975).
40. G. Richter Co., Ger. Patent 2,259,388 (1974); *Chem. Abstr.* **81**, 82369m (1974).
41. G. Richter Co., Neth. Patent Appl., 17,069 (1972); *Chem. Abstr.* **83**, 84848 (1975).
42. W. E. Jones, Ger. Patent 2,442,245 (1975); *Chem. Abstr.* **83**, 120821d (1975).
43. N. Neuss, M. Gorman, N. J. Cone, and L. L. Huckstep, *Tetrahedron Lett.*, 783 (1968).
44. S. S. Tafur, W. E. Jones, D. E. Dorman, E. E. Logsdon, and G. H. Svoboda, *J. Pharm. Sci.* **64**, 1953 (1975).
45. G. H. Svoboda, *Lloydia* **24**, 173 (1961).
46. S. Mukhopadhyay and G. A. Cordell, *J. Nat. Prod.* **44**, 611 (1981).
47. N. Neuss, M. Gorman, G. H. Svoboda, G. Maciak, and C. T. Beer, *J. Am. Chem. Soc.* **81**, 4754 (1959).
48. R. L. Noble, C. T. Beer, and J. H. Cutts, *Biochem. Pharmacol.* **1**, 347 (1958).
49. S. Kohlmunzer and H. Tomczyk, *Diss. Pharm. Pharmacol.* **19**, 403 (1967).
50. N. R. Farnsworth, *J. Pharm. Sci.* **61**, 1840 (1972).
51. M. Tin-Wa, N. R. Farnsworth, H. H. S. Fong, and J. Trojanek, *Lloydia* **33**, 261 (1970).
52. A. El-Sayed, G. A. Handy, and G. A. Cordell, *J. Nat. Prod.* **43**, 157 (1980).
53. A. El-Sayed, G. A. Handy, and G. A. Cordell, *J. Nat. Prod.* **46**, 517 (1983).
54. G. H. Svoboda, I. S. Johnson, M. Gorman, and N. Neuss, *J. Pharm. Sci.* **51**, 707 (1962).
55. R. Z. Andriamialisoa, N. Langlois, P. Potier, A. Chiaroni, and C. Riche, *Tetrahedron* **34**, 677 (1978).
56. W. E. Jones and G. J. Cullinan, U.S. Patent 3,887,565 (1973); *Chem. Abstr.* **83**, 97687 (1973).
57. R. L. Noble, C. T. Beer, and J. H. Cutts, *Ann. N.Y. Acad. Sci.* **76**, 882 (1958).
58. G. H. Svoboda, N. Neuss, and M. Gorman, *J. Am. Pharm. Assoc. Sci. Ed.* **48**, 659 (1959).
59. Atta-ur-Rahman, I. Ali, and M. Bashir, *J. Nat. Prod.* **47**, 554 (1984).
60. Atta-ur-Rahman, M. Bashir, M. Hafeez, N. Perveen, J. Fatima, and A. N. Mistry, *Planta Med.* **47**, 246 (1983).
61. S. Mukhopadhyay and G. A. Cordell, *J. Nat. Prod.* **44**, 335 (1981).
62. N. Langlois, R. Z. Andriamialisoa, and N. Neuss, *Helv. Chim. Acta* **63**, 793 (1980).
63. R. Z. Andriamialisoa, N. Langlois, and P. Potier, *Tetrahedron Lett.*, 2849 (1976).
64. G. H. Svoboda, M. Gorman, A. J. Barnes, Jr., and A. T. Oliver, *J. Pharm. Sci.* **51**, 518 (1962).
65. G. H. Svoboda, A. T. Oliver, and D. R. Bedwell, *Lloydia* **26**, 141 (1963).
66. G. H. Svoboda, *J. Pharm. Sci.* **52**, 407 (1963).
67. M. Gorman, G. H. Svoboda, and N. Neuss, *Lloydia* **28**, 269 (1965).
68. G. Richter Co., Can. Patent 948,625 (1975); *Chem. Abstr.* **82**, 103133 (1975).
69. G. Richter Co., Fr. Patent 2,210,392 (1975); *Chem. Abstr.* **82**, 116076 (1975).
70. G. Richter Co., Ger. Patent 2,259,388 (1974); *Chem. Abstr.* **81**, 82369 (1974).
71. G. Richter Co., Br. Patent 1,382,460 (1975); *Chem. Abstr.* **83**, 79459 (1975).
72. A. De Bruyn, L. De Taeye, and M. J. O. Anteunis, *Bull. Soc. Chim. Belg.* **89**, 629 (1980).
73. E. Wenkert, D. W. Cochran, E. W. Hagaman, F. M. Schell, N. Neuss, A. S. Katner,

P. Potier, C. Kan, M. Plat, M. Koch, H. Mehri, J. Poisson, N. Kunesch, and Y. Rolland, *J. Am. Chem. Soc.* **95**, 4990 (1973).
74. E. Wenkert, E. W. Hagaman, B. Lal, G. E. Gutowski, A. S. Katner, J. C. Miller, and N. Neuss, *Helv. Chim. Acta* **58**, 1560 (1975).
75. D. E. Dorman and J. W. Paschal, *Org. Magn. Res.* **8**, 413 (1986).
77. M. S. Morales-Rios, J. Espineira, and P. Joseph-Nathan, *Magn. Res. Chem.* **25**, 376 (1987).
78. K. Mislow and J. Siegel, *J. Am. Chem. Soc.* **106**, 3319 (1984).
79. J. W. Moncrief and W. N. Lipscomb, *J. Am. Chem. Soc.* **87**, 4963 (1965).
80. J. W. Moncrief and W. N. Lipscomb, *Acta Crystallogr.* **21**, 322 (1966).
81. J. P. Kutney, D. E. Gregonis, R. Imhof, I. Itoh, E. Jahngen, A. I. Scott, and W. K. Chan, *J. Am. Chem. Soc.* **97**, 5013 (1975).
82. N. Langlois, F. Guéritte, Y. Langlois, and P. Potier, *J. Am. Chem. Soc.* **98**, 7017 (1976).
83. J. P. Kutney, J. Beck, F. Bylsma, and W. J. Cretney, *J. Am. Chem. Soc.* **90**, 4504 (1968).
84. J. P. Kutney, J. Cook, K. Fuji, A. M. Treasurywala, J. Clardy, J. Fayos, and H. Wright, *Heterocycles* **3**, 205 (1975).
85. D. R. Brannon and N. Neuss, Ger. Patent 2,440,931 (1975); *Chem. Abstr.* **83**, 7184 (1975).
86. P. Potier, N. Langlois, Y. Langlois, and F. Guéritte, *J. Chem. Soc., Chem. Commun.* **1975**, 670 (1975).
87. A. I. Scott, F. Guéritte, and S.-L. Lee, *J. Am. Chem. Soc.* **100**, 6253 (1978).
88. N. Langlois and P. Potier, *J. Chem. Soc., Chem. Commun.*, 582 (1979).
89. A. De Bruyn, L. De Taeye, and R. Simonds, *Bull. Soc. Chim. Belg.* **90**, 185 (1981).
90. J. P. Kutney, J. Balsevich, and G. H. Bokelman, *Heterocycles* **4**, 1377 (1976).
91. A. De Bruyn, J. Sleechecker, J. P. De Jonghe, and J. Hannart, *Bull. Soc. Chim. Belg.* **92**, 485 (1983).
92. G. E. Gutowski, A. S. Katner, and J. C. Miller, unpublished results cited in Ref. *75*.
93. S. S. Tafur, W. E. Jones, D. E. Dorman, E. E. Logsdon, and G. H. Svoboda, unpublished results cited in Ref. *44*.
94. N. Neuss and S. S. Tafur, U.S. Patent 3,968,113 (1976); *Chem. Abstr.* **85**, 143352 (1976).
95. S. S. Tafur, J. L. Occolowitz, T. K. Elzey, J. W. Paschal, and D. E. Dorman, *J. Org. Chem.* **41**, 1001 (1976).
96. W. E. Jones and G. J. Cullinan, U.S. Patent 3,887,565 (1975); *Chem. Abstr.* **83**, 97687 (1975).
97. G. A. Cordell, *Lloydia* **37**, 219 (1974).
98. K. H. C. Baser, *Chim. Acta Turc.* **8**, 133 (1980).
99. A. I. Scott, S.-L. Lee, M. G. Culver, W. Wan, T. Hirata, F. Guéritte, R. L. Baxter, H. Nordlöv, C. A. Dorschel, H. Mizukami, and N. E. Mackenzie, *Heterocycles* **15**, 1257 (1981).
100. J. P. Kutney, *Heterocycles* **15**, 1405 (1981).
101. J. P. Kutney, *Pure Appl. Chem.* **54**, 2523 (1982).
102. J. P. Kutney, B. Aweryn, L. S. L. Choi, T. Honda, P. Kolodziejczyk, N. G. Lewis, T. Sato, S. K. Sleigh, K. L. Stuart, and B. R. Worth, *Tetrahedron* **39**, 3781 (1983).
103. R. Verpoorte, *Pharm. Wkbld.* **121**, 248 (1986).
104. J. P. Kutney, *Heterocycles* **25**, 617 (1987).
105. F. Reda, *Pharmazie* **33**, 233 (1978).
106. A. N. Masoud, L. A, Sciuchetti, N. R. Farnsworth, R. N. Blomster, and W. A. Meer, *J. Pharm. Sci.* **57**, 589 (1968).

107. U. Eilert, L. R. Nesbitt, and F. Constabel, *Can. J. Bot.* **63**, 1540 (1985).
108. R. Krishnan, M. V. Chandravadana, P. R. Ramachander, and H. B. Kumar, *Herb. Hung.* **22**, 47 (1983).
109. B. G. Mersey and A. J. Cutler, *Can. J. Bot.* **64**, 1039 (1986).
110. F. Constabel, *Int. Rev. Cytol. Suppl.* **16**, 209 (1983).
111. K. Esau, "Plant Anatomy," 2nd Ed., Wiley, New York, 1965.
112. J. Balsevich, V. DeLuca, and W. G. W. Kurz, *Heterocycles* **24**, 2415 (1986).
113. P. E. Dadonna, J. L. Wright, and C. R. Hutchinson, *Phytochemistry* **15**, 941 (1976).
114. P. A. Babcock and D. P. Carew, *Lloydia* **25**, 209 (1962).
115. A. Harris, H. B. Nylund, and D. P. Carew, *Lloydia* **27**, 322 (1964).
116. G. B. Boder, M. Gorman, I. S. Johnson, and P. J. Simpson, *Lloydia* **27**, 328 (1964).
117. I. Richter, K. Stolle, D. Gröger, and K. Mothes, *Naturwissenschaften* **52**, 305 (1965).
118. R. J. Krueger, D. P. Carew, J. H. C. Lui, and E. J. Staba, *Planta Med.* **45**, 56 (1982).
119. T. Endo, A. Goodbody, and M. Misawa, *Planta Med.* **53**, 479 (1987).
120. K.-H. Knobloch, G. Bast, and J. Berlin, *Phytochemistry* **21**, 591 (1982).
121. K. Hirata, A. Yamanaka, N. Kurano, K. Miyamoto, and Y. Miura, *Agric. Biol. Chem.* **51**, 1311 (1987).
122. Y. Miura, K. Hirata, and N. Kurano, *Agric. Biol. Chem.* **51**, 611 (1987).
123. Y. Miura, K. Hirata, N. Kurano, K. Miyamoto, and K. Uchida, *Planta Med.* **54**, 18 (1988).
124. B. D. Patterson and D. P. Carew, *Lloydia* **32**, 131 (1969).
125. D. P. Carew and R. J. Krueger, *Phytochemistry* **16**, 1461 (1977).
126. A. I. Scott, H. Mizukami, T. Hirata, and S.-L. Lee, *Phytochemistry* **19**, 488 (1980).
127. J. Stöckigt and H. J. Soll, *Planta Med.* **40**, 22 (1980).
128. V. Pétiard, F. Guéritte, N. Langlois, and P. Potier, *Physiol. Veg.* **18**, 711 (1980).
129. W. Kohl, B. Witte, and G. Höfle, *Z. Naturforsch.* **36B**, 1153 (1981).
130. W. Kohl, B. Witte, and G. Höfle, *Z. Naturforsch., B: Anorg. Chem., Org. Chem.* **37B**, 1346 (1982).
131. V. Pétiard, D. Courtois, F. Guéritte, N. Langlois, and B. Mompon, *Plant Tissue Cult.*, 309 (1982).
132. F. Guéritte, N. Langlois, and V. Pétiard, *J. Nat. Prod.* **46**, 144 (1983).
133. W. Kohl, B. Witte, W. S. Sheldrick, and G. Höfle, *Planta Med.* **50**, 242 (1984).
134. W. Kohl, B. Witte, and G. Höfle, *Planta Med.* **47**, 177 (1983).
135. P. Westerkemper, U. Wieczorek, F. Guéritte, N. Langlois, Y. Langlois, P. Potier, and M. H. Zenk, *Planta Med.* **39**, 24 (1980).
136. R. V. Schenk and A. C. Hildebrandt, *Can. J. Bot.* **50**, 199 (1972).
137. R. W. Raab and S.-L. Lee, *Phytochem. Bull.* **20**, 40 (1988).
138. K. H. Knobloch and J. Berlin, *Plant Cell, Tissue Organ Cult.* **2**, 333 (1983).
139. J. M. Mérillion, J. C. Chénieux, and M. Rideau, *Planta Med.* **47**, 169 (1983).
140. A. Stafford, L. Smith, and M. W. Fowler, *Plant Cell, Tissue Organ Cult.* **4**, 83 (1985).
141. R. J. Kreuger and D. P. Carew, *Lloydia* **41**, 327 (1978).
142. D. Courtois, A. Kurkdjian, and J. Guern, *Plant Sci. Letts.* **18**, 85 (1980).
143. J. P. Kutney, W. J. Cretney, J. R. Hadfield, E. S. Hall, V. R. Nelson, and D. C. Wigfield, *J. Am. Chem. Soc.* **90**, 3566 (1968).
144. A. R. Battersby, A. R. Burnett, and P. G. Parsons, *J. Chem. Soc.*, 1193 (1969).
145. M. H. Zenk, H. El-Shagi, H. Arens, J. Stöckigt, E. W. Weiler, and B. Deus, *in* "Plant Cell Culture and Its Biotechnological Applications," p. 27. Springer-Verlag, Berlin, 1977.
146. J. P. Kutney, B. Aweryn, K. B. Chatson, L. S. L. Choi, and W. G. W. Kurz, *Plant Cell Rep.* **4**, 259 (1985).
147. S.-L. Lee, K.-D. Cheng, and A. I. Scott, *Phytochemistry* **20**, 1841 (1981).

148. B. Deus-Neumann and M. H. Zenk, *Planta Med.*, 365 (1984).
149. D. Neumann, G. Krauss, M. Heike, and D. Gröger, *Planta Med.* **48**, 20 (1983).
150. E. Müller, D. Neumann, and A. Nelles, *Nova Acta Leopoldina, Suppl.* **7**, 133 (1976).
151. M. H. Zenk, 13th Congress der Pharmazeutischen Gesellschaft der DDR, Leipzig, November, 1980.
152. B. Deus-Neumann and M. H. Zenk, *Planta* **162**, 250 (1984).
153. W. G. W. Kurz, K. B. Chatson, F. Constabel, J. P. Kutney, L. S. L. Choi, P. Kolodziejczyk, S. K. Sleigh, K. L. Stuart, and B. R. Worth, *Phytochemistry* **19**, 2583 (1980).
154. J. P. Kutney, L. S. L. Choi, P. Kolodziejczyk, S. K. Sleigh, K. L. Stuart, B. R. Worth, W. G. W. Kurz, K. B. Chatson, and F. Constabel, *Phytochemistry* **19**, 2589 (1980).
155. J. P. Kutney, L. S. L. Choi, P. Kolodziejczyk, S. K. Sleigh, K. L. Stuart, B. R. Worth, W. G. W. Kurz, K. B. Chatson, and F. Constabel, *Heterocycles* **14**, 765 (1980).
156. W. G. W. Kurz, K. B. Chatson, F. Constabel, J. P. Kutney, L. S. L. Choi, P. Kolodziejczyk, S. K. Sleigh, K. L. Stuart, and B. R. Worth, *Helv. Chim. Acta* **63**, 1891 (1980).
157. A. I. Scott, P. B. Reichardt, M. B. Slaytor, and J. G. Sweeney, *Bioorg. Chem.* **1**, 157 (1971).
158. J. P. Kutney, L. S. L. Choi, P. Kolodziejczyk, S. K. Sleigh, K. L. Stuart, B. R. Worth, W. G. W. Kurz, K. B. Chatson, and F. Constabel, *Helv. Chim. Acta* **64**, 1837 (1981).
159. W. G. W. Kurz, K. B. Chatson, F. Constabel, J. P. Kutney, L. S. L. Choi, P. Kolodziejczyk, S. K. Sleigh, and K. L. Stuart, *Planta Med.* **39**, 284 (1980).
160. W. G. W. Kurz, K. B. Chatson, F. Constabel, J. P. Kutney, L. S. L. Choi, P. Kolodziejczyk, S. K. Sleigh, K. L. Stuart, and B. R. Worth, *Planta Med.* **42**, 22 (1981).
161. J. P. Kutney, L. S. L. Choi, P. Kolodziejczyk, S. K. Sleigh, K. L. Stuart, W. G. W. Kurz, K. B. Chatson, and F. Constabel, *J. Nat. Prod.* **44**, 536 (1981).
162. A. I. Scott, H. Mizukami, and S.-L. Lee, *Phytochemistry* **18**, 795 (1979).
163. J. Schallenberg and J. Berlin, *Z. Naturforsch.* **34C**, 541 (1979).
164. F. Sasse, M. Buchholz, and J. Berlin, *Z. Naturforsch., C: Biosci.* **38C**, 916 (1983).
165. A. I. Scott and S.-L. Lee, *J. Am. Chem. Soc.* **97**, 6906 (1975).
166. J. Stöckigt, J. Treimer, and M. H. Zenk, *FEBS Lett.* **70**, 267 (1976).
167. J. Stöckigt, H.-P. Husson, C. Kan-Fan, and M. H. Zenk, *J. Chem. Soc., Chem. Commun.*, 164 (1977).
168. H.-P. Husson, C. Kan-Fan, T. Sévenet, and J.-P. Vidal, *Tetrahedron Lett.*, 1889 (1977).
169. K. L. Stuart, J. P. Kutney, T. Honda, N. G. Lewis, and B. R. Worth, *Heterocycles* **9**, 647 (1978).
170. J. P. Kutney, L. S. L. Choi, T. Honda, N. G. Lewis, T. Sato, K. L. Stuart, and B. R. Worth, *Helv. Chim. Acta* **65**, 2088 (1982).
171. J. Stöckigt, H. Gundlach, and B. Deus-Neumann, *Helv. Chim. Acta* **68**, 315 (1985).
172. V. W. Rodwell, *in* "Metabolic Pathways" (D. M. Greenberg, ed.), 3rd Ed., Vol. 3, p. 191. Academic Press, New York, 1969.
173. T. Suga, T. Hirata, T. Shishibori, and K. Tange, *Chem. Lett.*, 189 (1974).
174. D. C. Wigfield and B. P. Wen, *Bioorg. Chem.* **6**, 511 (1977).
175. E. Leete, A. Ahmad, and I. Kompis, *J. Am. Chem. Soc.* **87**, 4168 (1965).
176. R. L. Baxter and W. R. McLauchlan, *J. Chem. Soc., Chem. Commun.*, 1245 (1985).
177. J. Balsevich, F. Constabel, and W. G. W. Kurz, *Planta Med.* **44**, 231 (1982).
178. J. Balsevich and W. G. W. Kurz, *Planta Med.* **49**, 79 (1983).
179. J. Balsevich, *Planta Med.* **51**, 128 (1985).
180. R. Guarnaccia and C. J. Coscia, *J. Am. Chem. Soc.* **93**, 6320 (1971).
181. A. R. Battersby, A. R. Burnett, and P. G. Parsons, *Chem. Commun.*, 1282 (1968).

182. A. R. Battersby, A. R. Burnett, and P. G. Parsons, *J. Chem. Soc. C*, 1193 (1969).
183. K. T. D. De Siva, G. N. Smith, and K. E. H. Warren, *Chem. Commun.*, 905 (1971).
184. W. P. Blackstock, R. T. Brown, and G. K. Lee, *Chem. Commun.*, 910 (1971).
185. A. R. Battersby and K. H. Gibson, *Chem. Commun.*, 902 (1971).
186. K. C. Mattes, C. R. Hutchinson, J .P. Springer, and J. Clardy, *J. Am. Chem. Soc.* **97**, 6720 (1975).
187. O. Kennard, P. J. Roberst, N. W. Isaacs, F. H. Allen, W. D. S. Motherwell, K. H. Gibson, and A. R. Battersby, *Chem. Commun.*, 899 (1971).
188. J. Stöckigt and M. H. Zenk, *J. Chem. Soc., Chem. Commun.*, 646 (1977).
189. J. Stöckigt and M. H. Zenk, *FEBS Lett.* **79**, 233 (1977).
190. R. T. Brown, J. Leonard, and S. K. Sleigh, *Phytochemistry* **17**, 899 (1978).
191. M. Rüffer, N. Nagakura, and M. H. Zenk, *Tetrahedron Lett.*, 1593 (1978).
192. N. Nagakura, M. Rüffer, and M. H. Zenk, *J. Chem. Soc., Perkin Trans. 1*, 2308 (1979).
193. A. R. Battersby, N. G. Lewis, and J. M. Tippett, *Tetrahedron Lett.*, 4849 (1977).
194. J. F. Treimer and M. H. Zenk, *FEBS Lett.* **97**, 159 (1979).
195. H. Mizukami, H. Nordlöv, S.-L. Lee, and A. I. Scott, *Biochemistry* **18**, 3760 (1979).
196. J. Treimer and M. H. Zenk, *Eur. J. Biochem.* **101**, 225 (1979).
197. U. Pfitzner and M. H. Zenk, *Planta Med.* **46**, 10 (1982).
198. A. I. Scott, S.-L. Lee, and W. Wan, *Biochem. Biophys. Res. Commun.* **75**, 1004 (1977).
199. A. I. Scott, S.-L. Lee, and W. Wan, *Heterocycles* **6**, 1552 (1977).
200. J. Stöckigt, M. Rüffer, M. H. Zenk, and G.-A. Hoyer, *Planta Med.* **33**, 188 (1978).
201. H. Arens, J. Stöckigt, E. W. Weiler, and M. H. Zenk, *Planta Med.* **34**, 37 (1978).
202. J. F. Treimer and M. H. Zenk, *Phytochemistry* **17**, 227 (1978).
203. T. Hemscheidt and M. H. Zenk, *FEBS Lett.* **110**, 187 (1980).
204. A. R. Battersby and E. S. Hall, *Chem. Commun.*, 793 (1969).
205. J. Stöckigt, *J. Chem. Soc., Chem. Commun.*, 1097 (1978).
206. J. Stöckigt, G. Höfle, and A. Pfitzner, *Tetrahedron Lett.* **21**, 1925 (1980).
207. C. Kan-Fan and H.-P. Husson, *J. Chem. Soc., Chem. Commun.*, 1015 (1979).
208. S.-L. Lee, T. Hirata, and A. I. Scott, *Tetrahedron Lett.*, 691 (1979).
209. P. Heinstein, J. Stöckigt, and M. H. Zenk, *Tetrahedron Lett.* **21**, 141 (1980).
210. T. Hirata, S.-L. Lee, and A. I. Scott, *J. Chem. Soc., Chem. Commun.*, 1081 (1979).
211. J. Stöckigt, *Phytochemistry* **18**, 965 (1979).
212. M. H. Zenk, *J. Nat. Prod.* **43**, 438 (1980).
213. P. Brodelius, B. Deus, K. Mosbach, and M. H. Zenk, *FEBS Lett.* **103**, 93 (1979).
214. W. Fahn, H. Gundlach, B. Deus-Neumann, and J. Stöckigt, *Plant Cell Rep.* **4**, 333 (1985).
215. W. Fahn, E. Laussermair, B. Deus-Neumann, and J. Stöckigt, *Plant Cell Rep.* **4**, 337 (1985).
216. V. DeLuca, J. Balsevich, R. T. Tyler, and W. G. W. Kurz, *Plant Cell Rep.* **6**, 458 (1987).
217. P. E. Dadonna and C. R. Hutchinson, *J. Am. Chem. Soc.* **96**, 6805 (1974).
218. N. Langlois and P. Potier, *J. Chem. Soc., Chem. Commun.*, 102 (1978).
219. K. L. Stuart, J. P. Kutney, T. Honda, and B. R. Worth, *Heterocycles* **9**, 1419 (1978).
220. J. P. Kutney, C. A. Boulet, L. S. L. Choi, W. Gustowski, M. McHugh, J. Nakano, T. Tamotsu, H. Tsukamoto, G. M. Hewitt, and R. Suen, *Heterocycles* **27**, 613 (1988).
221. M. Misawa, T. Endo, A. Goodbody, J. Vukovic, C. Chapple, L. Choi, and J. P. Kutney, *Phytochemistry* **27**, 1355 (1988).
222. A. Goodbody, T. Endo, J. Vukovic, J. P. Kutney, and L. S. L. Choi, *Planta Med.* **51**, 136 (1988).
223. A. I. Scott, *Bioorg. Chem.* **3**, 398 (1974).

224. S. B. Hassam and C. R. Hutchinson, *Tetrahedron Lett.*, 1681 (1978).
225. J. P. Kutney, B. Aweryn, L. S. L. Choi, P. Kolodziejczyk, W. G. W. Kurz, K. B. Chatson, and F. Constabel, *Heterocycles* **16**, 1169 (1981).
226. J. P. Kutney, B. Aweryn, L. S. L. Choi, P. Kolodziejczyk, W. G. W. Kurz, K. B. Chatson, and F. Constabel, *Helv. Chim. Acta* **65**, 1271 (1982).
227. K. L. Stuart, J. P. Kutney, T. Honda, and B. R. Worth, *Heterocycles* **9**, 1391 (1978).
228. R. L. Baxter, C. A. Dorschel, S.-L. Lee, and A. I. Scott, *J. Chem. Soc., Chem. Commun.*, 257 (1979).
229. K. L. Stuart, J. P. Kutney, and B. R. Worth, *Heterocycles* **9**, 1015 (1978).
230. F. Guéritte, N. V. Bac, Y. Langlois, and P. Potier, *J. Chem. Soc., Chem. Commun.*, 452 (1980).
231. R. L. Baxter, M. Hasan, N. E. Mackenzie, and A. I. Scott, *J. Chem. Soc., Chem. Commun.*, 791 (1982).
232. W. R. McLauchlan, M. Hasan, R. L. Baxter, and A. I. Scott, *Tetrahedron* **39**, 3777 (1983).
233. N. G. Lewis, Ph.D. Thesis, University of British Columbia, 1978, quoted in Ref. *219*.
234. B. Deus and M. H. Zenk, *Biotechnol. Bioeng.* **24**, 1965 (1982).

—Chapter 2—

SYNTHESES OF VINBLASTINE-TYPE ALKALOIDS

Martin E. Kuehne

Department of Chemistry
University of Vermont
Burlington, Vermont 05405

AND

István Markó

Department of Chemistry
The University of Sheffield
Sheffield S37 HF, England

I. Introduction

The binary* indole–indoline alkaloids vinblastine (VLB, **1**) and vincristine (VCR, **2**) (Fig. 1) have triggered considerable chemical investigation as a result of their extensive clinical use as antineoplastic agents (*2–4*), coupled with an arduous isolation from plant material,† and because of

* Though these alkaloids are not truly composed of two identical monomeric units, they are popularly named dimers or dimeric alkaloids. We prefer to avoid this incorrect nomenclature and would like to encourage the use of the more adequate binary terminology. In another consideration of nomenclature, we describe quaternary salts derived from an imine functionality as imonium salts, in accord with the descriptor for other onium salts (ammonium, oxonium, etc.), rather than by the frequently used iminium terminology. This nomenclature was suggested earlier (*1*).

† VLB and VCR are said to be present in only minute amounts in *Catharanthus roseus* (L.) G. Don (0.00025% of leaf dry weight), and their separation from other monomeric or binary alkaloids involves a complicated and tedious procedure based on their differential basicity followed by chromatography. Industrial production of VLB and VCR has therefore been a serious problem, and consequently these drugs are among the most expensive on the pharmaceutical market.

THE ALKALOIDS, VOL. 37
Copyright © 1990 by Academic Press, Inc.
All rights of reproduction in any form reserved.

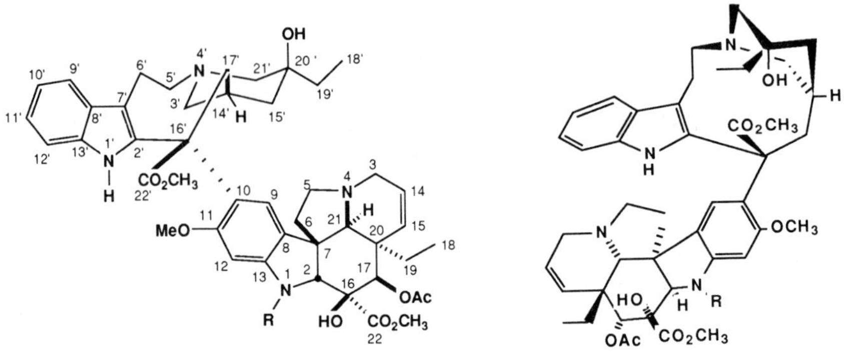

1 R = CH_3 : Vinblastine (VLB, Vincaleukoblastine)
2 R = CHO : Vincristine (VCR, Vincaleukocristine)

FIG. 1. (Left) Numbering of vinblastine-type alkaloids in this chapter according to the biogenetic scheme of LeMen and Taylor (*134*), with equivalent atoms in all synthetic intermediates equally labeled. (Right) Approximation of computer-generated, energy-minimized structure, obtained with the Clark Still MACROMODEL program.

their chemically and biogenetically intriguing structures. The fascinating story relating the isolation and structural determination of these natural products is described elsewhere (*5–11*) and is not covered in this chapter. Instead, we focus on the synthetic challenges that had to be overcome for an efficient preparation of vinblastine-type alkaloids.

One of the major difficulties in the synthesis of these binary indole–indoline alkaloids is the necessity of generating the natural PARF (priority antireflective) (*12*) relative stereochemistry between C-14′ and C-16′, as well as the requirement for controlling the absolute stereochemistry at C-16′, which must be (*S*). Other epimers at these positions lack the high cytotoxicity, with mitotic arrest at metaphase, that is the basis of the anticancer activity of these compounds (*13,14*).

Thus far all synthetic approaches have utilized a coupling of vindoline **3** to a precursor of the remaining half of the binary VLB structure, with the hope of obtaining the desired stereoselectivity. Three different strategies have now reached the goal of obtaining the C-16′–C-14′ PARF relative stereochemistry, and one of these provided an enantioselective total synthesis of vinblastine (Section V,B). Because of the importance of preparing a variety of synthetic congeners for pharmacological testing, any synthetic route toward VLB-type alkaloids should be as flexible as possible, and consequently the presently developed syntheses, and future synthetic strategies, must be judged on the basis of their relative versatility as well as on their efficacy and stereoselectivity.

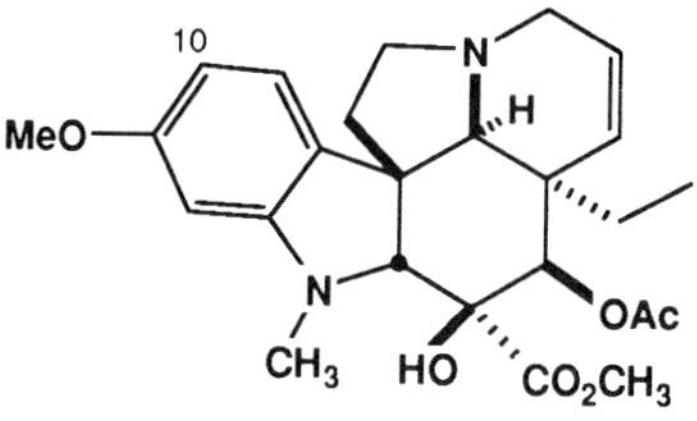

3: Vindoline

(10 - Vindolinyl : Vi)

II. Hydroxyalkylindole Route

The first reported preparation of a binary alkaloid of the vinblastine type came from the efforts of J. Harley-Mason and Atta-ur-Rahman (*15,16*). In accord with a proposal for the synthesis of vinblastine-type alkaloids (*17*), their synthetic strategy was based on a coupling reaction previously developed by G. Büchi *et al.* for the synthesis of *Voacanga* alkaloids (*18*), * as shown in Scheme 1. There, acid-catalyzed ionization

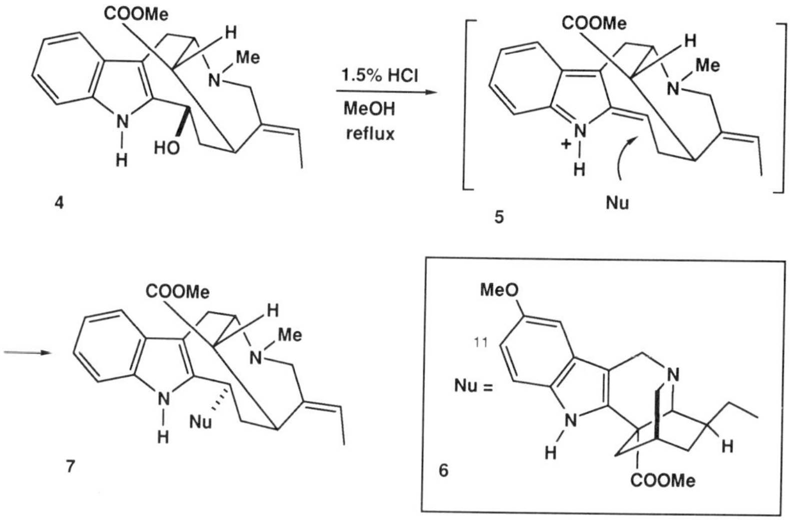

SCHEME 1

* A simple and earlier example of this coupling process can be found in the rapid self-condensation of 1-hydroxy-1,2,3,4-tetrahydrocarbazole in the presence of dilute acids (*19*).

of vobasinol (**4**) had produced a delocalized imonium ion **5**, which reacted with voacangine (**6**) at C-11 to give voacamine (**7**).

For the synthesis of VLB-type alkaloids (Schemes 2 and 3), the tetracyclic lactam **8**, prepared in three steps from the acetal **9**, was treated with phosphorus pentasulfide, and the resulting thiolactam was then desulfurized with Raney nickel to yield the amino acid **10** (Scheme 2). Rearrangement to a cleavamine skeleton was brought about by heating of **10** with acetic anhydride. This interesting reaction proceeds most probably by initial N-acylation, derived from a mixed anhydride **12**, to form the pentacyclic *N*-acyl ammonium intermediate **13**, which suffers ring opening to the imonium ion **14**. Trapping of the latter by acetate ion (or another

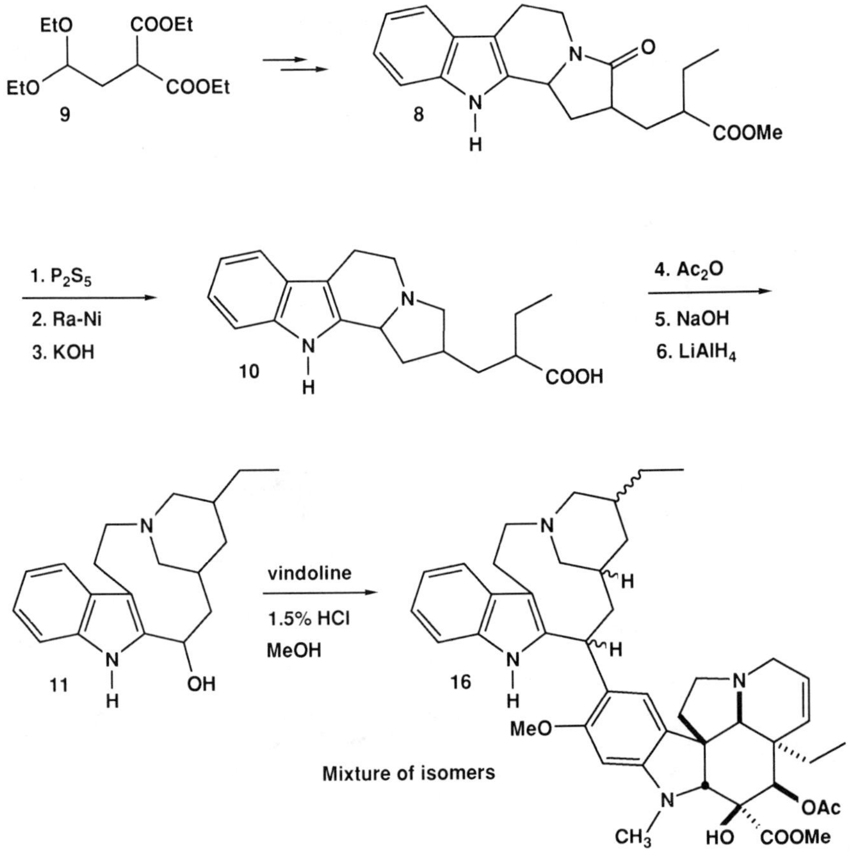

EtO COOEt EtO COOEt 9 → 8 N H N O COOMe

1. P_2S_5 2. Ra-Ni 3. KOH → 10 N H N COOH → 4. Ac_2O 5. NaOH 6. $LiAlH_4$

11 N H N OH → vindoline 1.5% HCl MeOH → 16 N H N H H MeO N CH_3 N OAc HO COOMe

Mixture of isomers

SCHEME 2

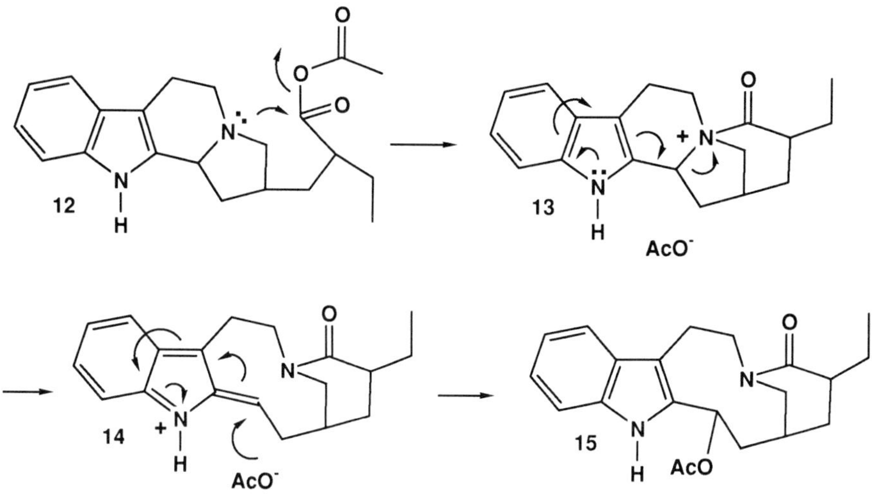

SCHEME 3

nucleophile) delivers the cleavamine lactam **15** (Scheme 3).* Cleavage of the acetate group and hydride reduction of the lactam function in **15** (Scheme 2) gave the key coupling precursor **11** as a mixture of isomers. This mixture of bridged indoloazonines, on treatment with 1% methanolic HCl and vindoline (**3**), afforded, in unspecified yield, the binary alkaloids **16** as a mixture of unspecified stereoisomers. Although these products lacked the C-16′ carbomethoxy and C-20′ hydroxy substituents of VLB, and proved later to possess the wrong C-16′–C-14′ relative stereochemistry, this was the first successful approach to the preparation of vinblastine-type binary alkaloids.

III. Chloroindolenine Route

A related approach, developed almost simultaneously by Neuss and co-workers at Eli Lilly (*25*) and by Kutney *et al.* in Vancouver (*26*), utilized as coupling substrate the chloroindoline alkene **17**, which, through equilibrium with the corresponding chloroindolenine, was readily prepared by chlorination of dihydrocleavamine **18**. Condensation with the 17-deacetyl-16-carboxyhydrazide derivative of vindoline (**19**) gave a binary indole–indoline product (**20**) (Scheme 4). The undesired C-16′ con-

* This reaction has been extensively applied by Kutney *et al.* and Takano *et al.* For some leading references, see Refs. *20–24*.

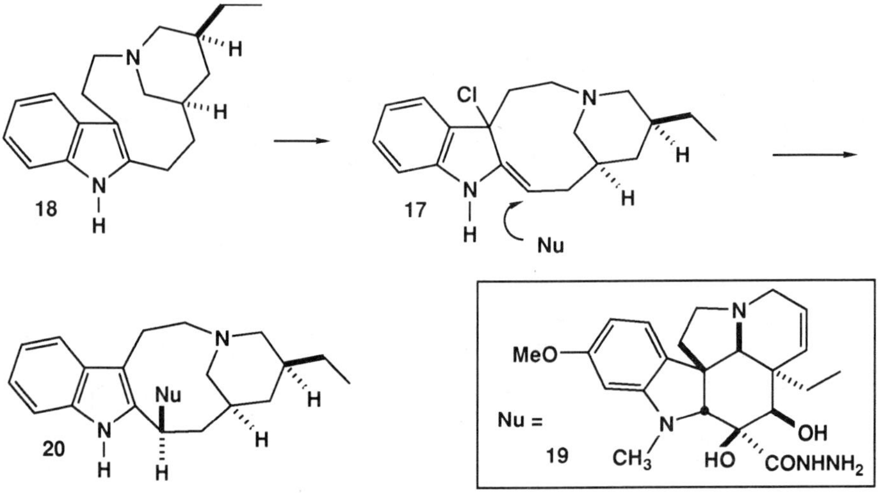

Scheme 4

figuration was assigned to this product by the Lilly group, which in turn also indicated that the compound **16**, previously prepared by Atta-ur-Rahman, has the VLB-unlike configuration at C-16′.

A similar strategy was reported soon after by Atta-ur-Rahman for the synthesis of 16′-epianhydrovinblastine (Scheme 5) (*27*). Reductive cleavage of catharanthine (**21**) provided a mixture of epimeric carbomethoxycleavamines (**22**) which, without separation, was transformed to a mix-

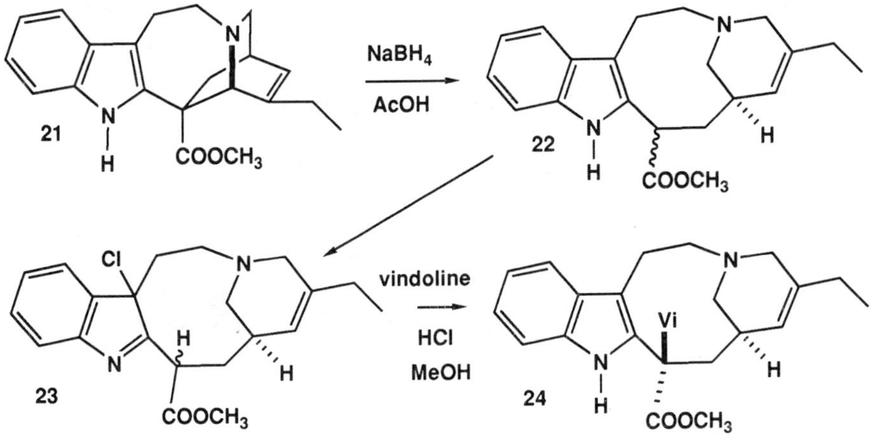

Scheme 5

ture of chloroindolenines (**23**). The coupling reaction was effected by reaction of vindoline with the tautomeric chloroindoline alkenes in the presence of methanolic hydrochloric acid. Once again, however, the binary alkaloid product **24** possessed the wrong C-16′ (*R*) configuration.

That only the wrong C-16′ diastereomer seemed to be produced in this reaction was then demonstrated by the Kutney group, who prepared a series of binary indole–indoline alkaloids using the chloroindolenine approach. The apparent simplicity of this coupling reaction and the rapidity in assembling such binary alkaloids prompted an extensive study of reaction conditions (*28*), with the desire to find a procedure suitable for generation of the C-16′ (*S*) isomer, required for anticancer activity. Despite the intensive effort involved in this in-depth study, no success could be realized, and it was therefore widely accepted that ". . . it is very unlikely that any natural dimer could be obtained in this way" (*7*). At this point it may be noted, however, that we were able to show subsequently that the desired C-16′–C-14′ PARF relative stereochemistry can be obtained as a preferential result, albeit only in very low yield [3.6% PARF versus 2.1% PREF (priority reflective)], when the chloroindolenine reaction with vindoline is initiated with silver tetrafluoroborate (*13*).

Although the ultimate goal could not be realized in the earlier studies of this coupling reaction, important information on the reaction mechanism was generated. It seems likely that the coupling proceeds through the intermediacy of a cation **25**, the fate of which depends on the reaction conditions. Trapping of the cation by vindoline resulted in the binary product **27**, whereas reaction with the solvent methanol produced the methyl ether **26** (Scheme 6). This methoxycleavamine was the sole product obtained when the chloroindolenine **17** was treated with methanolic HCl in the absence of vindoline, and it is a likely intermediate in the reaction with vindoline, since it can be anticipated that acid-promoted ionization of the ether **26** and equilibration with the precursor imonium ion **25** should be a facile process. Indeed, when the isolated methoxycleavamine **26** was subjected to the coupling conditions in the presence of vindoline, a good yield of binary alkaloid **27** could be produced (*28*).

Finally, this study excluded a reaction pathway involving the quaternary ammonium salt **28**. When this salt, prepared by intramolecular alkylation, was submitted to the conditions of the coupling reaction with vindoline, no binary alkaloid was obtained, thus ruling out participation of the ammonium salt as an intermediate on the reaction pathway (*28*). However, consideration of the quaternary salt **28** as a possible intermediate was reopened by a report from Atta-ur-Rahman on the production of binary alkaloids of both the natural and unnatural C-16′ configuration (*29*), in which the coupling reaction was effected by using **28** as the electro-

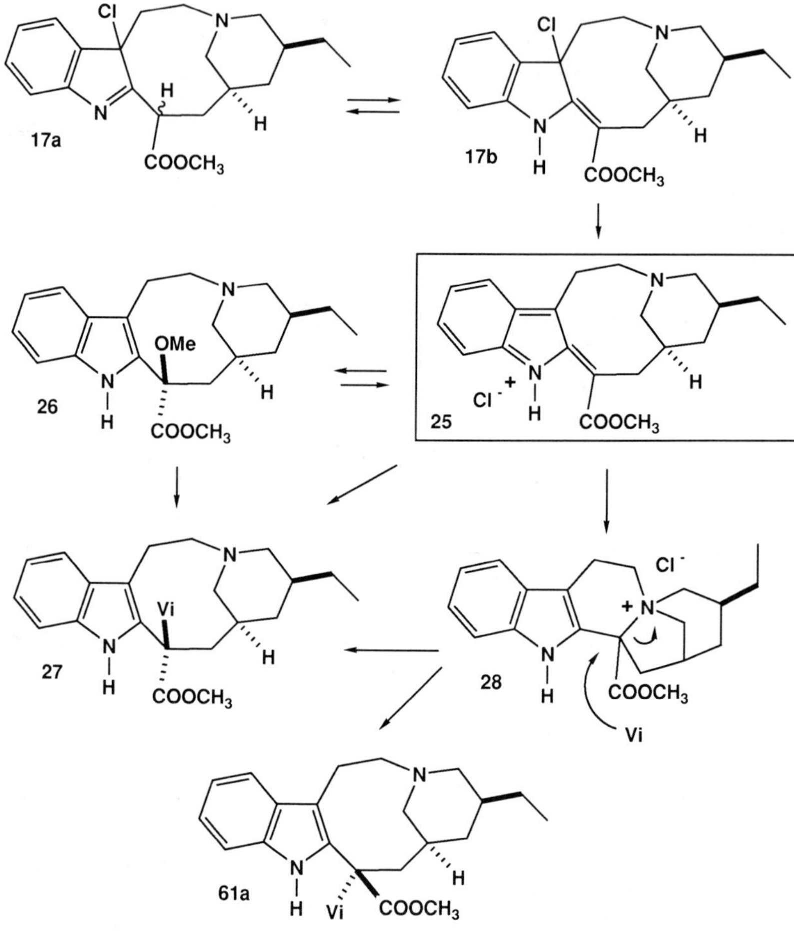

SCHEME 6

philic partner. Optimization of the reaction conditions, alas, gave only about a 10% yield of binary VLB-type alkaloids.

A rationalization of the generally selective formation of the undesired C-16′–C-14′ PREF relative stereochemistry on reaction of the chloroimine–chloroindoline alkene **17a,b** with vindoline (**3**) may be found in a preferred conformation, **25a** or **25b**, of the cationic intermediate **25**, where the C-3 methylene group hinders approach to one face of the nine-membered ring (Fig. 2). A conformationally more flexible nine-membered ring cationic intermediate **29a,b** would be expected to be formed from the chloro derivative of a D-secocleavamine **30**. From reaction of such a compound with vindoline (**3**), under the usual protic acid conditions for cou-

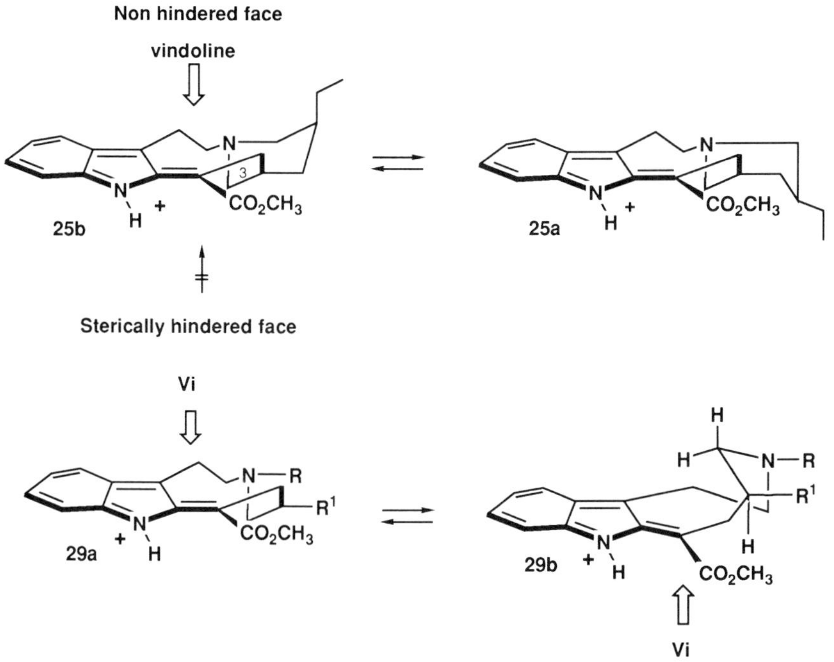

FIG. 2. Various conformations of cationic intermediates in the chloroindolenine approach to vinblastine-type alkaloids.

pling, followed by cyclization and debenzylation steps, the C-16′–C-14′ PARF and PREF products **31a,b** and **32a,b** were obtained in 16 and 40% yield, respectively (Scheme 7) (*13*). Coupling of a carbamate corresponding to the amine **30**, however, gave only C-16′–C-14′ PREF products (see Section **VI**) (*30–32*), indicating the importance of the character of N^b substitution on the conformation and charge stabilization in the cationic nine-membered ring intermediate. When the reaction of the chloroimine was initiated with silver tetrafluoroborate in dry acetone, the C-16′–C-14′ PARF and PREF products **31a,b** and **32a,b** were obtained in 6 and 7% yield, respectively (*13*).

In a synthesis of 14′-epileurosidine (**33**) by Kunesch *et al.* (*33*), using the general chloroindolenine coupling scheme and starting with an indoloazonine derived from reduction of pandoline (**34**) (*34*), the now anticipated C-14′–C-16′ PREF product was again obtained (Scheme 8). While this compound has the desired C-16′ (*S*) configuration, its unnatural C-14′ (*S*) configuration, derived from pandoline, results in a lack of the characteristic inhibition of tubulin polymerization that is found with the epimeric natural binary alkaloids and is required for anticancer cytotoxicity.

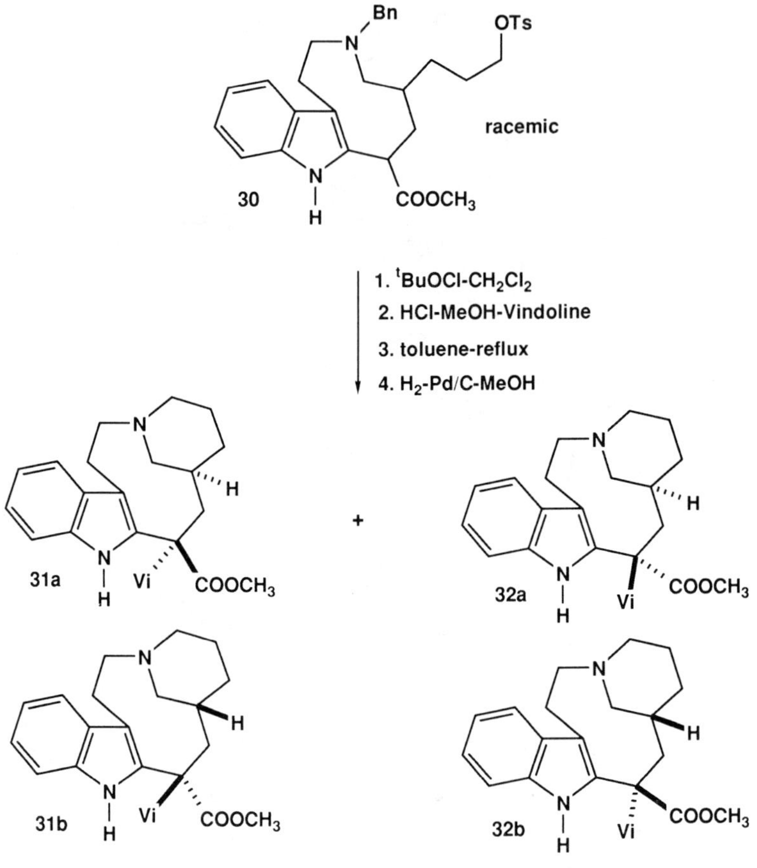

SCHEME 7

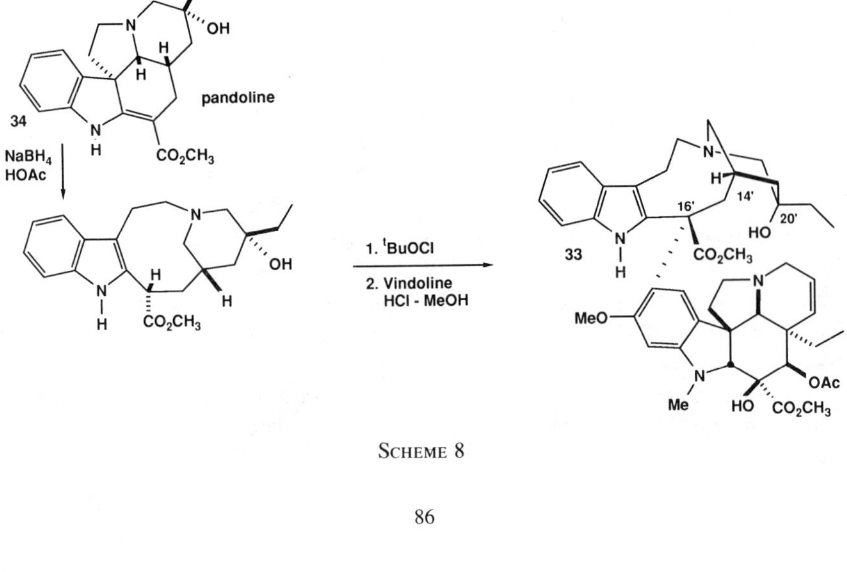

SCHEME 8

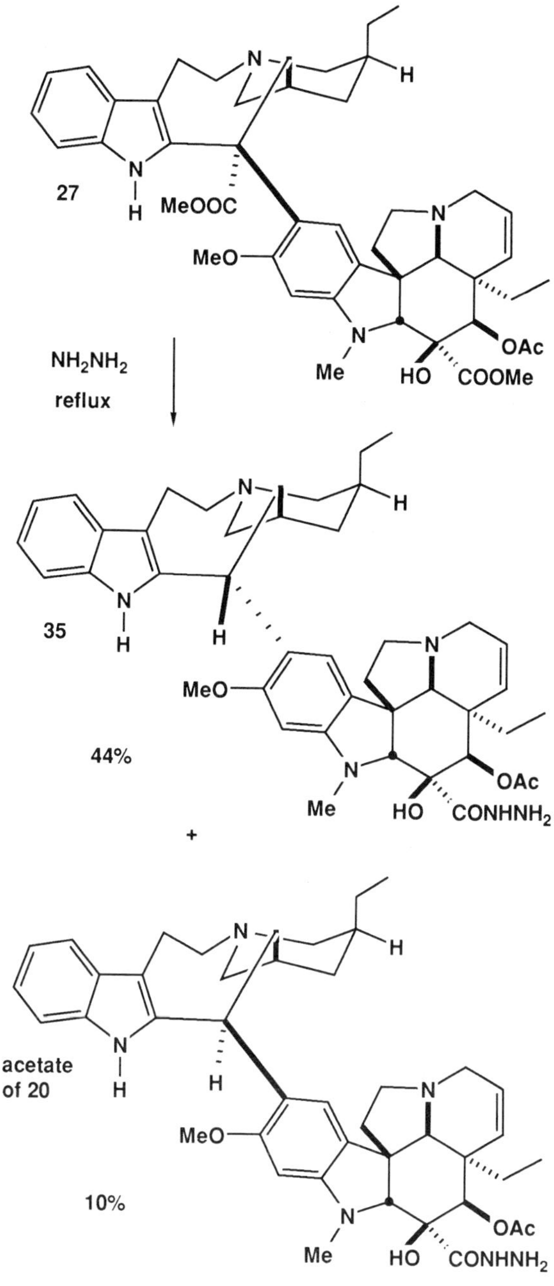

27

MeOOC

MeO

OAc

Me HO COOMe

NH_2NH_2
reflux

35

MeO

44%

OAc

Me HO $CONHNH_2$

+

acetate of 20

MeO

10%

OAc

Me HO $CONHNH_2$

Scheme 9

Studies directed toward an adjustment of the C-16′–C-14′ relative configuration, subsequent to the coupling reaction, showed that on decarbomethoxylation by heating with hydrazine, cleavamines with a 16′-α-vindolinyl substituent (i.e., **35**) were obtained from vinblastine as well as from the C-16′ epimeric series of coupling products (Scheme 9) (*35*). This suggests that a yet unreported subsequent carbomethoxylation process might provide the desired C-16′–C-14′ PARF stereochemistry. In an alternative approach, 15′-keto C-16′–C-14′ PREF binary compounds **36** and **37** were generated by the usual chloroindolenine coupling (Scheme 10), but an epimerization at C-14′ was not possible, presumably owing to preferential enolization toward the nonbridgehead C-20′ position (*36*).

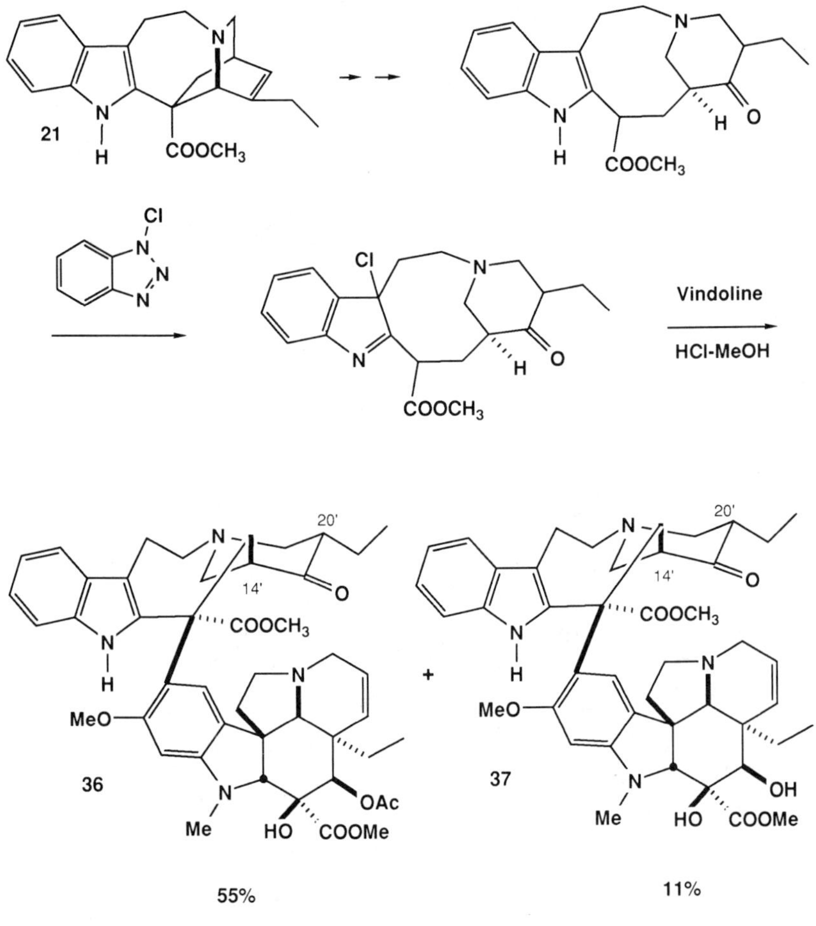

SCHEME 10

IV. β-Carboline Coupling

A variant of the Harley-Mason coupling reaction (Schemes 2 and 3) (*15,16*) with incorporation of the C-16 carbomethoxy substituent in the carboline ester **38** and intermolecular acylation, provided the model diastereomeric vindoline coupling products **39** in 70% yield (Scheme 11) (37a). Based on this sequence, it was proposed that $N^{b'}$-acylation with a synthetic vindoline precursor (**40a** or **40b**), with the correct absolute stereochemistry, might lead to an enantioselective intramolecular coupling reaction, with generation of the desired C-16′ (*S*) absolute stereochemistry (37a). Depending on the conformation of the nine-membered ring and on torsion about the sulfur bonds in the proposed imonium intermediate (**41a** or **41b**), the opposite result might, however, be more reasonably expected, unless the stereochemistry of the thioether is adjusted and/or an eventual C-14′ substituent could provide guidance for formation of the bond at C-16′ (see also Scheme 7).

A more promising approach to a synthesis of vinblastine-type compounds can be derived from the significant observation that reaction of one enantiomer of the carboline ester (+)-**38** with *p*-nitrobenzyl chloroformate and 3-methoxy-*N,N*-dimethylaniline, at 25°C, gave a model (+)-congener of **39** in 72% yield and 55% enantiomeric excess, thus indicating a conformational retention in the nine-membered cationic intermediate formed on acylation of the carboline ester **38** (37b).

V. Biomimetic Approach

The real breakthrough toward synthesis of vinblastine and, in fact, the first significant laboratory preparation of binary indole–indoline alkaloids with the natural C-16′–C14′ PARF configuration, was due to the work of the Potier–Langlois team at Gif (*38,39*; for reviews, see Refs. *40* and *41*), buttressed by results obtained by the Kutney group in Vancouver (*42,43,44*), and the efforts of Atta-ur-Rahman and associates in Karachi.* Their basic idea, which relied on the biogenetic consideration that binary indole–indoline alkaloids are formed in plants by the union of vindoline

* It is interesting to note that while vindoline and catharanthine are abundant in *Catharanthus roseus* (L.) G. Don, a carbomethoxycleavamine (the initially presumed biogenetic precursor of the binary alkaloids), could not be detected or isolated. This suggested to Atta-ur-Rahman (*45*) that the VLB-type alkaloids are formed by union of *Iboga* and *Aspidosperma* alkaloids, an idea reinforced by biochemical studies (*46–49*).

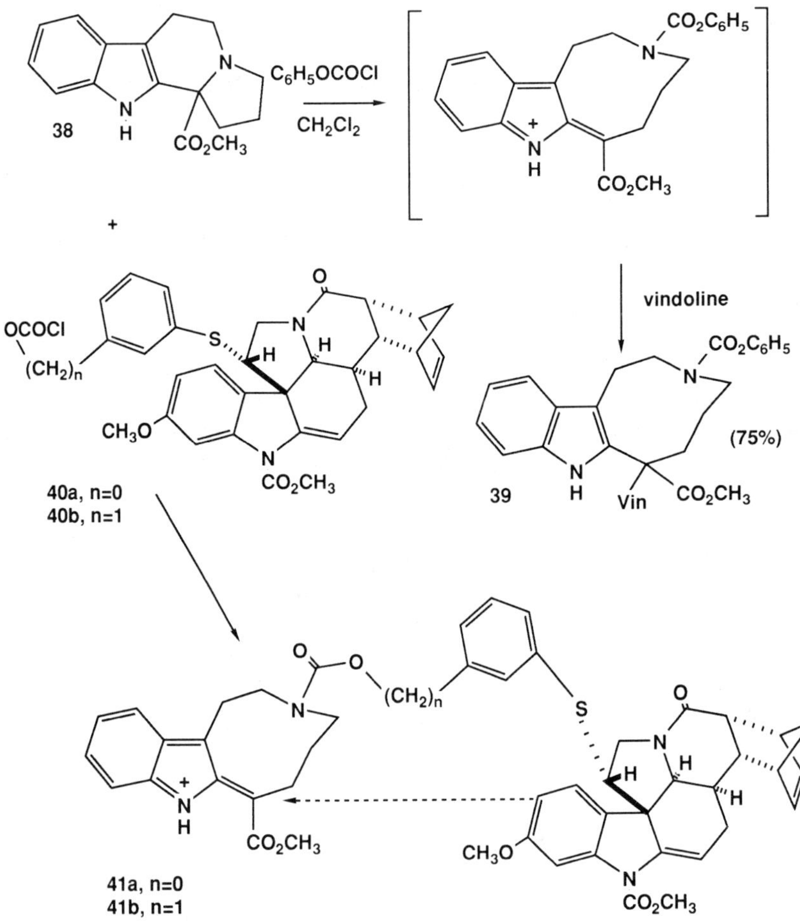

SCHEME 11

(**3**) with catharanthine (**21**) (two alkaloids abundant in the plant which produces vinblastine) (*42–49*), was to use a suitable catharanthine derivative instead of a chlorinated cleavamine for the coupling reaction (*50*). Particularly fruitful was the consideration that a synthesis of anhydrovinblastine (**42**) should be obtained from catharanthine with a leaving group on N^b (**43**). Such activation would allow rupture of the C-16–C-21 bond, leading to the imonium intermediate **44**, which, in turn, would be trapped by vindoline (Scheme 12). The modified Polonovski reaction (*51–53*), applied previously to the solution of other delicate problems in alkaloid

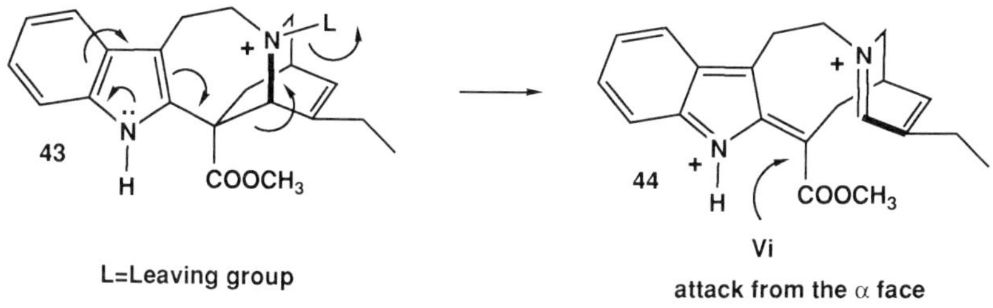

43
L=Leaving group
44
attack from the α face

SCHEME 12

chemistry (for reviews, see Refs. *54–56*), seemed well suited for achievement of this sequence.

A coupling of vindoline (**3**) with catharanthine *N*-oxide (**45**), mediated by trifluoroacetic anhydride, and subsequent reduction with sodium borohydride delivered anhydrovinblastine (**42**) in up to 40% yield, accompanied by minor amounts of the undesired C-16′ (*R*) isomer **24** as well as some 17-deacetylanhydrovinblastine (**46**) (Scheme 13) (*38–41*). The struc-

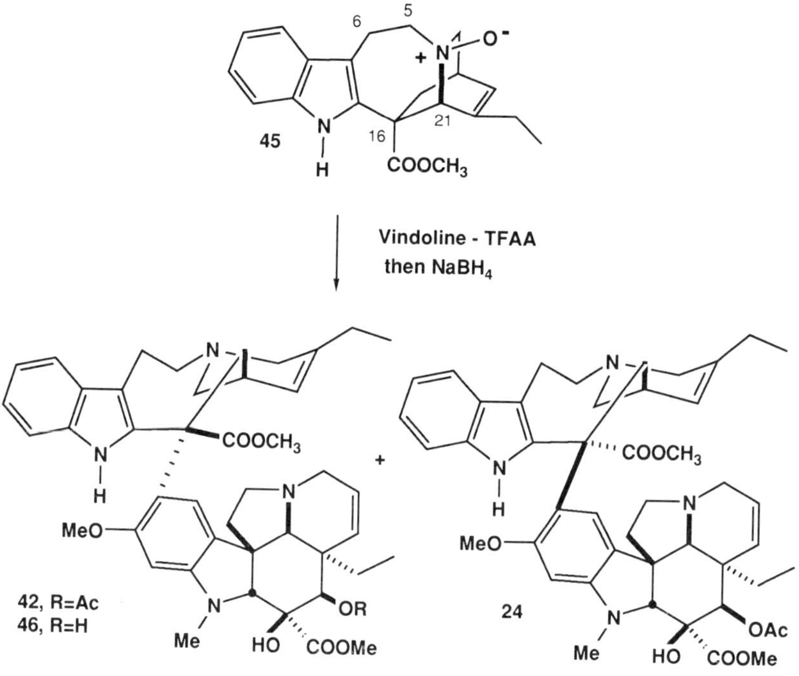

45
Vindoline - TFAA then $NaBH_4$
42, R=Ac
46, R=H
24

SCHEME 13

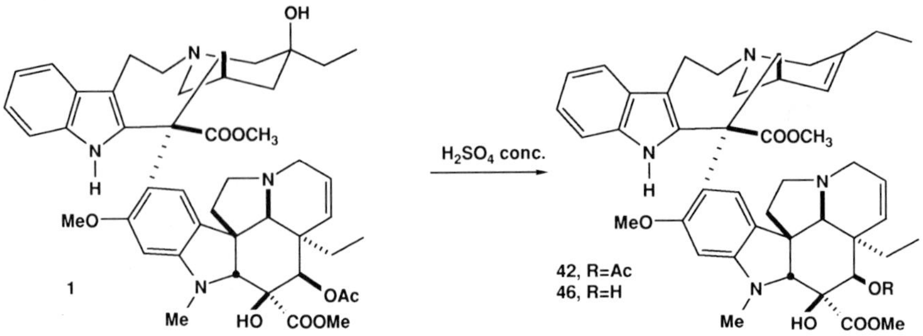

SCHEME 14

tures of the three binary alkaloids **42, 46,** and **24** were established by partial syntheses (*38–41*). Thus, vinblastine, on treatment with concentrated sulfuric acid, gave a mixture of anhydrovinblastine (**42**) and its 17-deacetyl congener **46**, which could be matched with the coupling products for establishment of their C-16′ stereochemistry (Scheme 14). The minor C-16′ epimer **24** had been obtained from the cleavamine **23**, following the chloroindolenine route (Scheme 5).

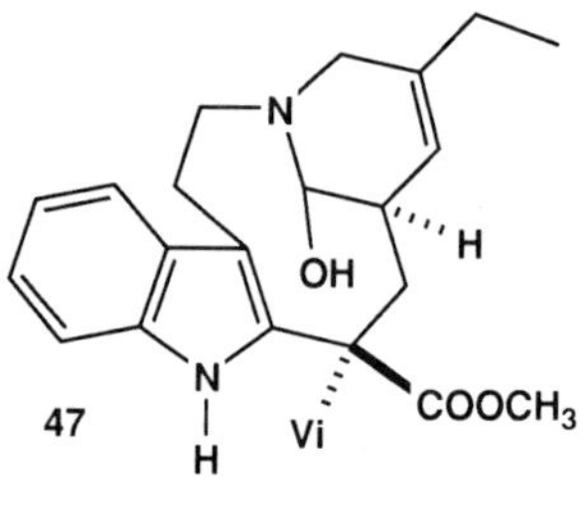

When the biomimetic coupling of vindoline to catharanthine *N*-oxide was performed at −50°C, only the desired (16′*S*) binary alkaloid was produced, in approximately 50% yield, in addition to a compound tentatively formulated as the carbinolamine **47** (*43*). It was found that generation of the C-16′ configuration of the coupling products is critically dependent on reaction temperature, with a gradual loss, at higher temperatures, of the C-16′ (*S*) configuration obtained at −50°C culminating in exclusive formation of the C-16′ (*R*) configuration at 62°C (34% yield) (*42–44*). These results were interpreted on the basis of two different intermediates and/or reaction processes. It was suggested that at low temperature a cation **48**, retaining the original conformation of catharanthine, was produced and nucleophilically attacked from the sterically less hindered α face, leading to the C-16′ (*S*) product. At higher temperature, a conformational change takes place to give the imonium ion **49** (identical to the one generated employing the chloroindolenine approach), in which the β face of the cation is now less hindered. Accordingly, attack by vindoline then leads to the undesired C-16′ (*R*) configuration of the binary product (Scheme 15).

The high regioselectivity of the catharanthine *N*-trifluoroacetoxy (**50**) fragmentation is quite remarkable and deserves some comment. Both the C-5–C-6 and the C-16–C-21 bonds of catharanthine are antiperiplanar with respect to the N—O bond, and therefore ring opening can occur in both directions (Scheme 16). The breaking of the C-5–C-6 bond (path **a**) is most commonly observed in saturated isoquinuclidines, and it has been used in an elegant preparation of vallesamine (*57*). The driving force for the alternative fragmentation (path **b**) then is presumably the gain in conjugation on formation of the vinyl imonium salt **48** (*58*). As seen below, compounds lacking the C-15–C-20 double bond of catharanthine yield negligible amounts of C-16′ coupling products, the main fragmentation now following the expected path **a**.

Analysis of the by-products obtained on coupling of catharanthine *N*-

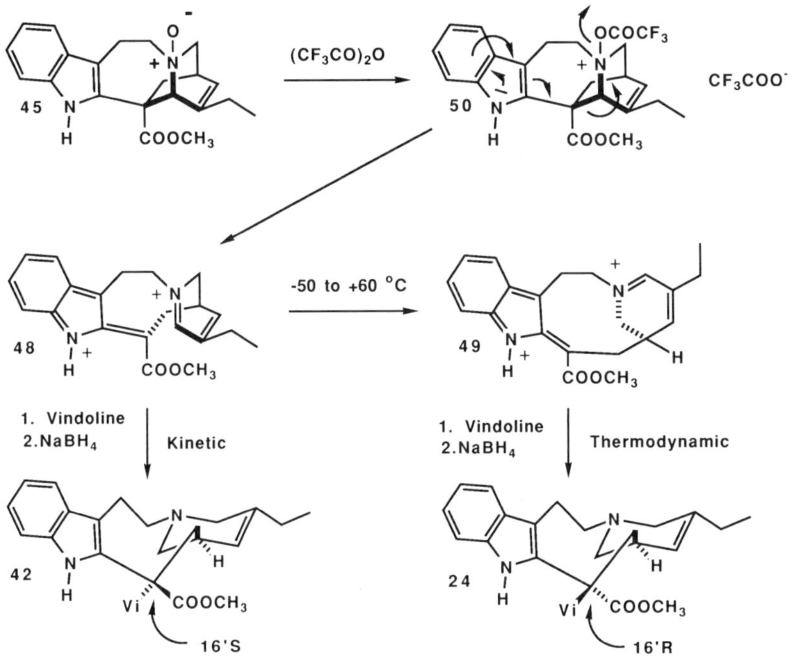

SCHEME 15

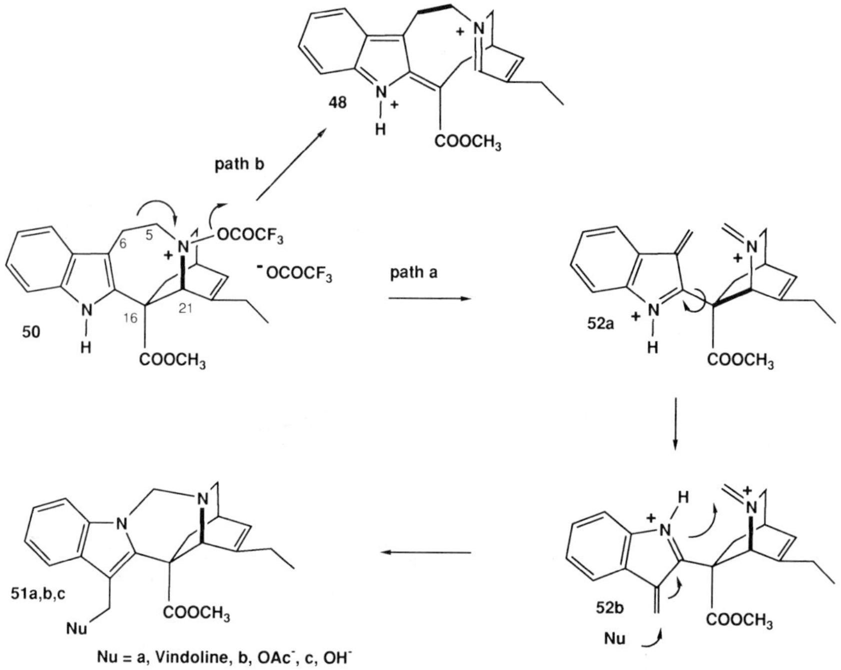

SCHEME 16

oxide with vindoline indicated a general structure **51** derived from rupture of the C-5–C-6 bond and addition of nucleophiles to C-6, with concomitant intramolecular trapping of the N^b imonium group in **52** by N^a (Scheme 16). When vindoline was omitted and catharanthine *N*-oxide was treated only with acetic anhydride, the acetate **51b**, resulting from addition of acetate ion to C-6, could be isolated as the sole product. With acetyl chloride–sodium hydroxide, addition of hydroxide and formation of the alcohol **51c** was found (*59*). A related skeletal rearrangement had been seen earlier, when ibogaine chloroindolenine (**53**) was heated with potassium cyanide (*60*). Fragmentation of the seven-membered ring, followed by a reaction of the resulting methylene imine **54** with cyanide, produced the nitrile **55** in good yield (Scheme 17).

For a synthesis of leurosidine (**56**), 15,20-dihydrocatharanthine *N*-oxide (**57**) was subjected to coupling with vindoline (**3**) under the modified Polonovski conditions. The initial adduct, imonium salt **58**, was converted to the enamine **59** in base. Oxidation of this product with osmium tetroxide proceeded chemo- and stereoselectively, without reaction of the

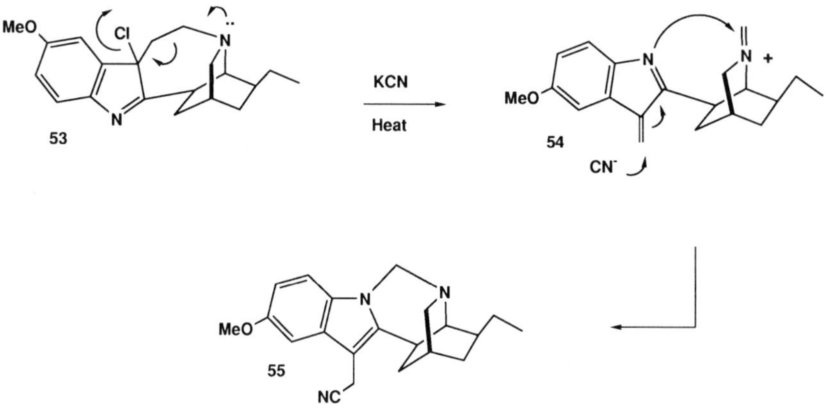

MeO Cl N N 53 KCN Heat MeO N N + 54 CN⁻ MeO N N 55 NC

SCHEME 17

C-14–C-15 double bond of the vindoline moiety (Scheme 18).* The stereochemistry of this hydroxylation can be assigned from the expectation that the bulky osmium reagent will attack the less hindered α face of the cleavamine segment of the binary alkaloid.† Without isolation, the crude dihydroxylation product **60** was subjected to reduction with sodium borohydride, and leurosidine (**56**) was obtained in about 5% yield (*62*). The low overall yield of this sequence most probably reflects the poor selectivity for cleavage of the C-16–C-21 bond of dihydrocatharanthine, as indicated above. A low yield was also obtained in another study of the coupling reaction of dihydrocatharanthine *N*-oxide (**57**) leading to 20′-deoxyleurosidine (**61a**) (*43*). From the same reaction, the C-20′ epimerization product 20′-deoxyvinblastine (**61b**), arising from enamine (**59**)–iminium ion (**58**) equilibration, could also be obtained (*43*).

In a parallel investigation, the Ban research group examined the coupling of a number of C-15 oxygenated derivatives of dihydrocatharanthine to vindoline (Scheme 19) (*63,64*). Here, too, C-20′ epimerization was

* Lack of reactivity of the $\Delta^{14,15}$ double bond of vindoline had been noticed earlier and might be the result of the particular conformation of that *Aspidosperma* alkaloid. Because of strong hydrogen bonding between the hydroxyl group at C-16 and N^b, the E ring adopts a boat conformation, in which attack of an electrophile from the β face is precluded. This conformation also positions the C-20 ethyl substituent axial, thereby shielding the α face of the molecule.

† Such a selectivity for attack at the α face of binary alkaloids is not uncommon (*61*).

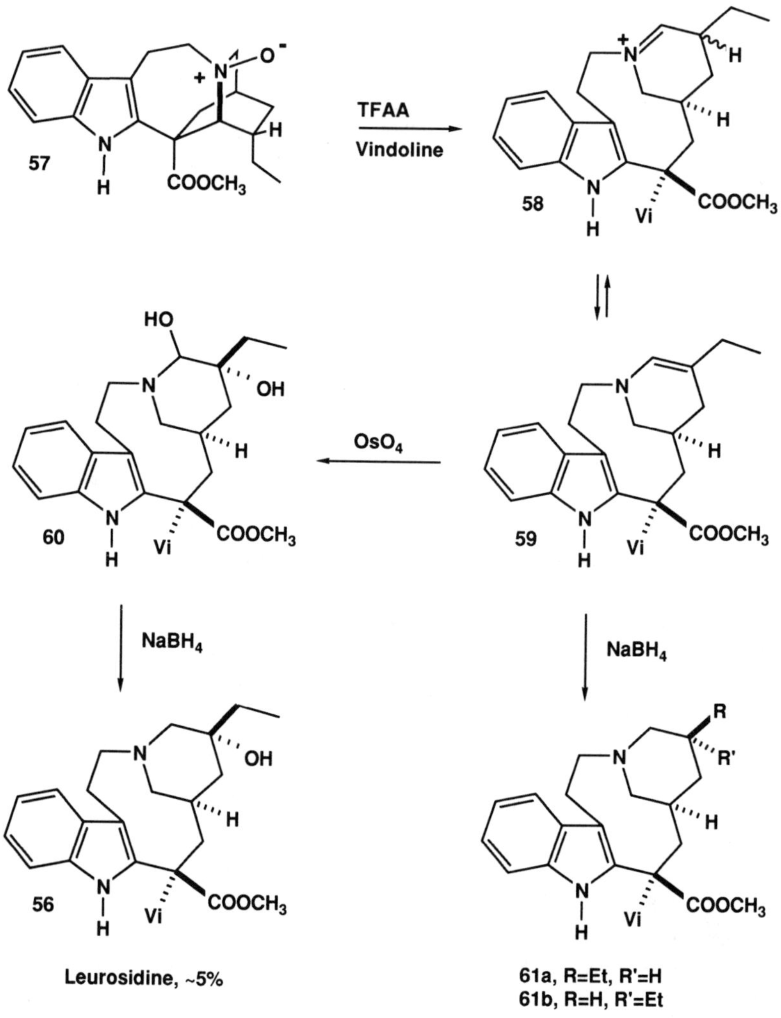

TFAA
Vindoline

57 58

OsO_4

60 59

$NaBH_4$ $NaBH_4$

56

Leurosidine, ~5%

61a, R=Et, R'=H
61b, R=H, R'=Et

SCHEME 18

found, with the 15-α-acetoxy-20-α-and-β-ethyl compounds **62** and **63** giving rise to the same 15-β-acetoxy-20-deoxyvinblastine (**64**) product.

Anhydrovinblastine (**42**) was produced from 15-β-acetoxy-20-α-ethyldihydrocatharanthine (**65**) in that study [*63,64*; an identical observation has been reported by Langlois et al. (*65,66*)], while the Kutney group reported the formation of 15-α-acetoxy-20-deoxyleurosidine (**66**) for this reaction (*67,68*). In all cases the major products of the reactions, **67–69**, were again

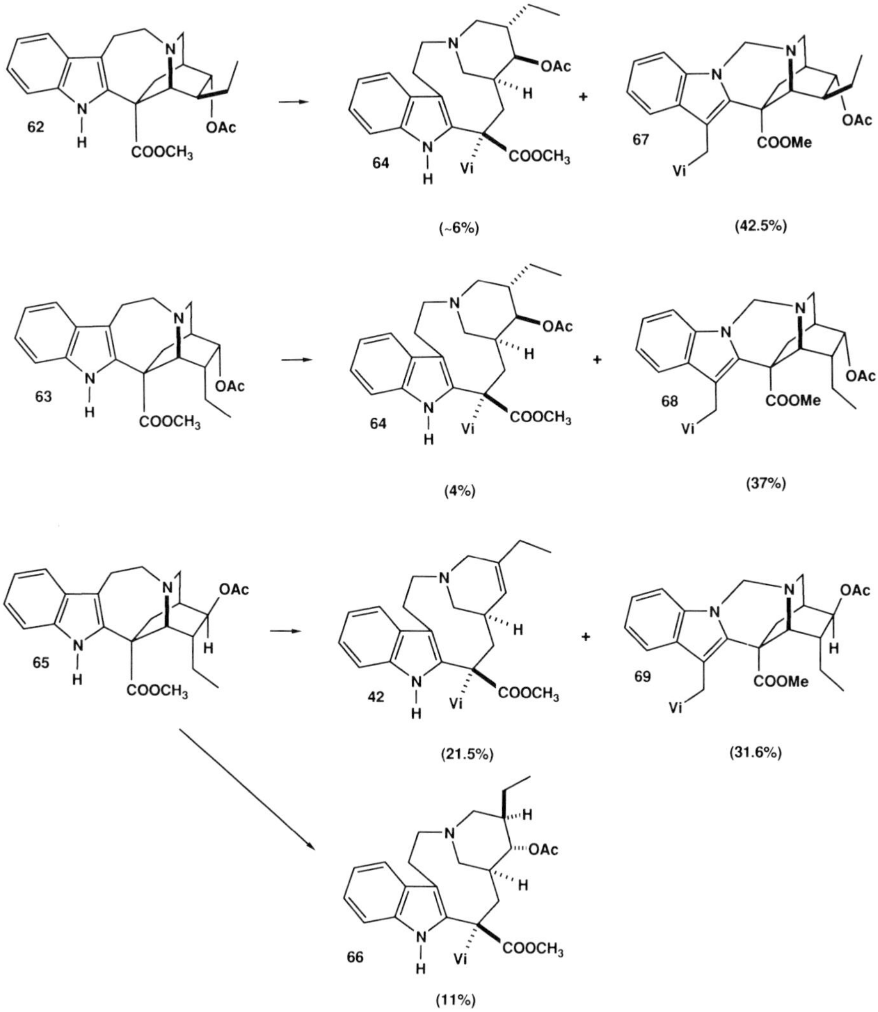

SCHEME 19

derived from C-5–C-6 bond rupture. This reaction pathway was also followed in the coupling of catharanthine lactone *N*-oxide (**70**), which led to the 1,3-piperazine lactone **71** and the corresponding lactol **72** (Scheme 20) (*63,64,69–74*).

Application of the modified Polonovski reaction to catharanthine β-epoxide (**73**) afforded leurosine (**74**) in 6% (*75*) or 20% yield (*70–74*), thus

SCHEME 20

securing the structure of that alkaloid as a (15′*R*,20′*S*) epoxide* (Scheme 21). The α-epoxide of catharanthine (**75**), on the other hand, provided only the product **76**, derived from C-5–C-6 cleavage.

Atta-ur-Rahman *et al.* reported the oxidation of catharanthine by a modified Prevost reaction to result in formation of the 20-α-acetoxy derivative. Coupling of its *N*-oxide (**77**) to vindoline, using trichloroacetic anhydride (rather than the usual trifluoro compound), and subsequent re-

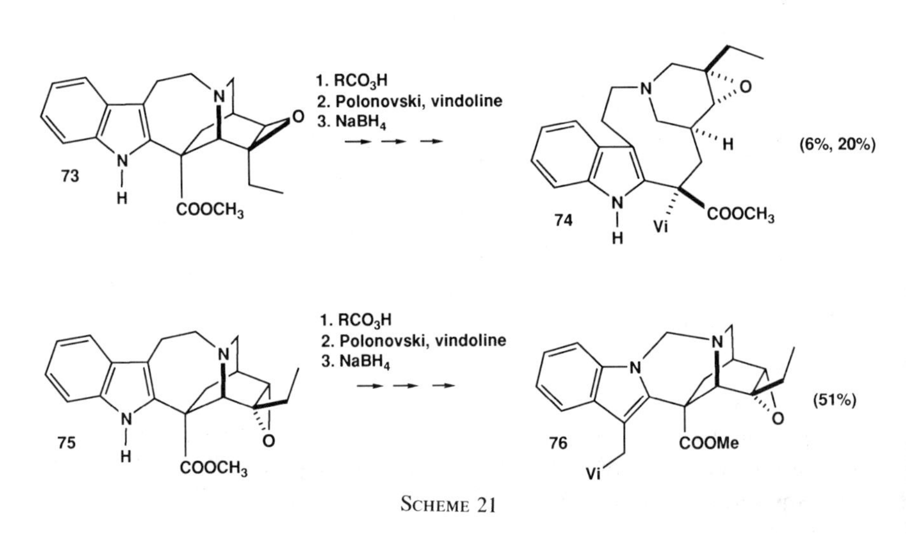

SCHEME 21

* The stereochemistry of the epoxide of anhydrovinblastine was misassigned in Ref. *71* but corrected in Ref. *72*.

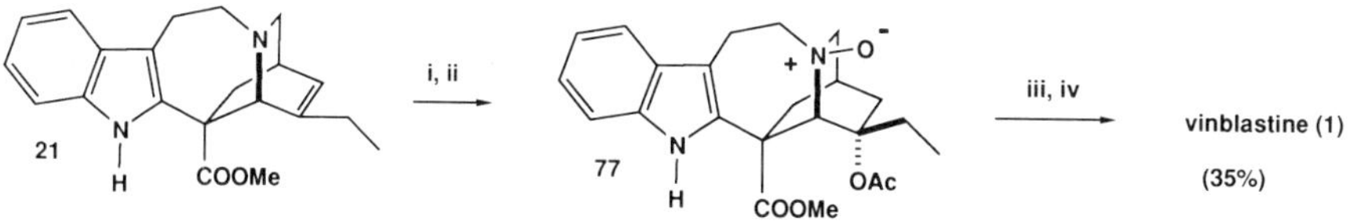

SCHEME 22. Reagents: i, AgOAc, I_2, AcOH, $NaBH_4$; ii, *m*-CPBA; iii, vindoline, $(CCl_3CO)_2O$; iv, $NaBH_4$ then Ac_2O, NaOAc.

duction were said to give vinblastine in as much as 35% yield (Scheme 22) (*76,77*). Unfortunately, the initial catharanthine oxidation to a 20-α-acetoxy product could not be repeated by others, thus casting doubt on the overall generation of vinblastine by this sequence (*78,79*).

The relatively facile formation of anhydrovinblastine (**42**) by the modified Polonovski reaction, and the poor yields experienced on coupling of other catharanthine derivatives, made anhydrovinblastine an attractive precursor for the preparation of additional binary alkaloids, including vinblastine itself (*80,81*).* Hydrogenation to 20′-deoxyleurosidine (**61a**), formation of its $N^{b'}$-oxide (**78**), and reaction with trifluoroacetic anhydride led to the enamine **59** (Scheme 23). Oxidation of this enamine with thal-

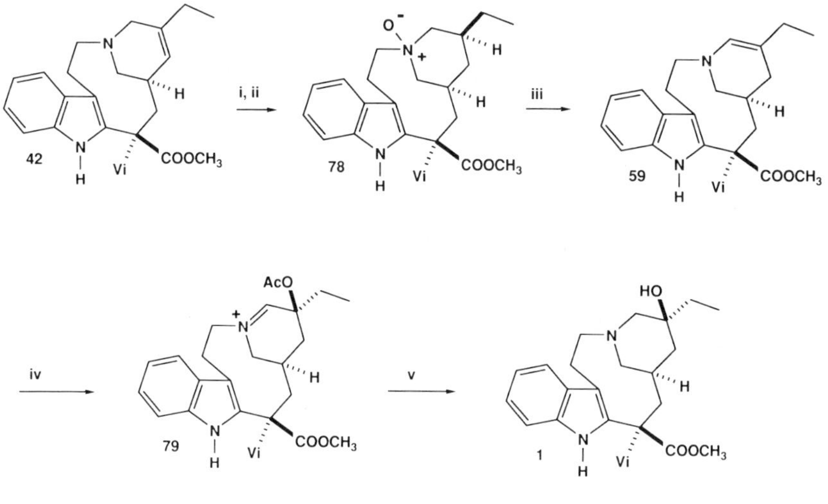

(30% overall)

SCHEME 23. Reagents: i, H_2/PtO_2; ii, *m*-CPBA; iii, $(CF_3CO)_2O$; iv, Tl $(OAc)_3$; v, $NaBH_4$.

* It is interesting to note that these experiments strongly suggested that anhydrovinblastine is the biogenetic precursor of the VLB-type alkaloids in the plant, a proposal recently substantiated (*47,48,82–88*).

lium triacetate (*89*) led to axial (β face) introduction of a 20′-acetoxy substituent. On reduction of the initially formed imonium product **79** with sodium borohydride and acetate hydrolysis, vinblastine was reportedly formed in 30% yield. The low overall yield in this sequence not only is derived from the thallium triacetate oxidation step, but it can also be expected as a consequence of failure to achieve the desired regioselectivity in the initial Polonovski oxidation of deoxyleurosidine (**61a**). The major result of that reaction proved to be cleavage of the C-5′–C-6′ bond (Scheme 24) and formation of seco derivatives **80** and **81**, which were obtained from trapping of the bisimonium intermediate **82** with various nucleophiles such as cyanide, hydroxide, or hydride (*90*). Similarly, when applied to anhydrovinblastine (**42**), the Polonovski oxidation sequence led to D′-seco derivatives **83–85,** as well as to the pharmacologically interesting 5′-nor compound **86**, derived from hydrolytic loss of formaldehyde and intramolecular trapping of the imonium function (Scheme 25) (*91–93*). This ring closure, unfortunately, seems to be dependent on the constitution of the piperidine ring. For example, 20′-deoxyleurosidine *N*-oxide (**78**), under the same conditions, produced only D′-seco derivatives, which resisted attempts at cyclization to a 5′-nor product (*90*).

While the coupling of vindoline to catharanthine *N*-oxide proceeds (particularly at low temperature) with high C-16′–C-14′ PARF stereoselectivity, this result can be influenced by structural modifications which either increase the stability of the intermediate, cleavamine-derived cation

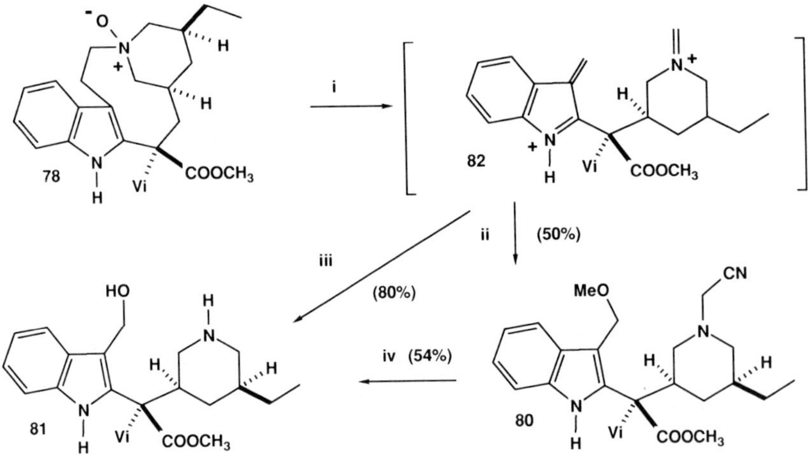

SCHEME 24. Reagents: i. $(CF_3CO)_2O$; ii, KCN–MeOH; iii, THF–H_2O–$NaHCO_3$; iv, $AgBF_4$–THF–H_2O.

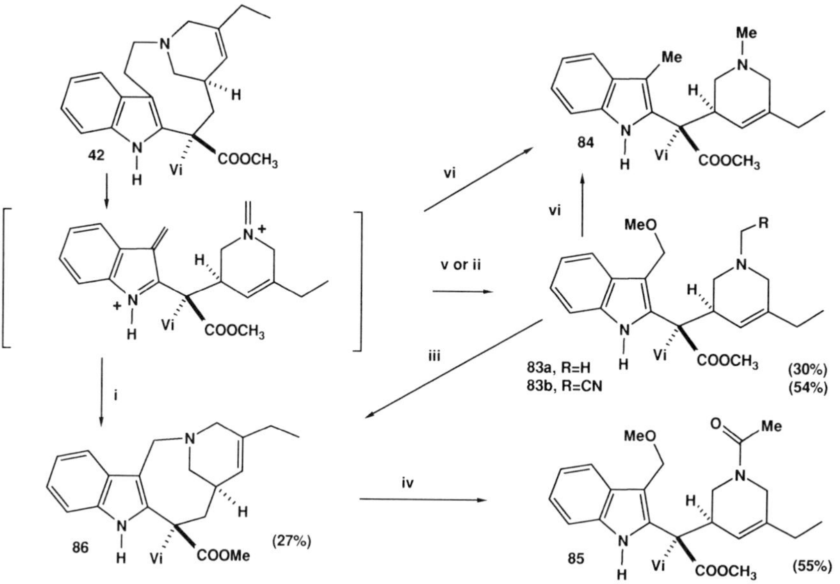

SCHEME 25. Reagents: i, THF–H_2O; ii, KCN–MeOH; iii, $AgBF_4$; iv, Ac_2O–MeOH; v, $NaBH_4$–MeOH; vi, $NaBH_3CN$.

or decrease the nucleophilicity of the vindoline reaction partner, thus allowing a conformational change of the cationic nine-membered ring system before formation of the vindoline to C-16′ bond. This effect could be seen in the coupling of decarbomethoxy catharanthine *N*-oxide **87** (*94*), where much of the stereoselectivity of introduction of the vindolinyl substituent was lost (**88**, **89** in Scheme 26) and, alternatively, in the coupling of vindorosine (demethoxyvindoline, **90**), where only the C-16′–C-14′ PREF diastereomer **91** (Scheme 27) was obtained in 18% yield on coupling to catharanthine *N*-oxide at 0°C (*39*).

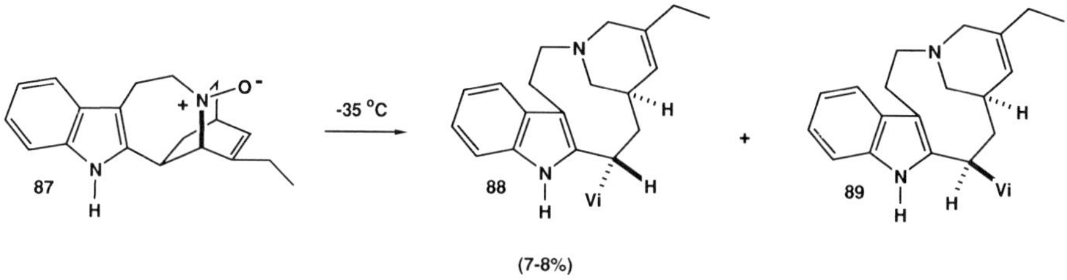

SCHEME 26

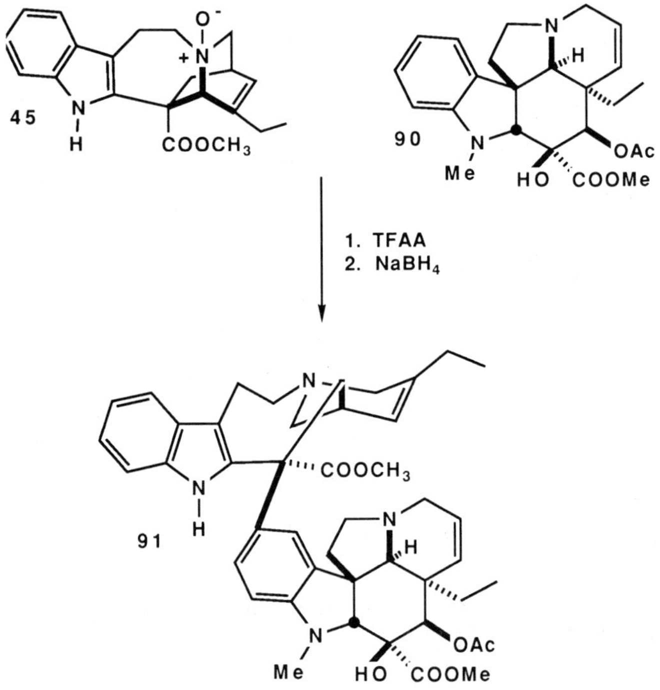

1. TFAA
2. $NaBH_4$

SCHEME 27

Coupling of other vindoline derivatives with ring D or E modified oxidation levels (**92–96**) to catharanthine *N*-oxide provided new binary products for biological evaluation (*39, 95–97*). The two diastereomeric C-16′–C-14′ PARF anhydrovinblastines **42** and **97** were obtained in a 46 : 54 ratio (50% yield) from racemic catharanthine (*98*), and the corresponding 20′-desethyl compounds **98** and **99** were generated at −20°C in a 1 : 1 ratio (16% yield each), and at −76°C in lower yields, together with the corre-

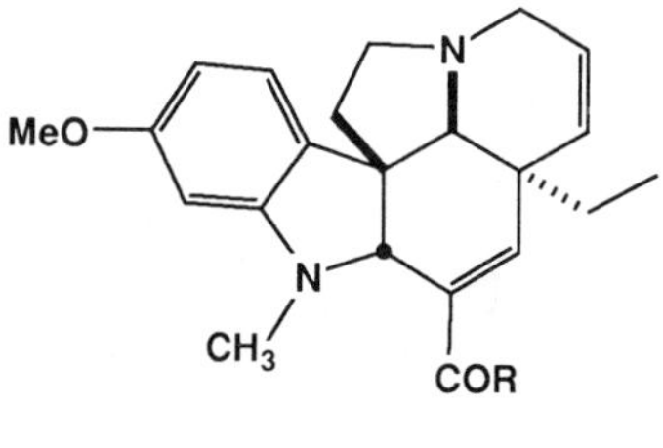

92, R = OMe (45%)
93, R = NHMe (36%)

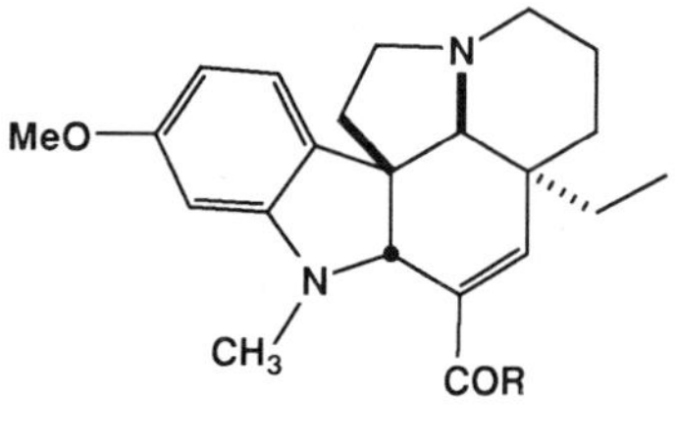

94, R = OMe (30%)
95, R = NHMe (22%)

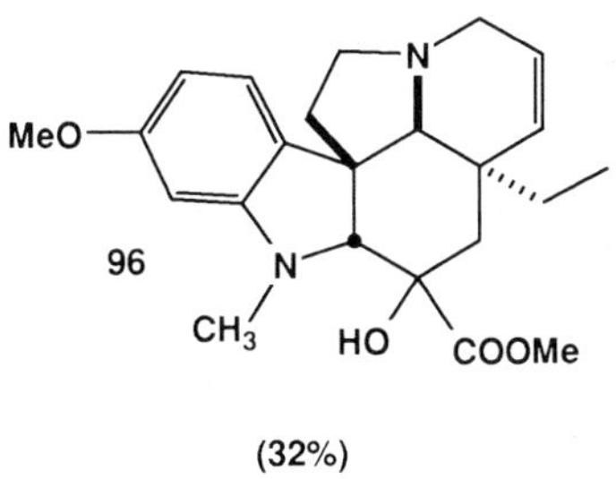

96

(32%)

sponding C-15′ methoxy substituted products **100** and **101** (Scheme 28) (*99*).

Initial unsuccessful attempts at biomimetic coupling of catharanthine and vindoline by the Kutney group had been directed at cleavage of the C-16–C-21 isoquinuclidine bond by ionization of a C-20 oxygenated pre-

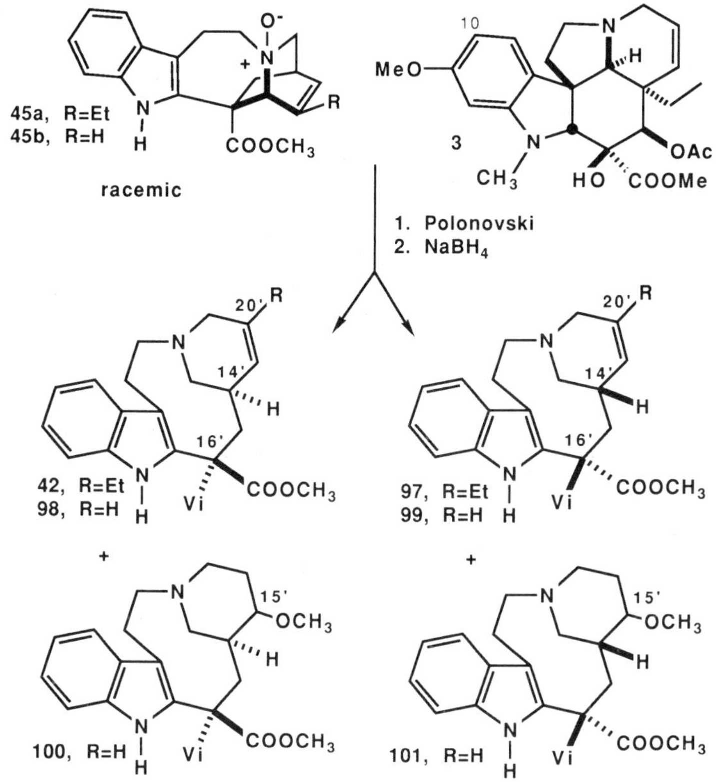

45a, R=Et
45b, R=H
racemic

3

1. Polonovski
2. $NaBH_4$

42, R=Et
98, R=H

97, R=Et
99, R=H

+

100, R=H

101, R=H

SCHEME 28

cursor **102** (Scheme 29) (*43, 100*). To this end, opening of a C-15–C-20 β-epoxide **103** would seem ideal, but it was found that catharanthine, on treatment with peracid, gave first the *N*-oxide **45**, while catharanthine hydrochloride on reaction with *m*-chloroperoxybenzoic acid was converted to a hydroxy indolenine **104**. A reaction of the latter with vindoline, followed by reduction with sodium borohydride, resulted in formation of a C-21 coupled carbomethoxycleavamine (**105**).

In another search for an alternative to Potier's modified Polonovski reaction of catharanthine *N*-oxide (**45**), it has now been found that anhydrovinblastine (**42**) can be generated directly, in 77% yield, from a reaction of catharanthine and vindoline in 0.01 *N* acid, promoted by ionized ferric salts, followed by reduction with sodium borohydride (Scheme 30) (*101*). Remarkably, the cation radical **106** generated by Fe(III), in accord with other simple amine oxidations by Lindsay Smith and Mead (*102*), resulted in isoquinuclidine fragmentation and coupling to vindoline at 0°C, without the conformational inversion observed in the modified Polonovski reaction at that temperature (see Scheme 15). Other metal oxidants or ligand-bound Fe(III) did not promote the coupling reaction. It will be of interest to see if the overwhelming competition of C-5–C-6 bond

102 15 16 21 20 X N H $COOCH_3$

H N $COOCH_3$ + vindoline

42 N H Vi $COOCH_3$

catharanthine (21) hydrochloride

103 N H $COOCH_3$ O

104 HO N $COOCH_3$

i. 1N HCl, vindoline
ii. $NaBH_4$

105 N H Vi $COOCH_3$

SCHEME 29

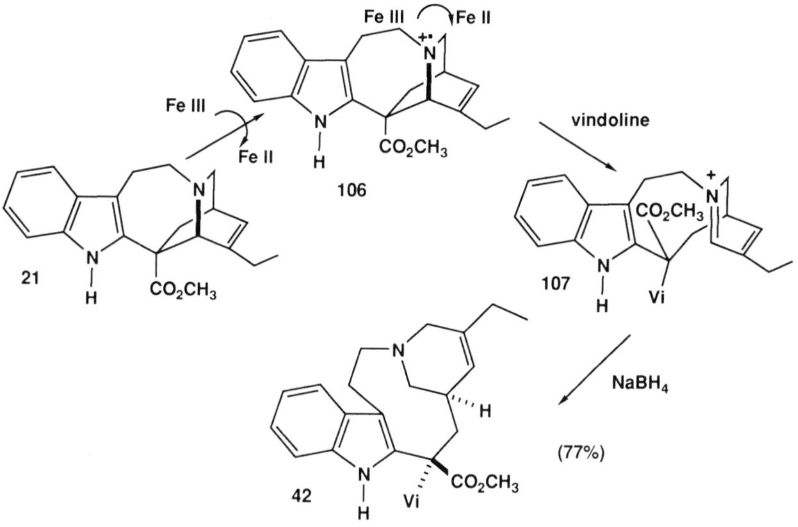

SCHEME 30

cleavage, which is found in compounds lacking the catharanthine double bond (see Scheme 16), will be influenced by this new variation of the biomimetic coupling concept.

VI. New Strategies for Generation of C-16′–C-14′ PARF Relative Stereochemistry

A. COUPLING OF INDOLOAZACYCLOUNDECANES

In an adaptation of the chloroindolenine approach studied previously with 16-carbomethoxycleavamines (e.g., Scheme 4), the Schill group generated a corresponding D-seco compound (**108**) (Scheme 31) for coupling to vindoline (*30–32*). Tryptamine was condensed with 3-(benzyloxymethyl)-5-carbomethoxypentanal (**109**) to form the tetrahydrocarbazole lactam **110**. After reduction of the lactam function with lithium aluminum hydride, a benzylchloroformate-induced cleavage of the benzylic C—N bond generated the eleven-membered hydroxy urethane **111** (compare Schemes 2 and 3). Following again a procedure developed by Harley-Mason and co-workers (*103–106*) and extended by Kutney *et al.* (*107*), this alcohol was homologated to ester **112**. Cleavage of its benzyl ether function by hydrogenolysis, formation of the corresponding mesylate **113**,

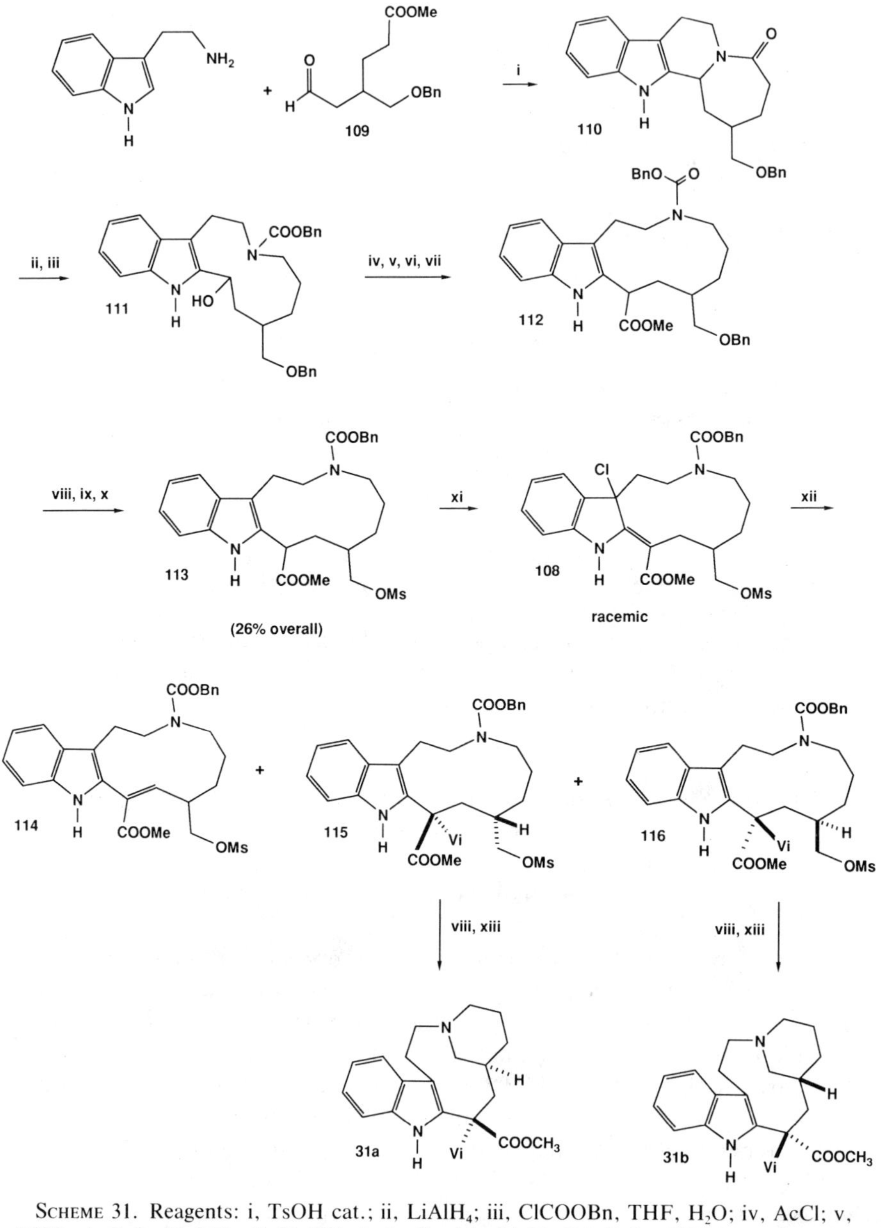

SCHEME 31. Reagents: i, TsOH cat.; ii, $LiAlH_4$; iii, ClCOOBn, THF, H_2O; iv, AcCl; v, KCN; vi, H_2O_2, NaOH; vii, MeOH, Amberlist 15; viii, H_2, Pd–C; ix, ClCOOBn; x, MsCl, Et_3N; xi, ${}^tBuOCl$; xii, vindoline, $BF_3 \cdot Et_2O$; xiii, Na_2CO_3.

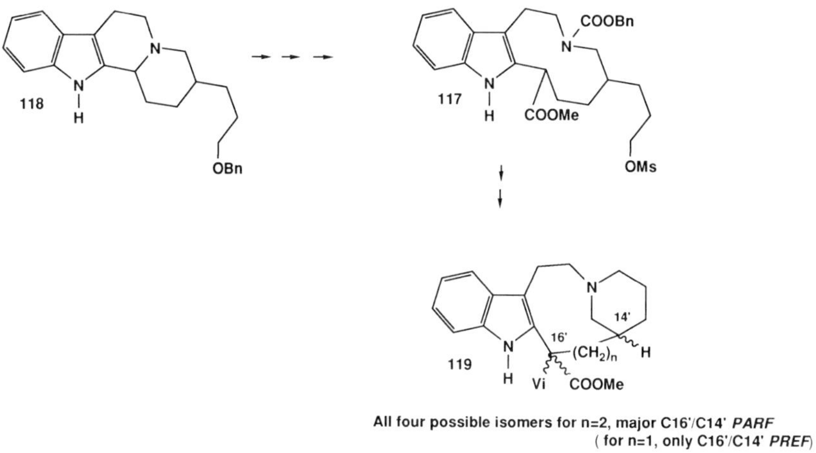

All four possible isomers for n=2, major C16'/C14' *PARF*
(for n=1, only C16'/C14' *PREF*)

SCHEME 32

and the usual chlorination then produced the racemic chloroindoline alkene **108** for the coupling reaction. A boron trifluoride-induced development of the derivative carbocation, in the presence of vindoline, led to the uncoupled ester product **114** and the binary azacycloundecanes **115** (30%) and **116** (30%) with the desired C-16′–C-14′ PARF relative stereochemistry. Chromatographic separation of the diastereomers and removal of the carbobenzyloxy protection from $N^{b'}$ allowed cyclization to 20′-desethyl-20′-deoxyvinblastine (**31a**) and 20′-desethyl-20′-deoxyvincovaline (**31b**).

A similar sequence was used for generation and coupling of the indoloazacyclodecane **117**, obtained from the tetrahydrocarbazole **118** (Scheme 32). Here, however, C-16′–C-14′ PREF as well as PARF products (**119**) were produced in the chloroindoline alkene coupling reaction, while the corresponding nine-membered ring D-seco cleavamine gave only the PREF product (see also Scheme 7).

B. Coupling of D-Seco ψ-vincadifformines

Intrigued by the hypothesis of a dehydrosecodine (**120**) as a key biogenetic intermediate in the natural generation of alkaloid structures of both the *Aspidosperma* (tabersonine, **121**) and the *Iboga* (catharanthine, **21**) types (Scheme 33) (*108, 109*),* we developed efficient biomimetic synthe-

* A quantitive, spontaneous cyclization of the 16-carbomethoxy C-20–C-21 unsaturated cleavamine to coronaridine (*110*) and the general failure of dehydrosecodine to serve as a synthetic precursor of catharanthine (see Ref. *111* for literature review) suggest, however, that the *Iboga* alkaloids may be preferably assigned a biogenetic origin based on cleavamine cyclizations rather than on a dehydrosecodine.

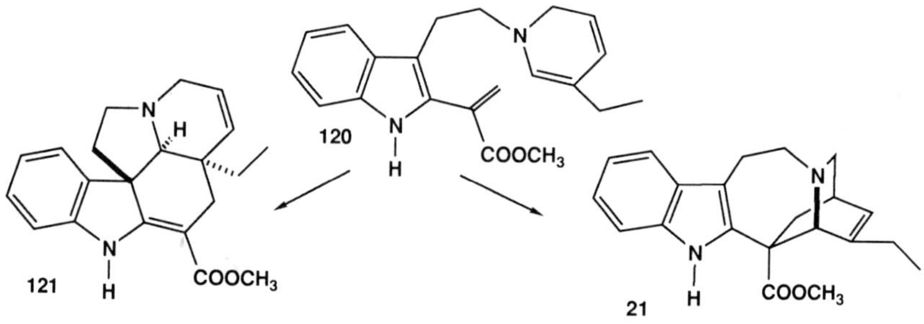

SCHEME 33

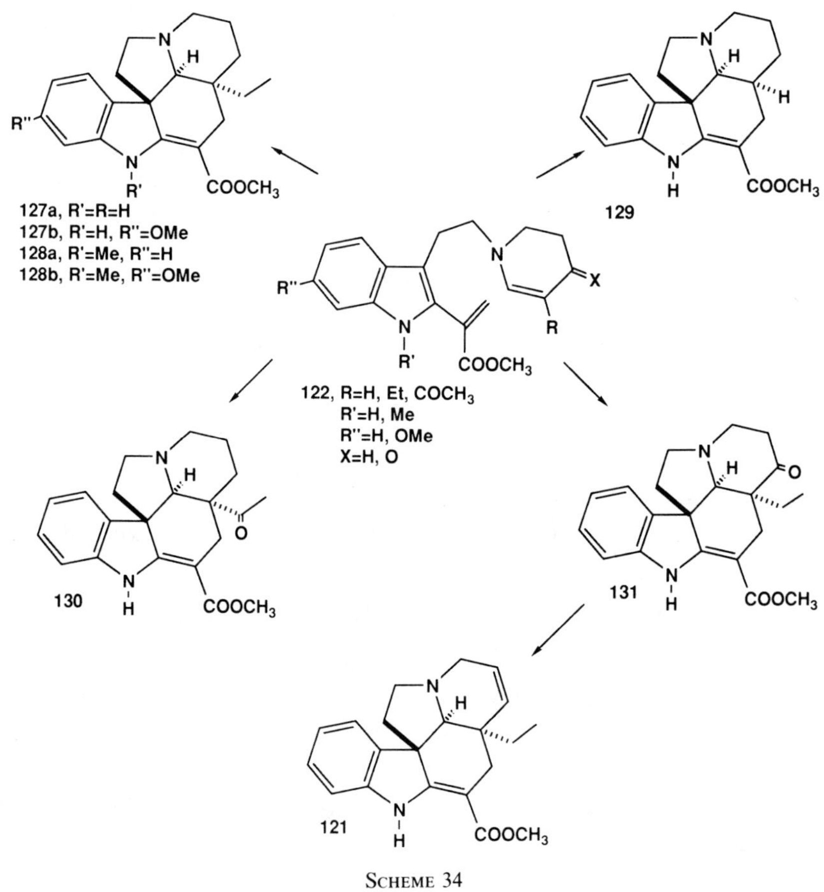

SCHEME 34

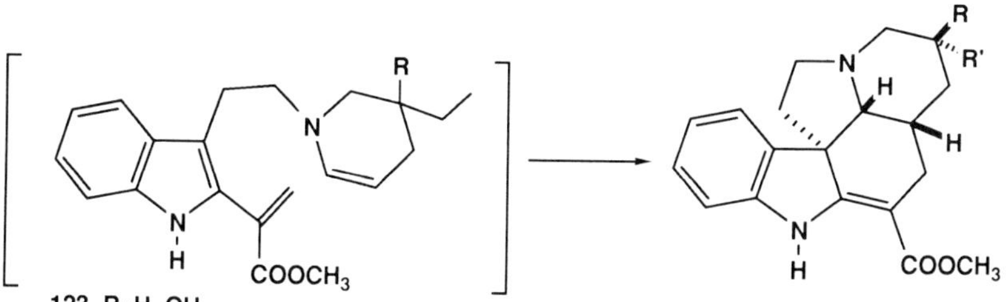

123, R=H, OH

34, R=Et, R'=OH
132, R=OH, R'=Et
133, R=H, R'=Et
134, R=Et, R'=H

SCHEME 35

ses which utilized the formal Diels–Alder potential of secodine- (**122–125**) and dehydrosecodine-type (**126**) intermediates. These reaction schemes (Schemes 34–39) provided total syntheses of vincadifformine (**127a**) (*112,113*), ervinceine (**127b**) (*113,114*), minovine (**128a**) (*113*), ψ- and 20-epi-ψ-vincadifformine (**133, 134**) (*115*), pandoline (**34**) (*115,116*), 20-epi-

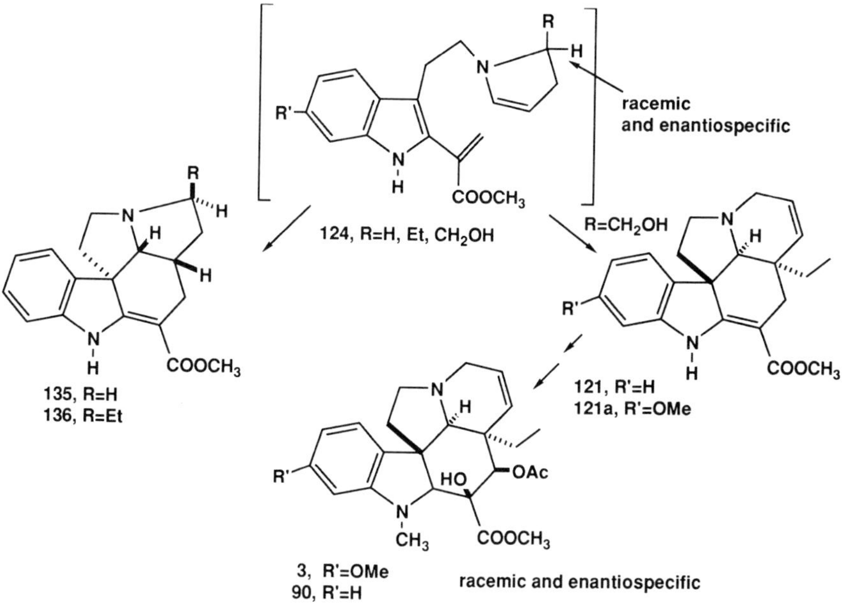

racemic and enantiospecific

124, R=H, Et, CH_2OH

R=CH_2OH

135, R=H
136, R=Et

121, R'=H
121a, R'=OMe

3, R'=OMe
90, R'=H

racemic and enantiospecific

SCHEME 36

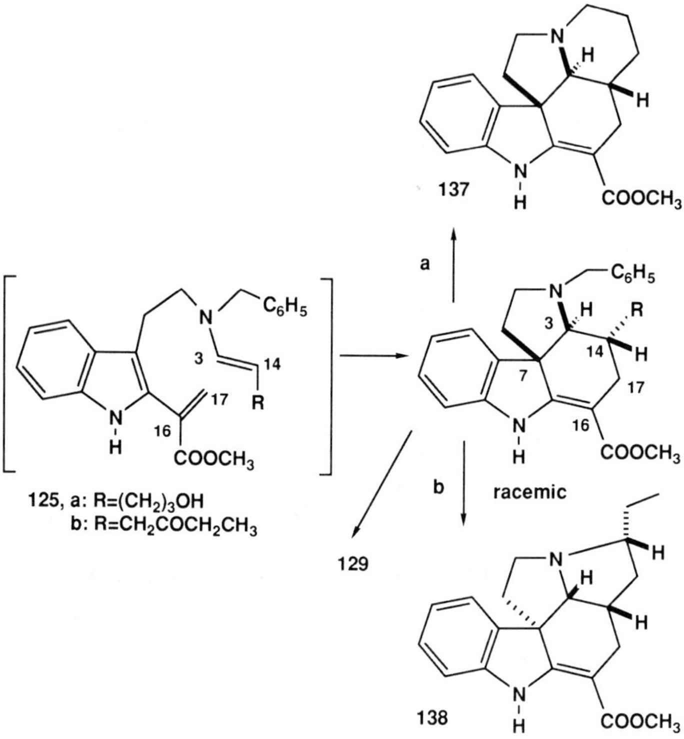

SCHEME 37

pandoline (**132**) (*115,116*), ibophyllidine (**138**) (*117*), 20-epiibophyllidine (**136**) (*117*), minovincine (**130**) (*118,119*), the D–E-*cis*- and -*trans*-desethylvincadifformines (**129, 137**) (*120,121*), tabersonine (**121**) (*111,116,122*), vindoline (**3**) (*123*), and catharanthine (**21**) (*111*), as well as 16-carbomethoxycleavamine-type products (**139–144**), in which the C-7–C-3 or C-7–C-21 bond of the corresponding preceding pentacyclic alkaloids had been broken reductively, to furnish the indoloazonine moiety inherent in half

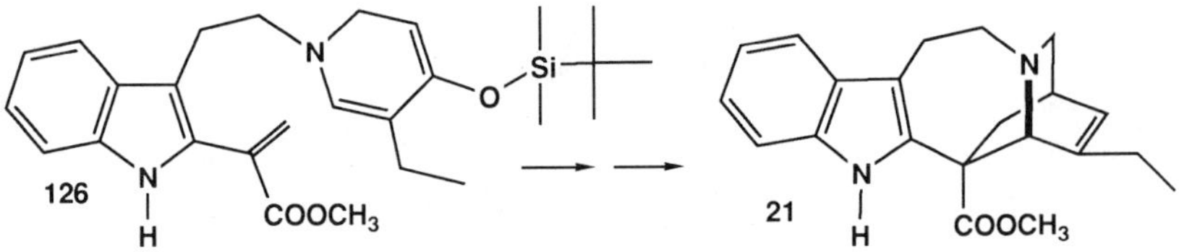

SCHEME 38

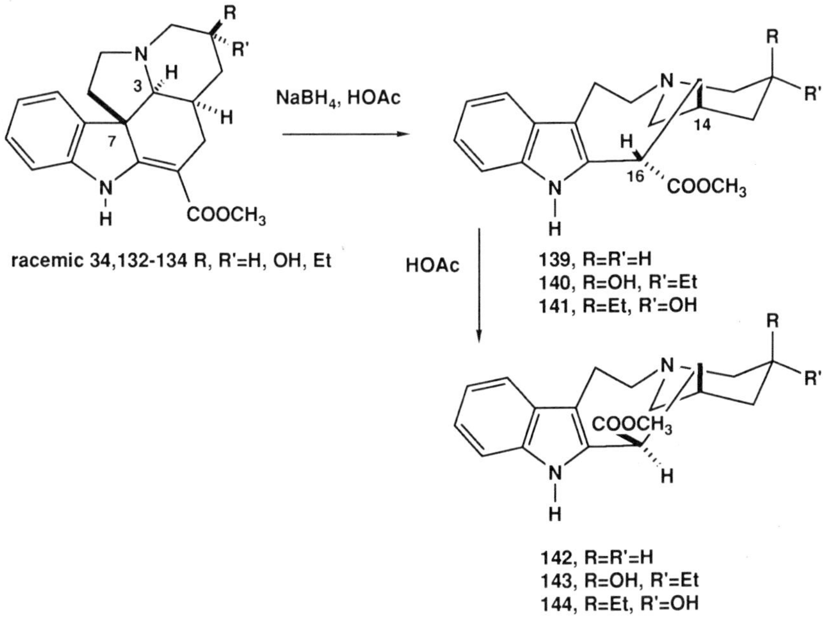

SCHEME 39

of the vinblastine (**1**) structure. Our syntheses of catharanthine (**21**) and vindoline (**3**), with high enantioselectivity in the latter case (*123*), thus provided us with precursors for a coupling reaction, which had been shown to lead to anhydrovinblastine (**42**) (*38–45*). The discovery of this reaction, enviable from the perspective of our interest in biomimetic syntheses, also revealed, however, its high specific structural requirement. This consideration, together with the modest yield obtained from the subsequent three-step conversion of anhydrovinblastine (**42**) to vinblastine (**1**), made it imperative to find a more general and practical route to the final product and analogous compounds that were desired for evaluation as anticancer agents. Such a route, which has now rewarded us with several alternative variants for the synthesis of vinblastine, as well as over 80 structural and stereochemical analogues of vinblastine (see Table I in Chapter 3, this volume, for some representative examples), could be obtained from reconsideration of our biomimetic secodine chemistry.

Reaction of vindoline (**3**) with the chloroindolenines derived from vincadifformine (**127a**), ψ-vincadifformine (**133**), ψ-tabersonine (**133a**), or synthetic pandoline (**34**) (*115, 116*), followed by reduction of the resulting imines **145** and **145a** (Scheme 40) with potassium borohydride, had given

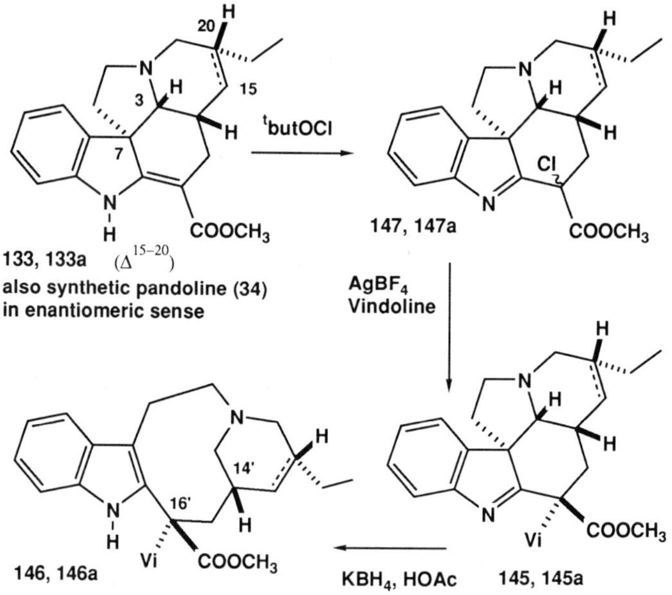

SCHEME 40

Lévy modest yields of indole–indoline products **146** and **146a**, but their relative stereochemistry at C-16′–C-14′ was exclusively PREF (*124, 125;* for the coupling methodology of Refs. *124* and *125*, see Ref. *126*), corresponding to the coupling products previously obtained from vindoline and the chloro derivatives of cleavamines (i.e., the unnatural C-16′–C-14′ relative stereochemistry, with resultant lack of vinblastine-like antitubulin activities). While a simple steric approach control can be invoked for explanation of a cleavamine to vindoline coupling reaction (Fig. 2), a more circumspect, stereoelectronics based analysis is required for the Lévy coupling, where one must consider that the species reacting with vindoline is not likely to be the initial pentacyclic chloro compound **147**, but rather a tetracyclic imonium salt **148** derived from rupture of the central C-7–C-3 bond, with placement of cationic charge onto $N^{b'}$ (Scheme 41). Support for this hypothesis was found in our observation that the relative ease or failure of coupling of such chloro imines correlates directly with the relative ability of the parent vinylogous urethanes to undergo reductive scission of the C-7–C-3 bond (e.g, the D–E-trans compound **137** did not give indoloazonines with sodium borohydride in hot acetic acid, and its chloro derivatives **150** do not undergo the coupling reaction with vindoline) (*13,121*).

If formation of the vindoline to C-16 bond is then concerted with refor-

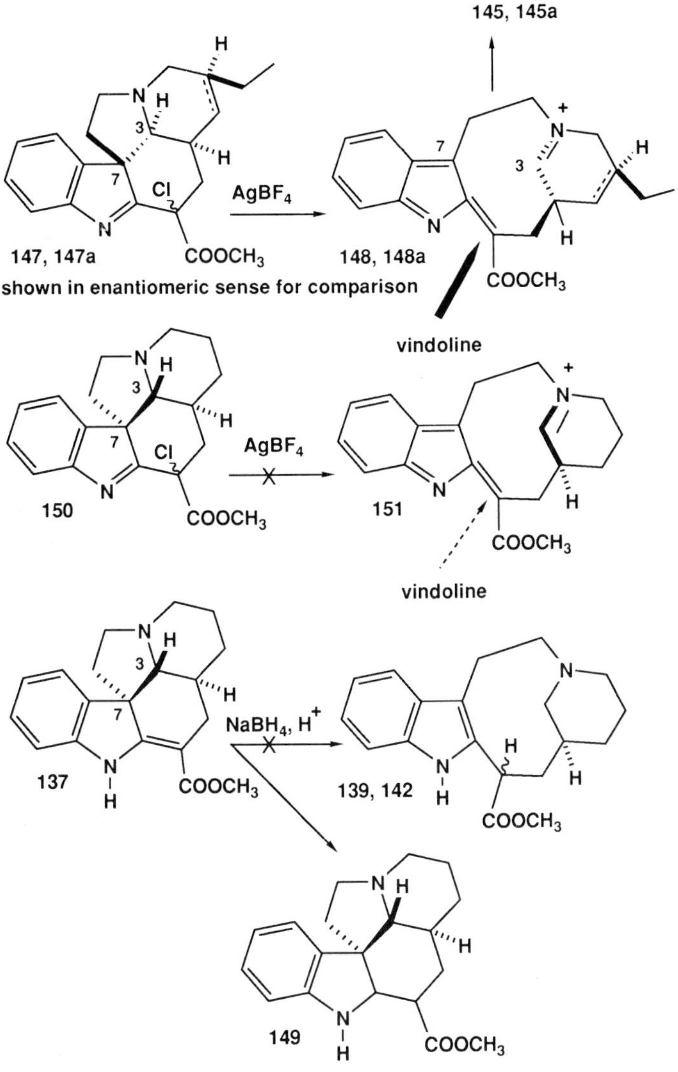

SCHEME 41

mation of the central C-7–C-3 bond of a ψ-vincadifformine, and results in the wrong C-16′–C-14′ relative stereochemistry, one could expect that a chloro imine precursor (**150**), with inverted stereochemistry at C-3 and C-7, should lead to a coupling product with the desired C-16′–C-14′ PARF relative stereochemistry. However, since such a D–E-trans-fused pentacyclic chloro compound (**150**) did not undergo the coupling reaction with vindoline, owing to lack of formation of an imonium intermediate (**151**) by rupture of the C-3–C-7 bond, an alternative coupling precursor was required. Such a compound had to allow for the stereoelectronic requirements for cleavage of the C-3–C-7 bond, i.e., a trans periplanar orientation of the C-3–C-7 bond and the N^b lone pair of electrons, while still maintaining the desired C-3–C-14 trans substitution. For such a system we turned to the D-seco compounds (**152**) as substrates for coupling to vindoline. These tetracyclic compounds were readily prepared through the generation of transient seco secodines **125a,c,d,** where the enamine function is formed exclusively as an *E*-substituted double bond (Scheme 42). The *E*-enamino acrylates **125** could be obtained by condensation of the indoloazepine **153** with appropriate aldehydes (**154**), benzylation of the resulting bridged indoloazepines (**155**) and fragmentation of their quaternary salts **156**. The subsequent intramolecular reaction of these seco secodines (**125**) then results in only the desired stereochemistry of the tetracyclic product **152**.

Chlorination of these vinylogous urethanes **152** with *tert*-butylhypochlorite, followed by reaction with silver tetrafluoroborate and protonated vindoline, at or below room temperature, provided almost exclusively coupling products **157** and **158** with the desired C-16′–C-14′ PARF rela-

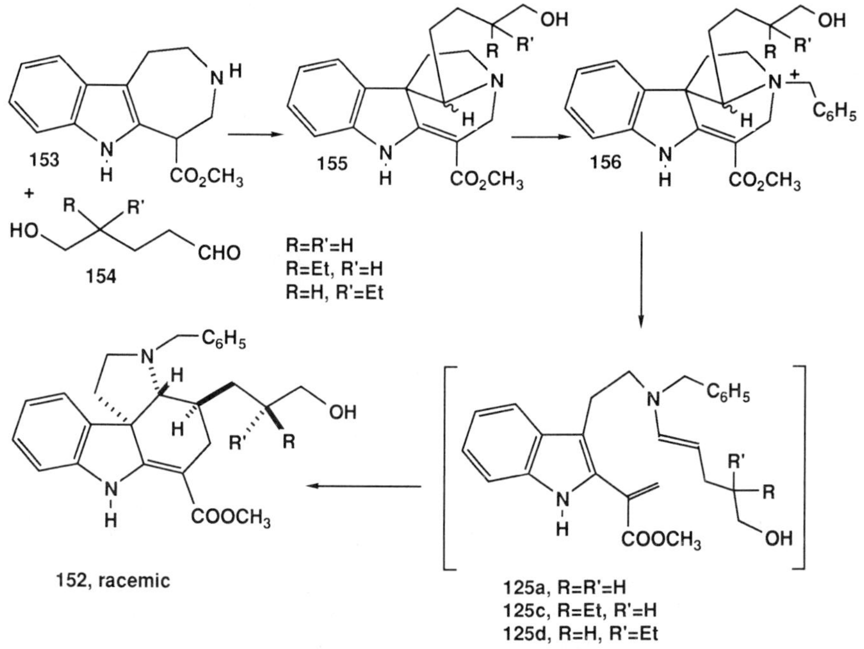

153 + 154 → 155 (R=R′=H; R=Et, R′=H; R=H, R′=Et) → 156 → [125a, R=R′=H; 125c, R=Et, R′=H; 125d, R=H, R′=Et] → 152, racemic

SCHEME 42

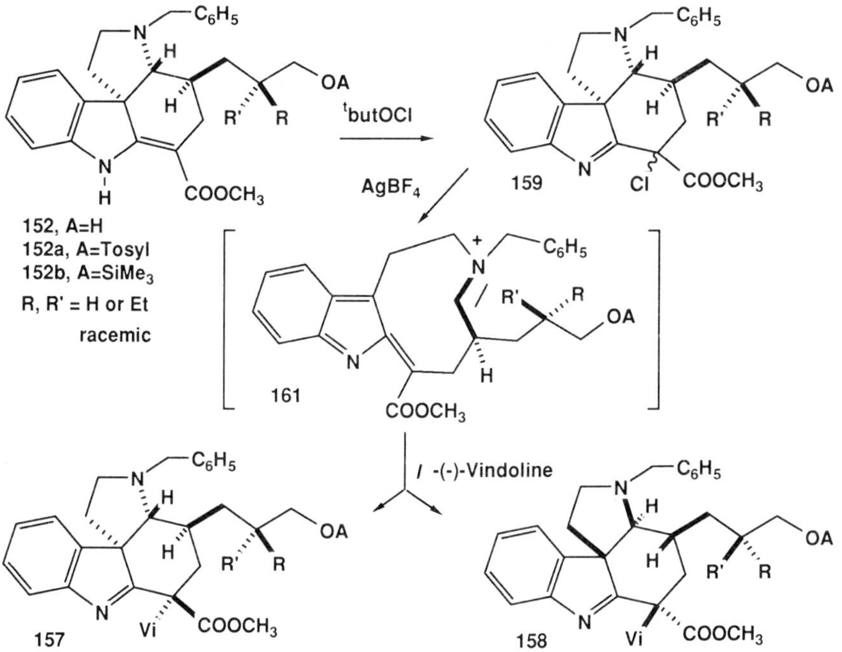

SCHEME 43

tive stereochemistry (Scheme 43). Coupling of various racemic chloro imines (**159**) with one enantiomer of vindoline (**3**) resulted, of course, in diastereomeric imine products **157** and **158** with modest or no diastereoselectivity, depending on the substitution pattern of the side chain of various chloroimines (**159**) examined, but essentially all products had the C-16′–C-14′ PARF relative stereochemistry. (A solution to the problem of asymmetric formation of the tetracyclic intermediates **152** is described below.)

In the chloroimine **159** to vindoline (**3**) coupling reaction, the C-16′–C-14′ PARF stereoselectivity was lost at elevated temperature, and a C-11′ methoxy substituent (but not a C-11′ bromo substituent) even favored formation of the C-16′–C-14′ PREF relative stereochemistry (**160**), while a C-10′ methoxy substituent on intermediate **159** resulted in a 9 : 1 PARF–PREF ratio (Scheme 44). These results, as well as the failure of a pentacyclic D–E-trans precursor **150** to undergo coupling, are consistent with a reversible cleavage of the C-7–C-3 bond during the reaction, with the resulting relative amount of C-16′–C-14′ PARF stereochemistry arising from a relative maintenance of the conformation of the initial tetracyclic chloro imine **159** in the imonium intermediate **161**. With increasing stabili-

161a

0 °C

55 °C X=Y=H

-5 °C X=OMe, Y=H

161b

157a, X=Y=H
157b, X=Br, Y=H
157c, X=H, Y=OMe
157d, X=OMe, Y=H

160a, X=Y=H
160b, X=OMe, Y=H

SCHEME 44

zation of the imonium intermediate **161**, or at elevated temperature, the C-16′–C-14′ PARF selectivity decreases because of diminished stereoelectronic control derived from the starting tetracyclic chloro imine conformation **161a**. Indeed, molecular modeling, using the Clark Still MACROMODEL program, showed that if only stereochemical control of the approach of vindoline to the nine-membered ring imonium intermediate **161a** is considered, then little, if any, C-16′–C-14′ PARF selectivity would be expected (*127*).

Reaction of the imine–indoline products **157** and **158** with potassium borohydride in acetic acid resulted in rupture of the C-3′–C-7′ bond and reduction of the resulting imonium function (Scheme 45). While the initial C-16′–C-14′ PARF imine–indolines **157** and **158** can be isolated, they are less stable on silica gel chromatography than the corresponding ψ-vincadifformine-derived C-16′–C-14′ PREF compounds **145**, and consequently purification of the vindoline coupling products and separation of diastereomers were best carried out at the indole–indoline stage (**162**, **163**).

The final phase of our syntheses of vinblastine and analogous compounds required closure of the piperidine ring in the seco cleavamine part of the indole–indoline compounds **162** and **163**. We had originally studied this reaction with the simple indolic C-16 and C-20 unsubstituted C-16 epimeric carbomethoxy seco cleavamines **164** and **165** and found that on cyclization they produced, through conformations **166** and **167**, almost exclusively quaternary salts **168** and **169**, which on debenzylation gave

SCHEME 45

stable, isolable atropisomers **170** and **171** of the C-20 unsubstituted carbomethoxycleavamines **172** and **173** (Scheme 46). On heating to 40°C (for **171**) or 100°C (for **170**), these compounds could be converted to their respective lower energy conformational isomers **172** and **173** (*121*).

When the corresponding seco cleavamines with a C-16′ 10-vindolinyl

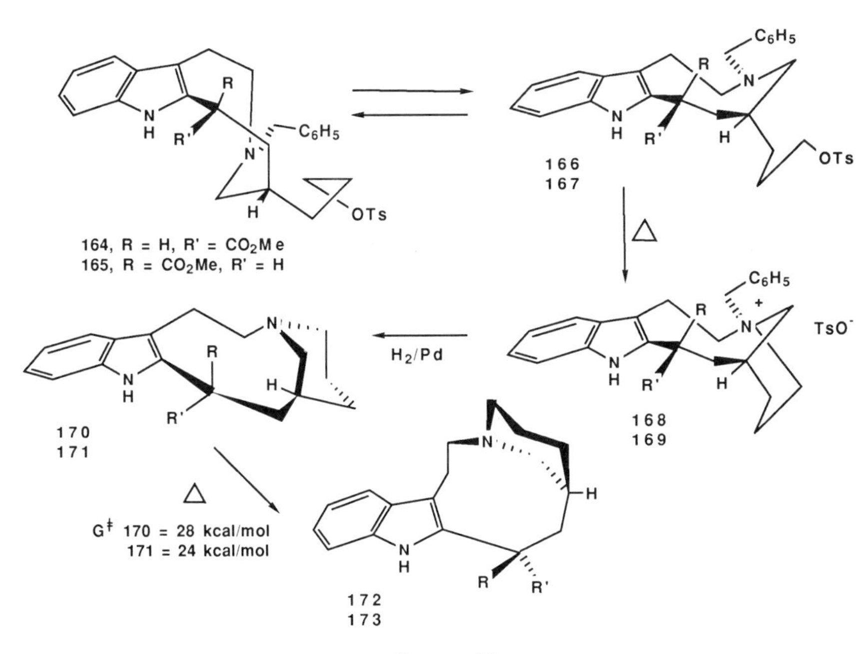

SCHEME 46

substituent (**162**, **163**; R = R′ = H) were subjected to the same cyclization and debenzylation (*13*), a 15 : 85 mixture of higher to lower energy atropisomers **174** and **175** was generated from the C-16′ (*S*) seco compound **162**, while the C-16′ (*R*) diastereomer **163** produced a 1 : 1 mixture of atropisomers **176** and **177** (Schemes 47 and 48). The presence of C-20′ α- or β-ethyl substituents changed these atropisomeric ratios in the direction expected from minimization of 1,3-diaxial repulsions, but such substitution did not drive the atropisomeric balance completely to the for-

162a → (110 °C; H_2 / Pd) → 174

R′ = Et, R = H → R = H, R′ = Et, 3%
R′ = H, R = Et → R = Et, R′ = H, 28%
R′ = R = H → R = R′ = H, 15%

162b → (110 °C; H_2 / Pd) → 175 (174 → 175: Δ)

R = H, R′ = Et → R = H, R′ = Et, 97%
R = Et, R′ = H → R = Et, R′ = H, 72%
R = R′ = H → R = R′ = H, 85%

SCHEME 47

163a

R' = Et, R = H
R' = H, R = Et
R = R' = H

110 °C, Pd/H_2

176

R' = Et, R = H, 43%
R' = H, R = Et, 60%
R = R' = H, 50%

163b

R' = Et, R = H
R' = H, R = Et
R' = R = H

110 °C, Pd/H_2

177

R' = Et, R = H, 57%
R' = H, R = Et, 40%
R' = R = H, 50%

SCHEME 48

mation of a unique product (*128*). Folding of the incipient piperidine ring into twist or boat conformations appears to accommodate the more severe diaxial interactions. Substitution on the nine-membered ring then outweighed the effect of substitution at C-20′.

Energy minimization calculations of the transition state of these cyclization reactions (*127*), using the Clark Still molecular modeling program and setting the bond length of the forming N^b to C-21′ bond at 1.5 times the bond length of the corresponding product bond, indicated that formation of the quaternary salt **169**, which leads to the higher energy atropisomer **171**, is favored by about 4 kcal/mol for cyclization of the C-16 β-carbomethoxy compound **165/167**, in reasonable agreement with the observed complete formation of that product lacking the C-16 vindolinyl substituent. From these calculations it can be seen not only that the transition states **166**, **167** versus **164**, **165** leading to the quaternary precursors of the higher and lower energy atropisomers **170**, **171** versus **172**, **173** differ with respect to C-17 of the nine-membered ring as an equatorial versus axial substituent on the incipient piperidine ring, but also that the nine-membered ring in these transition states assumes a chair versus tublike conformation. Thus, substitution on this nine-membered ring can be expected to result in modulation of the atropisomeric ratio obtained from these cyclization and debenzylation sequences. Calculations for the cyclizations with a C-16 α-phenyl substituent, as a model for our vindoline coupling products, indicated a drop of the energy difference to ~0.8 kcal/mol, now in favor of the lower energy atropisomeric product precursor, approximating our experimental results without consideration of long-range interaction effects found in the diastereomeric indole–vindoline compounds.

Biological evaluation of 20′-deoxyvinblastine and its 20′-desethyl congener showed a lack of vinblastine-like cytotoxicity with L1210 leukemia cells for the higher energy atropisomers (**174**) and generation of such cytotoxicity by thermal inversion to the lower energy conformers (**175**). With the thought that such a noncytotoxic, higher energy atropisomer might serve as a pro-drug for local site activation chemotherapy, given the proviso of substitution appropriate for lowering of the energy barrier of inversion and maintenance of high cytotoxicity in the product, we sought a method which would allow selective formation of only the higher energy atropisomer, irrespective of the remaining substitution pattern of the seco cleavamine bearing a vindoline substituent at C-16′. Such a method was found in our synthesis of vinblastine (**1**), where the intermediate higher energy atropisomer could be generated stereoselectively.

Once again considerations of stereoelectronic control provided the solution to achievement of the desired atropisomeric selectivity. Based on the well known trans-coplanar opening of epoxides, it could be expected that formation of the $N^{b\prime}$ to C-21′ bond from the C-21′–C-20′ epoxide **178**, with the C-20′ (*S*) absolute configuration, would result only in generation of the piperidine ring conformer **179**, where the resulting C-20′ hydroxyl substituent is equatorial. Less certain, however, was the realization of this cyclization reaction of a tertiary amine epoxide since the reverse process, namely, bond fragmentation and formation of tertiary amine epoxides from β-hydroxy ammonium salts, had been well documented with inter- and intramolecular examples (*129*). In the event, the reaction proceeded in 85% yield to furnish, after debenzylation, only the higher energy conformer of vinblastine (**180**) (*130*). On heating in toluene, this compound, which lacks the high cytotoxicity of vinblastine and which essentially does not inhibit microtubule assembly nor cause microtubule disassembly like vinblastine (see Chapter 3, this volume), was quantita-

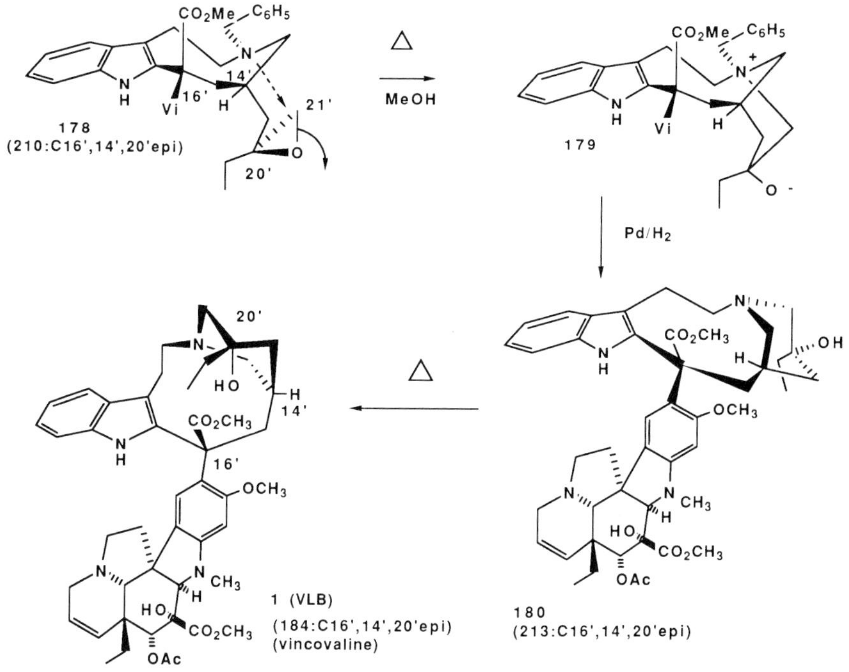

178
(210:C16',14',20'epi)

179

180
(213:C16',14',20'epi)

1 (VLB)
(184:C16',14',20'epi)
(vincovaline)

SCHEME 49

tively converted to the lower energy, natural conformer of vinblastine (**1**) (Scheme 49).

Cyclization of the corresponding C-20′ (*R*) epoxide (**181**) was also studied. Once again only the higher energy atropisomer of the product (**182**) was obtained after debenzylation. In this case, folding of the developing piperidine ring into a chair conformation (**183**), with a C-20′ pro-equatorial hydroxyl substituent, would entail generation of a 1,3,5-triaxial alkyl-substituted six-membered ring. This can be avoided by generation of a boat in the alternative cyclization conformation (shown for **181**), thus maintaining the required $N^{b\prime}$ C-21′ and C-20′ to OH coplanarity. Debenzylation then furnished the higher energy atropisomer of leurosidine (**182**), which on heating in toluene was converted to its natural, lower energy atropisomeric form (Scheme 50).

For an enantioselective synthesis of vinblastine (**1**) [or of leurosidine (**56**) and vincovaline (**184**, Scheme 49)], chirality at C-20′ was introduced at the outset through Sharpless (*131*) oxidations of 2-ethylpropenol (**185**). Subsequent elaboration to the substituted five-carbon aldehydes **186**

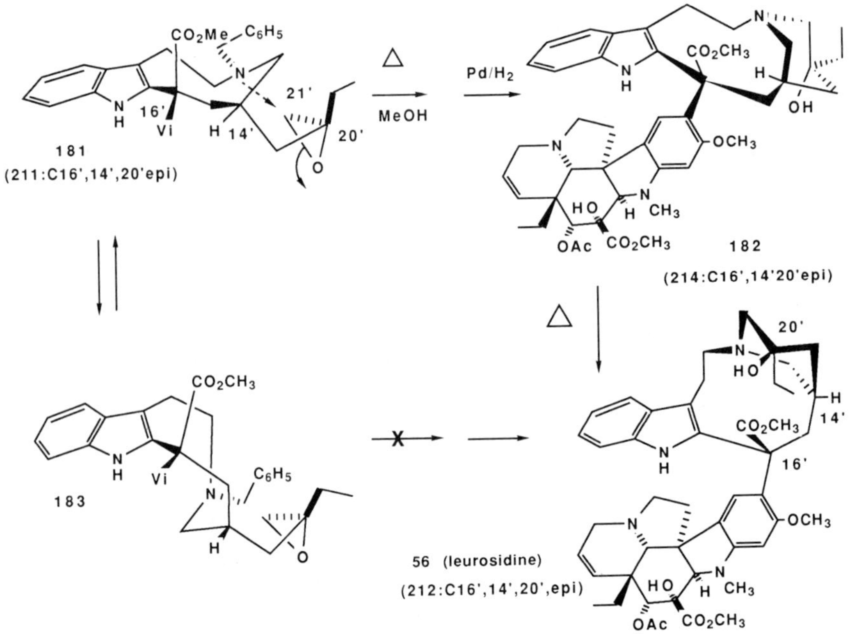

SCHEME 50

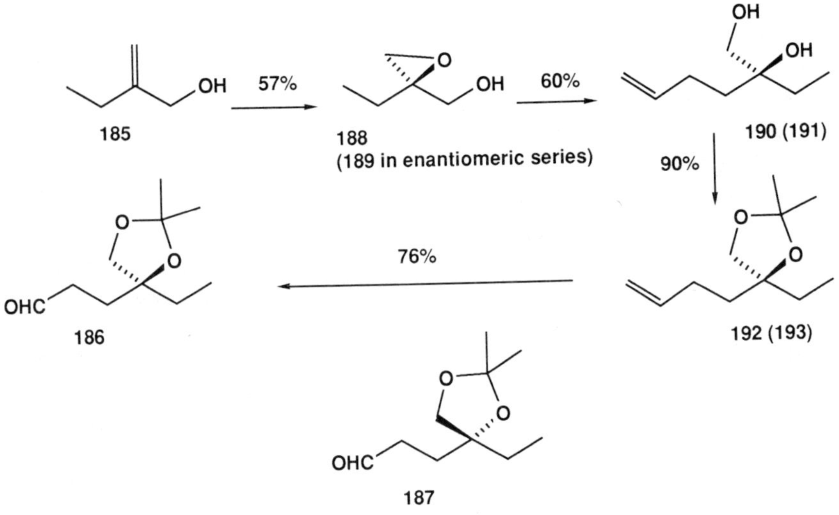

SCHEME 51

(**187**), required for our general synthetic strategy, was achieved by addition of allyl Grignard reagent to the epoxides **188** (**189**), masking of the resulting diols **190** (**191**) as their acetonides **192** (**193**), and ozonolysis (Scheme 51). Condensation of the aldehydes **186** (**187**) with the indoloazepine **153** and benzylation then provided the two series of key D-seco ψ-vincadifformines **194** versus **195** (and **196** versus **197**) in which the chiral centers at C-3, C-7, and C-14 were established (Scheme 52). No significant diastereoselectivity relative to C-20 could be expected in their formation, based on an intramolecular enamine–indoloacrylate reaction where

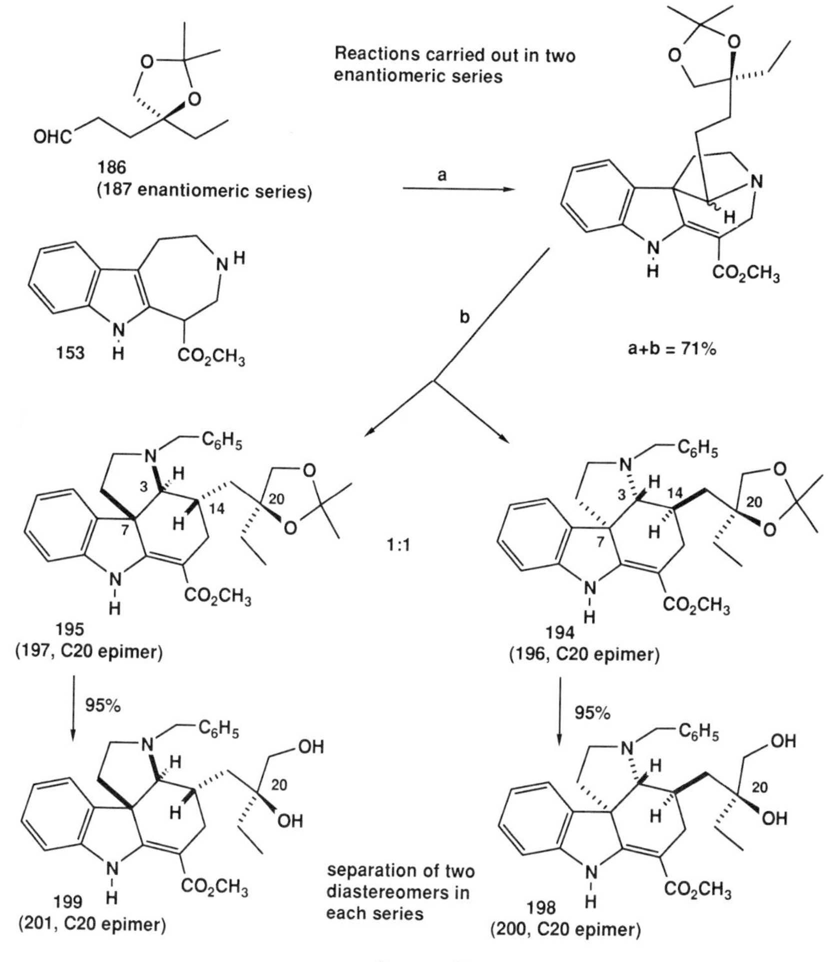

SCHEME 52

these centers are achiral. However, separation of the diastereomers **194/195** or **196/197**, for the purpose of generating stereoisomeric vinblastine congeners for biological evaluation, was readily achieved by chromatography after hydrolysis of the acetonide function.

Conversion of the resulting separate D-seco D–E trans ψ-vincadifformine diols **198–201** to their primary tosylates and tertiary trimethylsilyloxy derivatives **202–205** and coupling to vindoline by the chlorination–silver tetrafluoroborate–potassium borohydride sequence provided amino tosylates **206–209**, which could be directly subjected to cyclization or, alternatively, converted to the C-20′–C-21′ epoxides **178**, **181**, **210**, and **211** by reaction with tetrabutylammonium fluoride (Scheme 53). While cyclization of the tosylates **206–209** led essentially only to quaternary salts which could be debenzylated to provide the lower energy atropisomer of vinblastine (**1**), leurosidine (**56**), vincovaline (**184**), and its C-20′ epimer (**212**) respectively, cyclization of the epoxides **178**, **181**, **210**, and

198
199-201 diastereomeric series

Ts_2O 76%; TMSCl 92%

202
203-205 diastereomeric series

80%: i. ^{t}butOCl ii. vindoline, $AgBF_4$ iii. $NaBH_4$

206
207-209 diastereomeric series

but_4NF 85%

178
181 C20′ epimer
210,211 corresponding C16′-C14′ Parf diastereomers

SCHEME 53

211 gave only the corresponding higher energy atropisomers **180**, **182**, **213** and **214** (Schemes 49 and 50).

In order to obtain stereoselective formation of the chiral centers C-3, C-7, and C-14, we explored the use of chiral derivatives of our indoloazepine **153**. Earlier, we had already found that the *N*-benzylindoloazepine ester **215** rearranged on heating to an α-methylene lactam **216**, indicating the possibility of thermal generation of an intermediate secondary amine indoloacrylate **217**. It was also found that this intermediate could be trapped with a variety of aldehydes. Thus, D-seco D–E-trans vincadifformine congeners (**218**) could be obtained by condensation of the indoloazepine **215** with aldehydes at 100°C (Scheme 54) (*132*). Consequently, introduction of a chiral substituent onto N^b of the indoloazepine **153** prior to condensation with our 4-ethyl-4,5-dihydroxypentanal acetonides **186** and **187** appeared to be an option for chiral induction in the formation of cen-

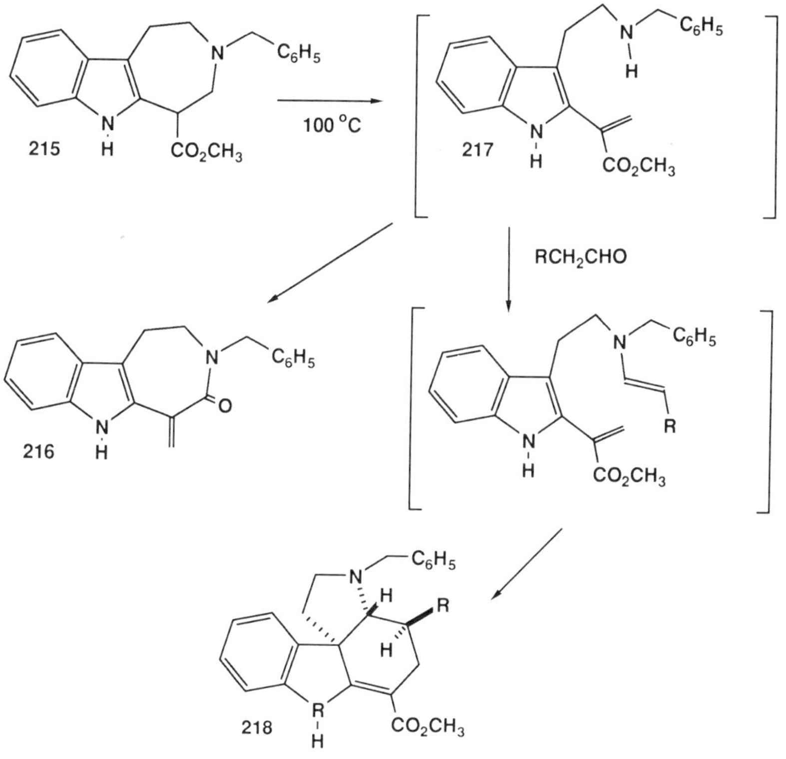

Scheme 54

ters C-3, C-7, and C-14.* A 1-(1-naphthyl) ethylamine (**219**) was chosen in order to maximize the π-stacking potential of the aryl, enamine, and acrylate functions in the intermediate **220** leading to the D-seco vincadifformine product. Thus, a 4 : 1 ratio of diastereomers **221** and **222** was formed (Scheme 55).

The required chirally substituted indoloazepine **219** was prepared by reaction of (*S*)-1-(1-naphthyl)ethylamine with the methiodide of 4-piperidone (**223**), followed by a Fischer indole reaction leading to the γ-carboline **224** (Scheme 56). Chlorination and reaction with thallium dimethylmalonate, followed by decarbomethoxylation of the resulting in-

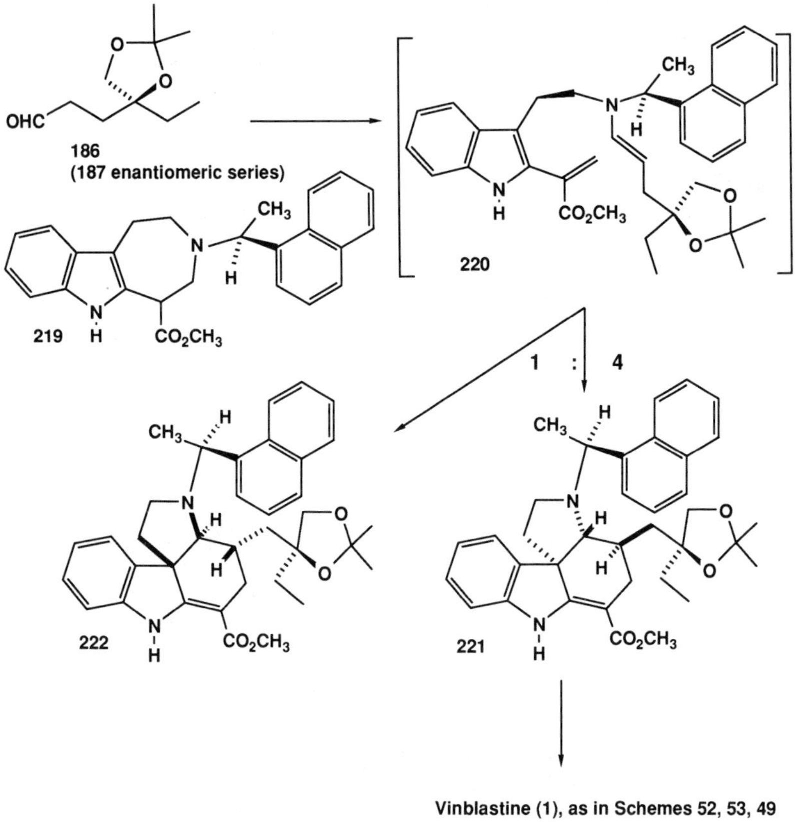

186 (187 enantiomeric series)

219

220

1 : 4

222

221

Vinblastine (1), as in Schemes 52, 53, 49

SCHEME 55

* An analogous strategy, using a chiral phenethylamine derivative, was followed by Legseir *et al.* (*133*) in a synthesis of tubotaiwin.

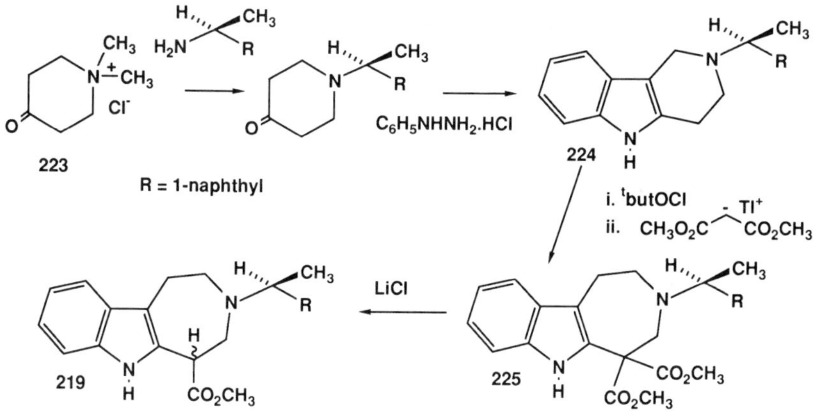

SCHEME 56

doloazepine diester **225**, provided the desired indoloazepine **219**, which was then carried through to vinblastine (**1**) and leurosidine (**56**), using the respective chiral aldehydes **186** and **187**. Similarly, vincovaline **184** and its C-20′-epimer **212** could be obtained from the (*R*)-1-(1-naphthyl) ethylamine derivative.

Acknowledgments

We thank Professor J. Lévy for providing us with copies of Refs. *124* and *125*. We also thank Professor Clark Still for a gift of the MACROMODEL computer program. Our research was supported by grant R01-12010 from the National Cancer Institute.

REFERENCES

1. A. G. Cook, ed., "Enamines," Chap. 8. Dekker, New York, 1968.
2. I. S. Johnson, J. G. Armstrong, M. Gorman, and J. P. Burnett, *Cancer Res.* **23**, 1390 (1963).
3. K. Gerzon, "Anticancer Agents Based on Natural Product Models" (J. M. Cassady and J. D. Douros, eds.), p. 271. Academic Press, New York, 1980.
4. H. Wisniewski, M. L. Shelanski, and R. D. Terry, *J. Cell. Biol.* **38**, 224 (1968).
5. W. I. Taylor and N. R. Farnsworth, eds., "The *Catharanthus* Alkaloids." Dekker, New York, 1975.
6. G. A. Cordell, *in* "Heterocyclic Compounds: The Monoterpenoid Indole Alkaloids" (J. E. Saxton, ed.), Vol. 25, Part 4, p. 539. Wiley, New York, 1983.
7. G. A. Cordell and J. E. Saxton, *in* "The Alkaloids" (R. H. F. Manske and R. G. A. Rodrigo, eds.), Vol. 20, p. 1. Academic Press, New York, 1981.
8. A. A. Gorman, M. Hesse, H. Schmid, P. G. Wasser, and W. H. Hopff, *in* "Alkaloids, A Specialist Periodical Report," Vol. 1, p. 200. The Chemical Society, London, 1971.

9. W. I. Taylor, *in* "The Alkaloids" (R. H. F. Manske and H. L. Holmes, eds.), Vol. 11, p. 99. Academic Press, New York, 1968.
10. W. I. Taylor, *in* "The Alkaloids" (R. H. F. Manske and H. L. Holmes, eds.), Vol. 8, p. 269. Academic Press, New York, 1965.
11. M. Lounasma and A. Nemes, *Tetrahedron* **38**, 227 (1982).
12. F. A. Carey and M. E. Kuehne, *J. Org. Chem.* **47**, 3811 (1982).
13. M. E. Kuehne, T. C. Zebovitz, W. G. Bornmann, and I. Markó, *J. Org. Chem.* **52**, 4340 (1987).
14. L. S. Borman, M. E. Kuehne, P. A. Matson, I. Markó, and T. Zebovitz *J. Biol. Chem.* **263**, 6945 (1988).
15. J. Harley-Mason and Atta-ur-Rahman, *J. Chem. Soc., Chem. Commun.*, 1048 (1967).
16. J. Harley-Mason and Atta-ur-Rahman, *Tetrahedron* **36**, 1057 (1980).
17. G. Büchi and R. E. Manning, *J. Am. Chem. Soc.* **88**, 2532 (1966).
18. G. Büchi, R. E. Manning, and S. A. Monti, *J. Am. Chem. Soc.* **86**, 4631 (1964).
19. S. G. P. Plant, R. Robinson, and M. Tomlinson, *Nature (London)* **165**, 928 (1950).
20. J. P. Kutney, W. J. Cretney, P. Le Quesne, B. McKague, and E. Piers, *J. Am. Chem. Soc.* **92**, 1712 (1970).
21. S. Takano, M. Hirama, and K. Ogasawara, *J. Org. Chem.* **45**, 3729 (1980).
22. S. Takano, C. Murakata, and K. Ogasawara, *Heterocycles* **14**, 1301 (1980).
23. S. Takano, M. Yonaga, K. Chiba, and K. Ogasawara, *Tetrahedron Lett.* **21**, 3697 (1980).
24. S. Takano, M. Yonaga, S. Yamada, S. Hatake-Yama, and K. Ogasawara, *Heterocycles* **15**, 309 (1981).
25. N. Neuss, M. Gorman, N. J. Cone, and L. L. Huckstep, *Tetrahedron Lett.*, 783 (1968).
26. J. P. Kutney, J. Beck, F. Bylsma, and W. J. Cretney, *J. Am. Chem. Soc.* **90**, 4504 (1968).
27. Atta-ur-Rahman, *Pak. J. Sci. Ind. Res.* **14**, 487 (1971).
28. J. P. Kutney, J. Beck, F. Bylsma, J. Cook, W. J. Cretney, K. Fuji, R. Imhof, and A. M. Treasurywala, *Helv. Chim. Acta* **58**, 1690, (1975).
29. Atta-ur-Rahman, *J. Chem. Soc. Pak.* **1**, 81 (1979).
30. G. Schill, C. V. Priester, U. F. Windhovel, and H. Fritz, *Helv. Chim. Acta* **69**, 438 (1986).
31. G. Schill, U. P. Ulrich, F. Udo, H. F. Windhovel, and L. Hartmut, *Tetrahedron* **43**, 3729 (1987).
32. G. Schill, U. P. Claus, F. Udo, and H. F. Windhovel, *Tetrahedron* **43**, 3747, 3765 (1987).
33. N. Kunesch, P.-L. Vaucamps, A. Cavé, J. Poisson, and E. Wenkert, *Tetrahedron Lett.*, 5073 (1979).
34. J. Bruneton, A. Cavé, E. W. Hagaman, N. Kunesch, and E. Wenkert, *Tetrahedron Lett.*, 3567 (1976).
35. J. P. Kutney, E. Jahngen, and T. Okutani, *Heterocycles* **5**, 59 (1976).
36. R. Z. Andriamialisoa, N. Langlois, and Y. Langlois, *Heterocycles* **15**, 245 (1981).
37a. P. Magnus, B. Hewitt, K. Cardwell, P. Cairns, and P. Boniface, *Stud. Org. Chem. (Amsterdam)* **25**, 203 (1986).
37b. P. Magnus, M. Ladlow, J. Elliott, and S.-K. Chung, *J. Chem. Soc., Chem. Commun.* 518 (1989).
38. P. Potier, N. Langlois, Y. Langlois, and F. Gueritte, *J. Chem. Soc., Chem. Commun.*, 670 (1975).
39. N. Langlois, F. Gueritte, Y. Langlois, and P. Potier, *J. Am. Chem. Soc.* **98**, 7017 (1976).

40. P. Potier, *Ann. Pharm. Fr.* **38**, 407 (1980).
41. P. Potier, *J. Nat. Prod.* **43**, 72 (1980).
42. J. P. Kutney, A. H. Ratcliffe, A. M. Treasurywala, and S. Wunderly, *Heterocycles* **3**, 639 (1975).
43. J. P. Kutney, T. Hibino, E. Jahngen, T. Okutani, A. H. Ratcliffe, A. M. Treasurywala, and S. Wunderly, *Helv. Chim. Acta* **59**, 2858 (1976).*
44. J. P. Kutney, *Lloydia* **40**, 107 (1977).
45. Atta-ur-Rahman, *Pak. J. Sci. Ind. Res.* **14**, 487 (1971).
46. S. B. Hassam and C. R. Hutchinson, *Tetrahedron Lett.*, 1681 (1978).
47. A. I. Scott, F. Gueritte, and S.-L. Lee, *J. Am. Chem. Soc.* **100**, 6253 (1978).
48. R. L. Baxter, C. A. Dorschel, S.-L. Lee, and A. I. Scott, *J. Chem. Soc., Chem. Commun.*, 257 (1979).
49. F. Gueritte, N. V. Bac, Y. Langlois, and P. Potier, *J. Chem. Soc., Chem. Commun.*, 452 (1980).
50. Atta-ur-Rahman, *J. Chem. Soc. Pak* **1**, 81 (1979).
51. M. Lounasmaa and A. Koskinen, *Heterocycles* **22**, 1591 (1984).
52. P. Potier, *Rev. Latinoam. Quim.* **9**, 47 (1978).
53. A. Cavé, C. Kan-Fan, P. Potier, and J. Le Men, *Tetrahedron* **23**, 4681 (1967).
54. A. Ahond, A. Cavé, C. Kan-Fan, Y. Langlois, and P. Potier, *J. Chem. Soc., Chem. Commun.*, 517 (1970).
55. J. D. Phillipson and S. S. Handa, *Lloydia* **41**, 385 (1978).
56. P. Potier *in* "Stereoselective Synthesis of Natural Products," proceedings of the Seventh Workshop Conference, Hoechst (W. Bartmann and E. Winterfeldt, eds). Excerpta Medica, Amsterdam, Oxford, 1978.
57. A. I. Scott, C. L. Yeh, and D. Greenslade, *J. Chem. Soc., Chem. Commun.*, 947 (1978).
58. P. Potier, *J. Nat. Prod.* **43**, 72 (1980).
59. N. Langlois, F. Gueritte, Y. Langlois, and P. Potier, *Tetrahedron Lett.*, 1487 (1976).
60. G. Büchi and R. E. Manning, *J. Am. Chem. Soc.* **88**, 2532 (1966).
61. A. De Bruyn, J. Sleeckx, J. P. De Jonghe, and J. Hannart, *Bull. Soc. Chim. Belg.* **92**, 485 (1983).
62. N. Langlois and P. Potier, *Tetrahedron Lett.*, 1099 (1976).
63. T. Honma and Y. Ban, *Tetrahedron Lett.*, 155 (1978).
64. T. Honma and Y. Ban, *Heterocycles* **6**, 285, 291 (1977).
65. N. Langlois, R. Z. Andriamialisoa, and Y. Langlois, *Tetrahedron* **37**, 1951 (1981).
66. Y. Langlois, N. Langlois, and P. Potier, *C. R. Acad. Sci. Paris, Ser. C* **284**, 809 (1977).
67. J. P. Kutney and B. R. Worth, *Heterocycles* **6**, 905 (1977).
68. J. P. Kutney, T. Honda, P. M. Kazmaier, N. J. Lewis, and B. R. Worth, *Helv. Chim. Acta* **63**, 366 (1980).
69. P. Mangeney, R. Costa, Y. Langlois, and P. Potier, *C. R. Acad. Sci. Paris, Ser. C* **284**, 701 (1977).
70. J. P. Kutney and B. R. Worth, *Heterocycles* **6**, 297 (1977).
71. J. P. Kutney, J. Balsevich, G. H. Bokelman, T. Hibino, I. Itoh, and A. H. Ratcliffe, *Heterocycles* **4**, 997 (1976).
72. J. P. Kutney, J. Balsevich, G. H. Bokelman, T. Hibino, T. Honda, I. Itoh, A. H. Ratcliffe, and B. Worth, *Can. J. Chem.* **56**, 62 (1978).

*The stereochemistry of hydrogenation of anhydrovinblastine was initially misassigned. See also ref. 70–74.

73. J. P. Kutney, A. V. Joshua, P.-H. Liao, and B. R. Worth, *Can. J. Chem.* **55**, 3235 (1977).
74. J. P. Kutney, A. V. Joshua, and P.-H. Liao, *Heterocycles* **6**, 297 (1977).
75. Y. Langlois, N. Langlois, P. Mangeney, and P. Potier, *Tetrahedron Lett.*, 3945 (1976).
76. Atta-ur-Rahman, A. Basha, and M. Ghazala, *Tetrahedron Lett.*, 2351 (1976).
77. Atta-ur-Rahman, N. Waheed, and M. Ghazala, *Z. Naturforsch., Teil B* **31**, 264 (1976).
78. J. P. Kutney, G. H. Bokelman, M. Ichikawa, E. Jahngen, A. V. Joshua, P.-H. Liao, and B. R. Worth, *Can. J. Chem.* **55**, 3227 (1977).
79. J. P. Kutney, T. Honda, A. V. Joshua, N. G. Lewis, and B. R. Worth, *Helv. Chim. Acta* **61**, 690 (1978).
80. P. Mangeney, R. Z. Andriamialisoa, N. Langlois, Y. Langlois, and P. Potier, *C. R. Acad. Sci. Paris, Ser. C* **288**, 129 (1979).
81. P. Mangeney, R. Z. Andriamialisoa, N. Langlois, Y. Langlois, and P. Potier, *J. Am. Chem. Soc.* **101**, 2243 (1979).
82. N. Langlois and P. Potier, *J. Chem. Soc., Chem. Commun.*, 102 (1978).
83. F. Gueritte, N. V. Bac, Y. Langlois, and P. Potier, *J. Chem. Soc., Chem. Commun.*, 452 (1980).
84. K. L. Stuart, J. P. Kutney, T. Honda, and R. B. Worth, *Heterocycles* **9**, 1391, 1419 (1978).
85. N. Langlois and P. Potier, *J. Chem. Soc., Chem Commun.*, 582, (1979).
86. J. P. Kutney, B. Aweryn, L. S. L. Choi, P. Kolodziejczyk, W. G. W. Kurz, K. B. Chatson, and F. Constabel, *Helv. Chim. Acta* **65**, 1271 (1982).
87. J. P. Kutney, L. S. L. Choi, T. Honda, N. G. Lewis, T. Sato, K. L. Stuart, and B. R. Worth, *Helv. Chim. Acta* **65**, 2088 (1982).
88. J. P. Kutney, *Heterocycles* **25**, 617 (1987).
89. M. E. Kuehne and T. J. Giacobbe, *J. Org. Chem.* **33**, 3359 (1968).
90. P. Mangeney, R. Z. Andriamialisoa, N. Langlois, Y. Langlois, and P. Potier, *J. Org. Chem.* **44**, 3765 (1979).
91. R. Z. Andriamialisoa, N. Langlois, Y. Langlois, and P. Potier, *Tetrahedron* **36**, 3053 (1980).
92. P. Mangeney, R. Z. Andriamialisoa, J.-Y, Lallemand, N. Langlois, Y. Langlois, and P. Potier, *Tetrahedron* **35**, 2175 (1979).
93. R. Z. Andriamialisoa, N. Langlois, Y. Langlois, P. Potier, and P. Bladon, *Can. J. Chem.* **57**, 2572 (1979).
94. R. Z. Andriamialisoa, Y. Langlois, N. Langlois, and P. Potier, *C. R. Acad. Sci. Paris, Ser. C* **284**, 751 (1977).
95. J. P. Kutney, J. Balsevich, T. Honda, P.-H. Liao, H. P. M. Thiellier, and B. R. Worth, *Can. J. Chem.* **56**, 2560 (1978).
96. J. P. Kutney, K. K. Chan, W. B. Evans, Y. Fujise, T. Honda, F. K. Klein, and J. P. Souza, *Heterocycles* **6**, 453 (1977).
97. J. P. Kutney, W. B. Evans, and T. Honda, *Heterocycles* **6**, 443 (1977).
98. S. Raucher, B. L. Bray, and R. F. Lawrence, *J. Am. Chem. Soc.* **109**, 442 (1987).
99. F. Gueritte, N. Langlois, Y. Langlois, R. J. Sundberg, and J. D. Bloom, *J. Org. Chem.* **46**, 5393 (1981).
100. A. M. Treasurywala, Doctoral dissertation, University of British Columbia, Vancouver (1973).
101. J. Vucovic, A. E. Goodbody, J. P. Kutney, and M. Misawa, *Tetrahedron* **44,** 325 (1988).
102. J. R. Lindsay Smith and L. A. V. Mead, *J. Chem. Soc., Perkin Trans. 2*, 206 (1973), and references cited therein.

103. J. Harley-Mason, Atta-ur-Rahman, and J. A. Beisler, *J. Chem. Soc., Chem. Commun.*, 743 (1966).
104. J. Harley-Mason and Atta-ur-Rahman, *J. Chem. Soc., Chem. Commun.*, 208 (1967).
105. J. Harley-Mason and Atta-ur-Rahman, *Chem. Ind. (London)*, 1845 (1968).
106. Atta-ur-Rahman, J. A. Beisler, and J. Harley-Mason, *Tetrahedron* **36**, 1063 (1980).
107. J. P. Kutney, W. J. Cretney, P. Le Quesne, B. McKague, and E. Piers, *J. Am. Chem. Soc.* **92**, 1712 (1970).
108. E. Wenkert, *J. Am. Chem. Soc.* **84,** 98 (1962).
109. A. I. Scott, *Acc. Chem. Res.* **3**, 151 (1970), and references cited therein.
110. M. E. Kuehne and W. G. Bornmann, unpublished results.
111. M. E. Kuehne, W. G. Bornmann, W. G. Earley, and I. Markó, *J. Org. Chem.* **51**, 2913 (1986).
112. M. E. Kuehne, D. M. Roland, and R. Hafter, *J. Org. Chem.* **43**, 3705 (1978).
113. M. E. Kuehne, J. A. Huebner, and T. H. Matsko, *J. Org. Chem.* **44**, 2477 (1979).
114. M. E. Kuehne, T. H. Matsko, J. C. Bohnert, and C. L. Kirkemo, *J. Org. Chem.* **44**, 1063 (1979).
115. M. E. Kuehne, C. L. Kirkemo, J. C. Bohnert, and T. H. Matsko, *J. Org. Chem.* **45**, 3259 (1980).
116. M. E. Kuehne, F. J. Okuniewics, C. L. Kirkemo, and J. C. Bohnert, *J. Org. Chem.* **47**, 1335 (1982).
117. M. E. Kuehne and J. C. Bohnert, *J. Org. Chem.* **46**, 3443 (1981).
118. M. E. Kuehne and W. G. Earley, *Tetrahedron* **39**, 3707 (1983).
119. M. E. Kuehne and W. G. Earley, *Tetrahedron* **39**, 3713 (1983).
120. M. E. Kuehne, T. H. Matsko, J. C. Bohnert, L. Motyka, and D. Oliver-Smith, *J. Org. Chem.* **46**, 2002 (1981).
121. M. E. Kuehne and T. C. Zebovitz, *J. Org. Chem.* **52**, 4331 (1987).
122. M. E. Kuehne and D. E. Podhorez, *J. Org. Chem.* **50**, 924 (1985).
123. M. E. Kuehne, D. E. Podhorez, T. Mulamba, and W. G. Bornmann, *J. Org. Chem.* **52**, 347 (1987).
124. D. Royer, Doctoral dissertation, University of Reims (1980).
125. N. de Moraes Britto Filho, Doctoral dissertation, University of Reims (1981).
126. M. D. De Maindreville and J. Lévy, *Bull. Soc. Chim. Fr.*, II-179 (1981).
127. Calculations by T. D. Spitzer of our group at the University of Vermont using the MACROMODEL program.
128. M. E. Kuehne and W. G. Bornmann, *J. Org. Chem.* **54,** 3407 (1989).
129. A. Cope, *Organic Reactions* **11**, 317 (1960), see pp. 352, 390, 391.
130. P. A. Matson, Doctoral dissertation, University of Vermont, Burlington (1988).
131. R. M. Hanson and K. B. Sharpless, *J. Org. Chem.* **51**, 1922 (1986).
132. Results obtained with S. E. Kuehne in our laboratories in 1980.
133. B. Legseir, G. Massiot, J. Gulhem, and J. Vercauteren, *Tetrahedron Lett.* **28**, 3573 (1987).
134. J. LeMen and W. I. Taylor, *Experientia* **21**, 508 (1965).

—Chapter 3—

FUNCTIONAL HOT SPOT AT THE C-20′ POSITION OF VINBLASTINE

Linda S. Borman

Department of Pharmacology, and
Vermont Regional Cancer Center
University of Vermont
Burlington, Vermont 05405

AND

Martin E. Kuehne

Department of Chemistry, and
Vermont Regional Cancer Center
University of Vermont
Burlington, Vermont 05405

I. Introduction

Meaningful progress in a study of structure–activity relationships of the indole–indoline binary alkaloids has awaited the recent achievement of their total and stereoselective syntheses (see Chapter 2, this volume). Now, with such methodologies in hand, it is feasible to create the types of congeners required for serious consideration of the structure and function of such a complex molecule. Specifically, one can now undertake the directed syntheses of series of single-site modified congeners, which encompass the epimeric pairs for each substitution, as well as generate diastereomeric pairs in which the enantiomeric sense of the binary segments is varied. While most of the work toward this formidable task is still in progress in our laboratories, we describe in this chapter our evi-

THE ALKALOIDS, VOL. 37
Copyright © 1990 by Academic Press, Inc.
All rights of reproduction in any form reserved.

dence for a function-sensitive site at the C-20′ position of the binary alkaloid vinblastine (VBL).

The cellular and biochemical parameters that define the activity of VBL in this work were chosen from the perspective of new drug development for cancer chemotherapy. VBL and vincristine (VCR) are known spindle poisons, and their efficacy against rapidly growing tumors, such as leukemias and lymphomas, by attack at the cellular division cycle, is demonstrated by work with patients (*1–4*) and by cytotoxicity studies with cultured tumor cell lines (*5–8*). The "Vinca" binary alkaloid drugs also afflict a nongrowing cell type, the neuron. This drug action presents in the clinic as peripheral neuropathy and is dose-limiting for VCR. Patients (*9*) and animals (*10,11*) with Vinca binary alkaloid-induced neurotoxicity exhibit gross abnormalities of various CNS structures, including the dissolution of neurotubules and the formation of intracellular tubulin paracrystals. Although the evidence for structural perturbation of peripheral neurons is not strong, recent work has detected dynamic properties of neurotubules (*12*) that are consistent with the notion that the functional integrity of such neurons is drug sensitive. Therefore, our evaluation of the VBL congeners in this study comprised measurements of cytotoxicity and mitotic arrest in tumor cell lines and of antimicrotubule actions *in vitro*. The latter studies used purified microtubule protein (MTP) from calf brain (*13*) to assess the reactions of the VBL congeners with soluble MTP and steady-state microtubules.

II. C-20′ Modifications

In 1985, an evaluation of an initial series of synthetic C-20′-deoxy derivatives of VBL (*14*) detected the first significant modification of the drug's potency as an inhibitor of microtubule assembly *in vitro* at this molecular locus (Fig. 1). Whereas the singular deletion of the C-20′ hydroxyl group in either epimeric sense was uneventful, the additional loss of the ethyl moiety resulted in a 10-fold increase in the IC_{50} for microtubule assembly. The effect of these congeners on the growth of cultured murine leukemia (L1210) and rat colon adenocarcinoma (RCC-2) cells displayed a similarly distinctive profile (Fig. 2). Deoxydesethyl VBL inhibited the population growth of both cell lines with an IC_{50} of 2×10^{-7} *M*, a value 100- (RCC-2) or 300-fold (L1210) greater than the IC_{50} of VBL. In contrast, the two 20′-deoxy VBL derivatives were 10-fold less potent than VBL in each cell line.

Over the next three years, we synthesized a variety of new 20′-alkyl

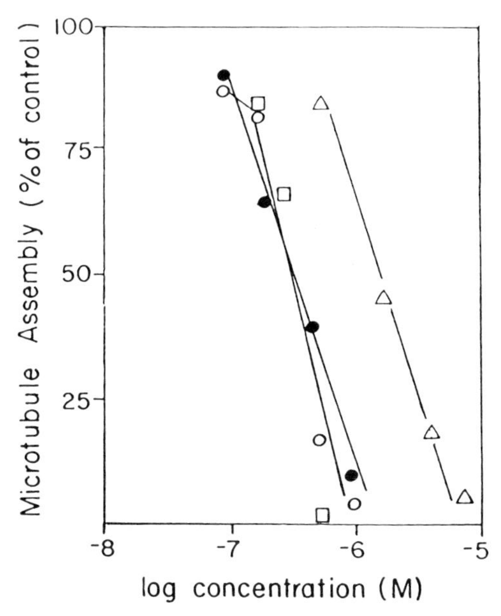

FIG. 1. Concentration-dependent inhibition of microtubule assembly by VBL (●), deoxy VBL (○), epideoxy VBL (□), and deoxydesethyl VBL (△).

congeners of VBL (Fig. 3) and expanded the evaluation of antimicrotubulet activity to include spiral aggregate formation from either the disassembly of microtubules or directly from soluble MTP by high concentrations ($>10^{-6}$ *M*) of the drug. The activity profiles of the alkyl congeners in the microtubule assays and the cellular properties of population growth inhibition and mitotic arrest are presented in Table I.

A. Antimicrotubule Activites

A perusal of Table I makes clear the general conclusion that alkyl modification at the C-20′ position of the VBL molecule is function sensitive for all and each of the three basic reactions of the drug with microtubules *in vitro*. If we begin the discussion with deoxydesethyl VBL (**5**) as the prototype compound for this series of congeners, we find that the 10-fold increase in the IC_{50} for microtubule assembly observed in our earlier work was accompanied by an alteration in the structure of the assembled microtubules. As exemplified in Fig. 4a, there are opened areas along the length of the tubule, and in some cases the linear array of protofilaments

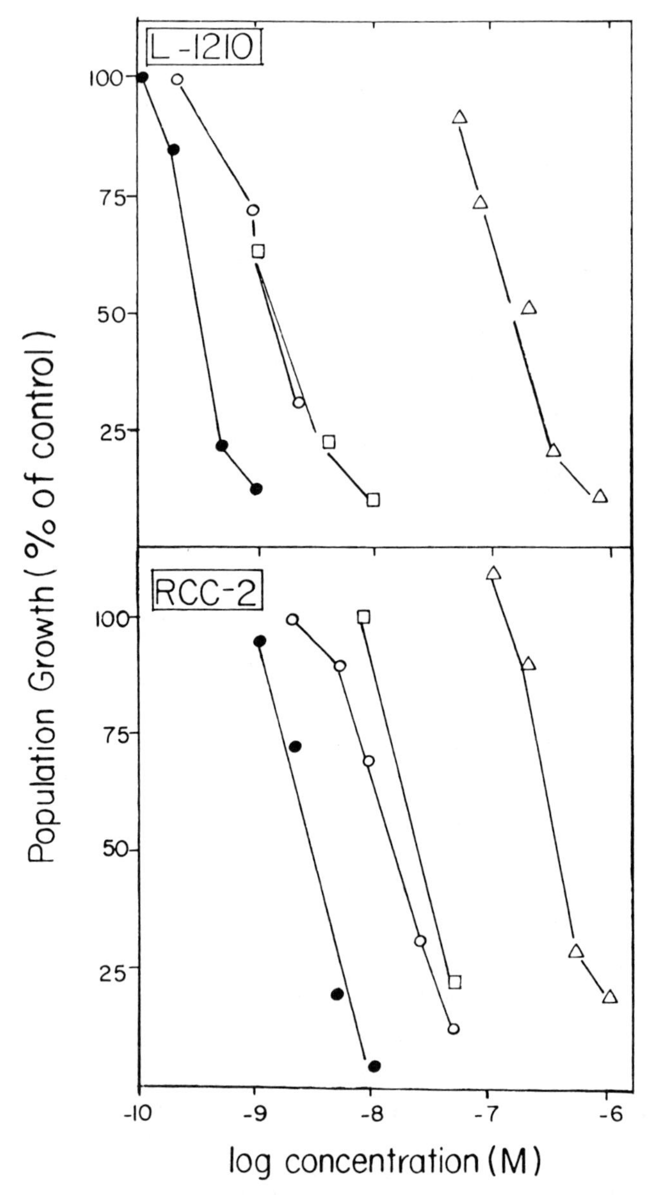

FIG. 2. Concentration-dependent inhibition of the growth of murine L1210 leukemia and rat RCC-2 colon adenocarcinoma cell populations by VBL (●), deoxy VBL (○), epideoxy VBL (□), and deoxydesethyl VBL (△).

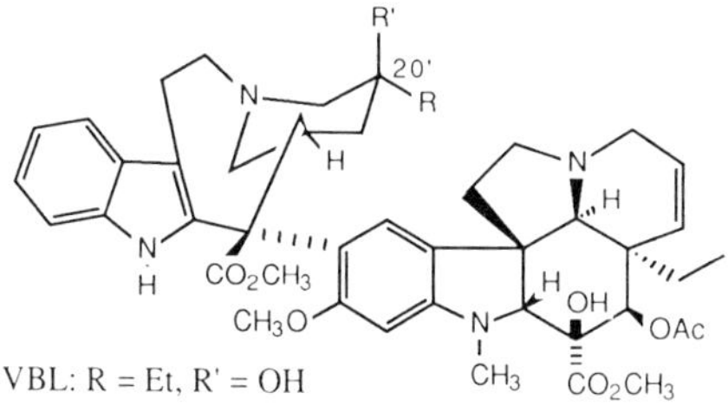

VBL: R = Et, R′ = OH

	R′	R
1. 20′-deoxy VBL	H	C_2H_5
2. 20′-*epi*-20′-deoxy VBL	C_2H_5	H
3. 20′-deoxy-20′-desethyl-20′-α-methyl VBL	H	CH_3
4. 20′-deoxy-20′-desethyl-20′-β-methyl VBL	CH_3	H
5. 20′-deoxy-20′-desethyl VBL	H	H
6. 20′-deoxy-20′-desethyl-20′-dimethyl VBL	CH_3	CH_3
7. 20′-desethyl VBL	OH	H
8. 20′-*epi*-20′-desethyl VBL	H	OH
9. 20′-deoxy-20′-desethyl-20′-α-*n*-propyl VBL	H	C_3H_7
10. 20′-deoxy-20′-desethyl-20′-β-*n*-propyl VBL	C_3H_7	H

FIG. 3. Structure of the parent molecule vinblastine (VBL) and C-20′ congeners **1–10**.

can be discerned in these regions. In addition to this effect, the exposure of soluble MTP to 10^{-4} M deoxydesethyl VBL resulted in single spirals (Fig. 4b) rather than in the formation of spiral aggregates, as is typical of the parent compound VBL (Fig. 4c). Completing the profile of this congener, we made the interesting observation that its reaction with steady-state microtubules at a high concentration formed normal spiral aggregates. These modified reactions of deoxydesethyl VBL with MTP are not primarily due to the loss of the hydroxyl group from the molecule. As we had surmised from earlier work, the reactions of deoxy VBL (**1**) and 20′-epideoxy VBL (**2**) with MTP/microtubules were indistinguishable from those of the parent compound VBL: short, structurally normal microtubules were produced in the presence of the IC_{50} concentration for assembly, and spiral aggregate formation was obtained from MTP or from steady-state microtubules with 10^{-4} M of these compounds. Analogous results were obtained with 20′-deoxy VBL congeners derived from natural sources (15).

TABLE I

C-20′ VINBLASTINE CONGENERS

Compound	Tubulin/Microtubule activities[a]			Activities with cultured cells[b]			
	MTP polymerization, IC_{50} ($M \times 10^7$)	Product of 10^{-4} M compound added to		Cell growth inhibition, IC_{50} ($M \times 10^9$)		Mitotic index (%)	
		MTP	Microtubules	L1210	RCC-2	L1210	RCC-2
Vinblastine (VBL)	2.0, 2.4	SA	SA	0.5 ± 0.2	4.8 ± 2.0	29.9	65.9
20′-Deoxy VBL (**1**)	2.3, 2.7	SA	SA	2.8 ± 0.9	34.0 ± 5.0	23.2	71.3
20′-Epideoxy VBL (**2**)	3.5, 5.5	SA	SA	2.9*	36.0*	25.9	N.D.
Alkyl congeners							
20′-Methyldeoxydesethyl VBL (**3**)	3.2, 1.5 [S]	SA	SA	7.5 ± 2.2	36.0*	28.7	N.D.
20′-Epimethyldeoxydesethyl VBL (**4**)	3.2, 2.8 [S]	SA	SA	3.1 ± 1.5	48.0*	20.1	N.D.
20′-Deoxydesethyl VBL (**5**)	26, 25	S	SA	104 ± 1	593 ± 83	24.0	70.5
20′-Dimethyldeoxydesethyl VBL (**6**)	29, 36	SA	MT	26.0 ± 5.2	716 ± 164	28.9	75.9
20′-Desethyl VBL (**7**)	32, 30	Am	MT	1150 ± 340	1000	26.2	N.D.
20′-Epidesethyl VBL (**8**)	No effect	MT	MT	6800*	1000	14.3	N.D.
20′-Propyldeoxydesethyl VBL (**9**)	25, 17	Am	N.D.	50.0*	580*	19.0	N.D.
20′-Epipropyldeoxydesethyl VBL (**10**)	25, 21	Am	MT	80.5 ± 1.5	630*	20.0	N.D.

[a]The IC_{50} values of the vinca binary alkaloids for microtubule assembly were measured from their concentration-dependent effects on steady-state turbidity levels. Values are presented from two separate experiments. The products induced by a high concentration (10^{-4} M) of each compound with MTP or steady-state microtubules assembled from MTP were determined by transmission electron microscopy of steady-state solutions from at least two separate preparations of protein with similar results. SA, Spiral aggregates; S, single spirals; Am, amorphous aggregates; MT, microtubules; N.D., not determined.

[b]The IC_{50} values for tumor cell population growth were determined from probability plots of the concentration of compound versus the percent control cell number after 48 hr (L1210 cells) or 72 hr (RCC-2 cells) of treatment. The values presented were averaged from at least three separate experiments, ±S.D. An asterisk denotes a value averaged from the duplicate samples of a single experiment. The mitotic index of the populations was measured after a 6-hr (L1210 cells) or a 21-hr (RCC-2 cells) incubation with each compound at 10-fold its IC_{70} concentration. The values presented are from a single, typical experiment. Control values for mitotic index are 3.0 and 1.0% for L1210 and RCC-2 cells, respectively.

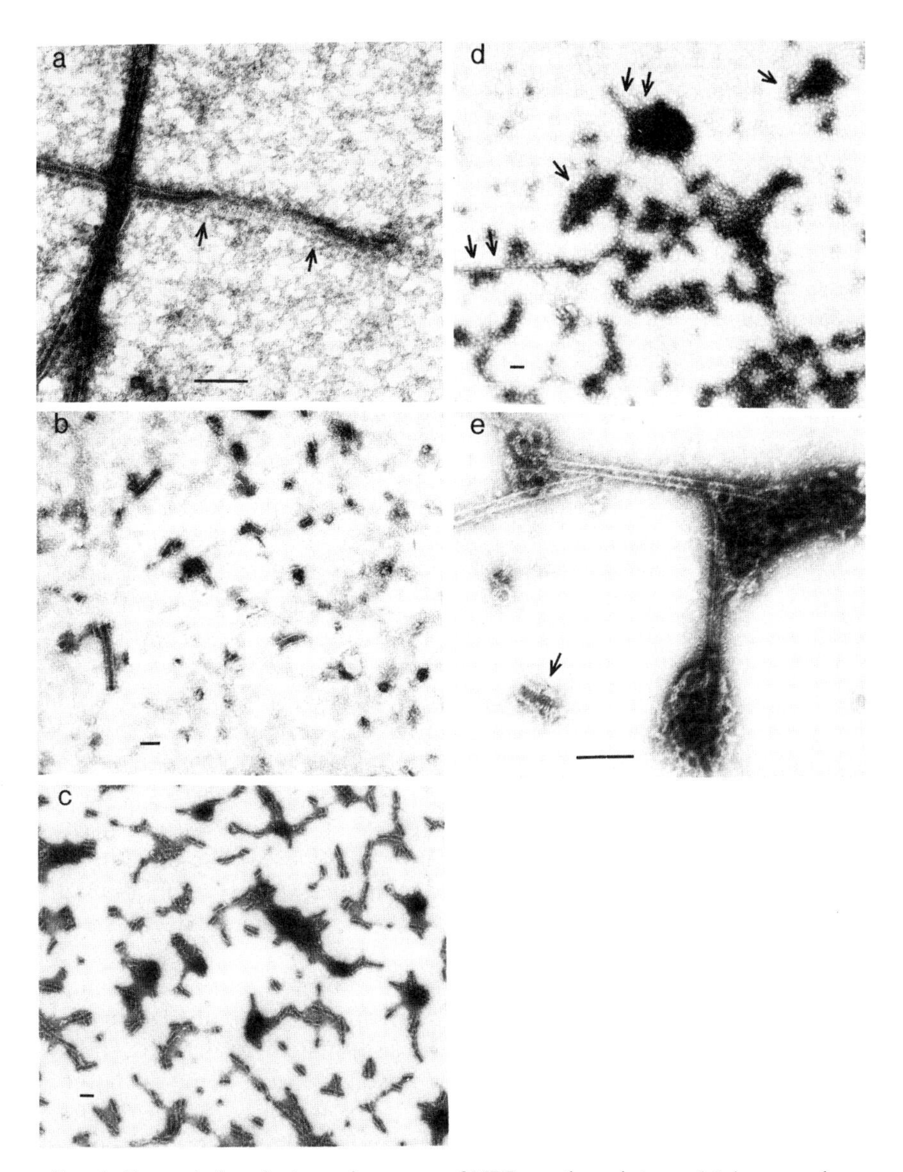

FIG. 4. Transmission electron microscopy of MTP reaction mixtures. (a) An opened area is seen in a microtubule polymerized in the presence of an IC_{50} concentration (3×10^{-6} *M*) of deoxydesethyl VBL (**5**). (b) Single spirals are formed from MTP and 10^{-4} *M* **5**. (c) Spiral aggregates are formed from MTP and 10^{-4} *M* VBL. (d) MTP incubated with an IC_{50} concentration (2×10^{-7} *M*) of epimethyldeoxydesethyl VBL (**4**) formed spiral aggregates both free in solution (single arrows) and associated with microtubules (double arrows). (e) Greater magnification of MTP incubated with an IC_{50} concentration (2×10^{-7} *M*) of methyldeoxydesethyl VBL (**3**) displays a free spiral (arrow) and spiralized material on the microtubules. Bar, 0.1 μm.

The incorporation of a single C-20′ methyl group into 20′-deoxy-20′-desethyl VBL produced congeners **3** and **4**, which displayed VBL-like potency as inhibitors of microtubule assembly. Also, spiral aggregates were formed from either steady-state microtubules or MTP at a high concentration of congener. However, total normalcy of reaction with the microtubule system *in vitro* did not occur with the monomethyl derivatives since we observed the formation of spiralized material in preparations of microtubules assembled in the presence of an IC_{50} concentration of each congener (Fig. 4d,e). This is the first report of spiral formation by a vinca binary alkaloid below a concentration of 10^{-6} *M* and constitutes a potentiation of reactivity with MTP compared to VBL that is specific to the monomethyl substitution of deoxydesethyl VBL.

This caveat is evident from the activity profiles of the other alkyl congeners of deoxydesethyl VBL in Table I: the dimethyl (**6**) and propyl (**9** and **10**) derivatives were an order of magnitude less inhibitory of microtubule assembly than VBL, and the high concentration effects on MTP and on microtubules were modified or lost. In this group of congeners, all of which lacked the ability to disassemble microtubules at 10^{-4} *M* concentration, only the dimethyl derivative (**6**) retained VBL-like activity with MTP at a high concentration and produced normal spiral aggregates. This observation links, once again, the methyl substitution at the C-20′ position with the reactivity of the molecule with MTP.

These data also clearly establish that reactions of C-20′ alkyl derivatives of deoxydesethyl VBL with soluble MTP or with its polymerized form as microtubules are dissociable events. Whereas the dimethyl- (**6**) and propyl- (**10**) substituted compounds inhibited the assembly of microtubules with an IC_{50} of about 3×10^{-6} *M*, they were inactive with steady-state microtubules at 10^{-4} *M*.

The series of C-20′ alkyl-modified VBL congeners of Table I contains one type of molecule which retains the hydroxyl group at that position: desethyl VBL (**7**) and epidesethyl VBL (desethylleurosidine, **8**). The former compound inhibited microtubule assembly at a reduced potency compared to VBL but lost completely the high-concentration reactions with MTP and microtubules. A comparison of this activity profile with that of deoxydesethyl VBL makes evident that the retention of the hydroxyl group at the C-20′ position of VBL, when the alkyl moiety is lacking, causes specific loss of the high-concentration functions of the molecule. We examined synthetic leurosidine and found it to be essentially inactive with the microtubule system *in vitro*, except that detectable inhibition of microtubule assembly was observed at 1×10^{-5} *M*. The removal of the ethyl moiety from leurosidine (congener **8**) eliminated this low-potency reaction with MTP, to produce the only C-20′ alkyl congener in our series that was totally inactive with the microtubule system *in vitro*.

Therefore, for the intact VBL molecule, it is the presence of the ethyl group and not the hydroxyl moiety at the C-20′ position that is required for normal potency and function as an antimicrotubule agent. The fact that simultaneous removal of both substituents at this position (congener deoxydesethyl VBL, **5**) restored the high-concentration-dependent activities of the molecule which were lost to the congener desethyl VBL (**7**) suggests that the singular presence of the hydroxyl group, in the natural epimeric sense, acts to inhibit the proper binding to tubulin as a dimer (MTP) or polymer (microtubule) that is required for spiral aggregate formation and microtubule disassembly. In the absence of the hydroxyl group at the C-20′ position of VBL, we found that the activity profile of the molecule at high concentrations with MTP or microtubules was dependent on the specific alkyl function present in the general order of methyl ⩾ ethyl > dimethyl > propyl. In addition, the normal reactivity of VBL with MTP under conditions of assembly is dependent on the presence of the ethyl moiety. Overall, these results negate the recent proposal (*16*) of a mechanism of anticancer activity for binary VBL-type alkaloids that is based on rupture of the binary structure by nucleophilic substitution at C-16′ by protein thiol groups.

These data from the C-20′ congeners make several new contributions to our knowledge concerning the reactivity of VBL-like binary alkaloids with the microtubule system *in vitro*. First, it is unlikely that the abberant, stoichiometric inhibition of MTP polymerization that is displayed by most of the alkyl congeners is a reflection of the same mechanism used to explain substoichiometric assembly inhibition by VBL, namely, the binding of the ligand to microtubule ends (*17*). This conclusion raises the interesting possibility that direct inactivation of tubulin by bound congener leads to assembly inhibition. Likewise, the polymerization IC_{50} values of the monomethyl derivatives (**3** and **4**) of deoxydesethyl VBL are apparently equivalent to VBL, and yet the former quantitations are manifested from unique reactions that contain spirals and spiralized material. Future studies are required to address the mechanistic details of the C-20′ congener specificity of action with tubulin and microtubules *in vitro*.

B. Cellular Properties

The activities of the various C-20′ alkyl congeners of VBL in cellular assays of growth inhibition and mitotic arrest are presented in Table I. Only one type of substitution, the 20′-methyl derivatives (**3** and **4**) of deoxydesethyl VBL, exhibited a level of cytotoxicity to leukemic and colon cancer cell lines that was in the nanomolar range and similar to the parent compounds. The remaining congeners fell into two groups with respect to potency: those with IC_{50} values for cell growth about an order of magnitude greater than the parent deoxy congeners (dimethyldeoxydesethyl VBL, **6**, propyldeoxydesethyl VBL, **9** and **10**, and deoxydesethyl VBL, **5**) and the very poorly cytotoxic desethyl compounds **7** and **8** with IC_{50} values at the micromolar level. Although the antimicrotubule activity profiles of the congeners and their respective abilities to inhibit cell growth do not present a clear correlation for the congeners as a group, it is apparent that methyl derivatives **3** and **4** have coupled parental potencies for both the microtubule system *in vitro* and for cell growth inhibition. Also, at the other extreme of activity, we found epidesethyl VBL (**8**), to be inactive in the microtubule assays and to be 10,000-fold less cytotoxic to cultured tumor cells than VBL.

The mitotic indices observed after treatment of both cell lines with 10-fold the IC_{70} concentration of each congener establish that the action of the vinca binary alkaloids with this cellular target *in vivo* is correlated with the ability of the molecule to inhibit microtubule assembly and not with its high-concentration-dependent activities with MTP or microtubules. Based on the current understanding that neurotoxicity and neurotubule damage are concerted events, the results of our structure–activity studies support the optimistic expectation that the development of a non-neurotoxic vinblastine–vincristine congener can be achieved.

III. Molecular Constraints of Congener Activity

As mentioned in Section I, molecules as complex as the indole–indoline binary alkaloids warrant an investigation of their critical stereochemistry, including the binary character, with respect to function. The activity profile of the epi C-16′, C-14′, C-20′-diastereomer of 20′-deoxy VBL is presented in Fig. 5a. The compound inhibits population growth only at micromolar potency, and it lacks detectable antimicrotubule activity. This profile is also typical of the diastereomers of VBL (vincovaline), and of the C-20′ congeners examined to date, epideoxy VBL (**2**), and deoxydesethyl VBL (**3**).

The sensitivity of the three types of antimicrotubule activity by VBL to modification at a single site on the cleavamine moiety poses an obvious question: Can the cleavamine moiety act alone to exert a functional attack at this target? The answer is no. The 20′-deoxy and epideoxy carbometh-

a 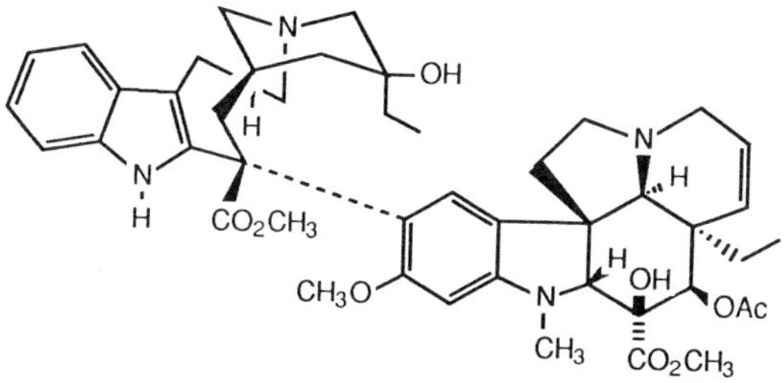 b

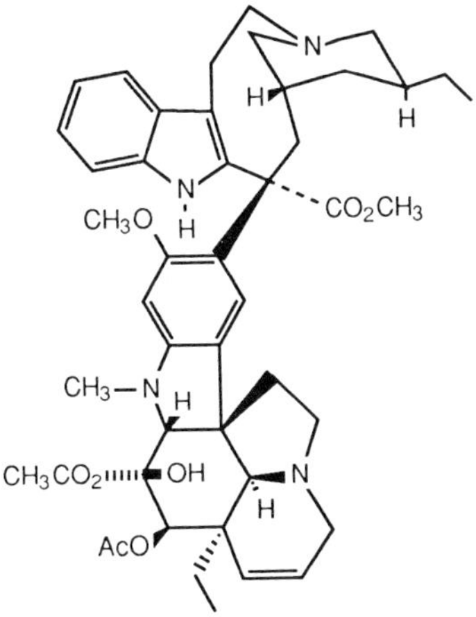

FIG. 5. Structure and activity profiles of the epi C-16′, C-14′, C-20′ diastereomer of 20′-deoxy VBL (**a**) and of the vinblastine unnatural piperidine conformational isomer (**b**). The parameters of activity, determined as described in Table I, were as follows:

Compound	Cell growth IC_{50} M ($\times 10^6$) L1210	Cell growth IC_{50} M ($\times 10^6$) RCC-2	Mitotic index L1210	MTP polymerization IC_{50} M ($\times 10^6$)	Product of 10^{-4} M added to MTP	Product of 10^{-4} M added to Microtubules
a	1.6 ± 0.2	1.4 ± 0.2	0.8	Inactive	MT	MT
b	4.1 ± 0.2	N.D.	N.D.	Inactive	MT	MT

oxydihydro cleavamines are poorly cytotoxic (IC_{50} >5 × $10^{-5}M$) and are inactive with MTP or steady-state microtubules. Also of note, these cleavamines do not affect the reactivity of VBL, deoxy VBL, or epideoxy VBL with the microtubule system *in vitro*.

In the course of our synthetic efforts, we discovered a new, unnatural conformational isomer of the VBL piperidine ring (see Chapter 2, this volume, for details). This compound is inactive with the microtubule system *in vitro* and is poorly cytotoxic to cultured tumor cells (Fig. 5b). Therefore, the functional determinism of the C-20′ position of VBL-like molecules mediates reactions, as yet unknown, that are dependent on the binary alkaloid structure and on the natural stereochemical configuration at C-16′ and C-14′ as well as on the conformation of the cleavamine moiety.

IV. Conclusions

The structure–function relationship of the indole–indoline binary alkaloids was relegated to obscurity until the recent achievement of methodologies for their complete syntheses (see Chapter 2, this volume). Our work with C-20′ congeners of VBL has established that the complex interactions between this molecule and tubulin or microtubules can be modified by structural alteration. The various, concentration-dependent reactions of VBL with the microtubule system *in vitro* are sensitive to subtle modifications at a single molecular locus. In addition, these reactions are distinctive on a mechanistic level as seen from the unique activity profiles of most of our C-20′ alkyl congeners. At first light, we can look toward the future with secured optimism.

Acknowledgments

Supported by Grants CH-424 and RD-228 from the American Cancer Society and Grant PO1 CA24543 to the Vermont Regional Cancer Center from the National Cancer Institute.

References

1. J. H. Cutts, *Cancer Res.* **21**, 168 (1961).
2. V. K. Vaitkevicius, R. W. Talley, J. L. Tucker, and M. J. Brennan, *Cancer* **15**, 294 (1962).
3. G. Cardineli, G. Cardineli, and T. N. Mehorta, *Proc. Am. Assoc. Cancer Res.* **4**, 10 (1963).
4. E. Frei III, J. Whang, R. B. Scoggins, E. J. Van Scott, D. P. Rall, and M. Ben, *Cancer Res.* **24**, 1918 (1964).
5. C. G. Palmer, D. Livengood, A. Warner, P. J. Simpson, and I. S. Johnson, *Exp. Cell. Res.* **20**, 198 (1960).
6. N. Burchovsky, A. A. Owen, A. J. Becker, and J. W. Till, *Cancer Res.* **25**, 1232 (1965).
7. R. S. Camplejohn, B. Schultze, and W. Maurer, *Cell Tissue Kinet.* **13**, 239 (1980).
8. A. M. Lengsfeld, B. Schultze, and W. Maurer, *Eur. J. Cancer* **17**, 307 (1981).
9. S. S. Schochet, Jr., P. W. Lampert, and K. M. Earle, *J. Neuropathol. Exp. Neurol.* **27**, 645 (1968).
10. W. A. Wisniewski, M. L. Shelanski, and R. D. Terry, *J. Cell Biol.* **38**, 224 (1968).
11. M. L. Schelanski and H. Wisniewski, *Arch. Neurol.* **20**, 199 (1969).
12. S. Okabe and N. Kirokawa, *J. Cell Biol.* **107**, 651 (1988).
13. M. L. Shelanski, F. Gaskin, and C. Cantor, *Proc. Natl. Acad. Sci. U.S.A.* **70**, 765 (1973).
14. L. S. Borman and M. E. Kuehne, *Proc. Am. Assoc. Cancer Res.* **26**, 993a (1985).
15. F. Zavala, D. Guénard, and P. Potier, *Experimentia* **34**, 1497 (1978).
16. P. Magnus, M. Ludlow, and J. Elliott, *J. Am. Chem. Soc.* **109**, 7929 (1987).
17. M. A. Jordan, R. C. Margolis, R. H. Himes, and L. Wilson, *J. Mol. Biol.* **187**, 61 (1986).

—CHAPTER 4—

MEDICINAL CHEMISTRY OF BISINDOLE ALKALOIDS FROM *CATHARANTHUS*

H. L. PEARCE

Lilly Research Laboratories
Indianapolis, Indiana 46285

I. Introduction

From the perspective of organic chemistry, the medicinal chemistry of the bisindole alkaloids derived from *Catharanthus roseus* (L.) G. Don circumscribes a large and exceedingly diverse collection of structurally complex molecules. There can be no doubt that the breadth and diversity of these compounds is due, in large part, to the remarkable biological

THE ALKALOIDS, VOL. 37
Copyright © 1990 by Academic Press, Inc.
All rights of reproduction in any form reserved.

activity inherent in the oldest members of the group, namely, vinblastine (**1**) and vincristine (**2**). The discovery of the experimental antitumor properties of these compounds was soon extended to the treatment of several human malignancies at a time that was coincident with both the recognition of cancer as menace to public health and the developing maturity of synthetic organic chemistry. These factors established a powerful incentive for medicinal chemistry research in the area, and the unique biochemical mechanism of action of **1** and **2** coupled with the increasing clinical reliance on chemotherapeutic regimes comprised of several drugs with nonoverlapping toxicities insured that "Vinca" chemistry would be pursued by research groups for decades.

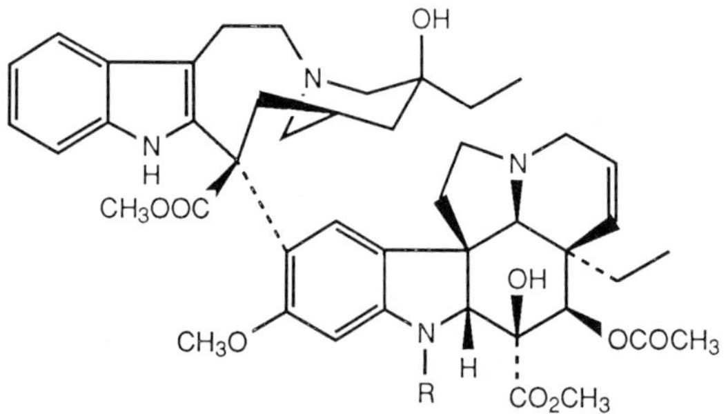

1, R = CH_3
2, R = CHO

The structural modification of natural products is useful in several ways. The known pharmacology of bisindole alkaloids is enriched by the diversity of chemical structures that are made available by structure modification and total synthesis. These molecules have served as biochemical probes in several areas of biology, especially in those of microtubule assembly and drug resistance. The most elusive prize, however, has remained the discovery of new compounds with clinical activity. In recent years several compounds have been evaluated in clinical trials, but vinblastine and vincristine remain the only bisindole alkaloids approved for the treatment of cancer in the United States. These compounds are joined by vindesine in Europe, and at least two new derivatives are the subject of ongoing clinical trials. Considering the breadth of chemical research in this area, the overall yield as measured by new compounds with clinical activity has been relatively low, but this observation is not unique in history of analog development in cancer research. Nevertheless, the search continues, and this chapter details the chemical endeavors to discover a new bisindole alkaloid with clinical activity.

In order to present an orderly and systematic overview, this chapter focuses on synthetic or semisynthetic derivatives of the naturally occurring bisindole alkaloids. The pharmacology of these compounds is described in detail elsewhere in this volume (Chapter 5), but in discussing the medicinal chemistry of this series it is necessary to introduce the essential features of their biochemical, cellular, and antitumor activities. Compounds are grouped according to the type of structure modification that each compound represents, with the largest group of chemical structures being composed of compounds formally derived by atomic or functional group substitutions or deletions. This group is followed by a smaller group of compounds that have been prepared by bond rearrangement of the bisindole carbon framework, followed by an even smaller group that have been prepared by multistep synthesis.

Because of the volume of material that must be included in describing the medicinal chemistry of these alkaloids, it is necessary to limit the scope of this chapter. The preparation of immunogens used in the production of antibodies for radioimmunoassay is not described, nor is the chemistry associated with the individual "monomer" units, vindoline or velbanamine (except as requisite starting materials for the preparation of "dimers").

II. Pharmacology

A detailed discussion of the pharmacology of the bisindole alkaloids derived from *Catharanthus* is presented elsewhere in this volume (Chapter 5); however, in order to discuss the medicinal chemistry of these compounds it is necessary to introduce the terms that will establish these agents in the context of new drug discovery. Vinca alkaloids have been reported to produce a variety of pharmacologic effects ranging from hypoglycemia to anti-inflammatory (*1–7*). In some instances these descriptions are based on purely anecdotal data, but in the majority of cases, the described effects are real, at least in an absolute sense. The primary impetus for new compound synthesis has been on the discovery of new antitumor agents, and the pharmacologic data presented in this chapter deal entirely with this aspect.

The antiproliferative effects of bisindole alkaloids are well established. In general, these compounds are extremely toxic to a wide variety of cells in culture. The cytotoxic effects require a minimal exposure time approximately equal to the doubling time of the cells under test, usually 12 to 72 hr, at a drug concentration of 10 n*M* to 1 μ*M* in order to achieve an ED_{50} (*8,9*). Cell types that are characterized by short doubling times (e.g., lymphocytes) tend to be more sensitive than those cell types that divide more slowly (e.g., epithelial cells). In contrast, neuronal cell preparations are extremely sensitive (*10,11*).

Dividing cells in culture exposed to vinblastine or vincristine are arrested from further growth during mitosis (*12,13*). In fact, the antimitotic effects of this class of compounds is ubiquitous. These effects are observed at relatively low concentrations ($<1\ \mu M$), and are reversible when drug is removed from the media prior to lysis of the arrested cells. The concentration of drug required to elicit an antimitotic effect is usually comparable to that required to produce a cytotoxic effect in the same cell type (*14,15*). Originally, this type of analysis was exceedingly laborious, but the introduction of laser- and computer-based fluoresence activated cell sorters (FACS) has rendered this type of analysis routine. Nevertheless, a cytotoxic, non-cell cycle-specific bisindole alkaloid has yet to be discovered.

The cause of the cell cycle specificity of the bisindole alkaloids may be associated with the ability of these compounds to interact with the protein tubulin and thereby inhibit the polymerization (and depolymerization) of microtubules (*16,17*). In this respect the cellular pharmacology of vinca alkaloids is similar to that of other cytotoxic natural products such as colchicine or podophyllotoxin. On closer inspection, however, Wilson determined that the specific binding site on tubulin occupied by vinblastine or vincristine is chemically distinct from the site occupied by the other natural products (*18*). Subsequent experiments have determined that the maytansinoids, a class of ansa-macrocycles structurally distinct from the bisindoles, may bind to tubulin at an adjacent (or overlapping) site (*19*). A partial correlation of the antimitotic activity of these compounds with their tubulin binding properties has been made, but discrepancies in cellular uptake probably preclude any quantitative relationship of these effects (*20*).

With the exception of the effect on microtubules described in the foregoing paragraph, the bisindole alkaloids have little or no effect on macromolecular synthesis at subtoxic concentrations (*21,22*). In experiments utilizing radiolabeled precursors ([^{3}H]leucine, -uridine, or -thymidine) cells cultured in the presence of vinblastine showed no differential incorporation of radioactivity. Furthermore, there is no indication that treatment of cells with vinblastine or vincristine produces alterations in cellular DNA (*23,24*).

The primary clinical application of the antiproliferative effects of the bisindole alkaloids has been in the treatment of human neoplasms. A variety of experimental antitumor systems have been used to study the antitumor properties of these alkaloids in animal models, and much of the information presented for specific compounds in this chapter highlights these effects. Early studies employed syngeneic mice bearing hematopoietic cells in the intraperitoneal space. Antitumor activity is usually noted as the effect on survival of the treated animals compared with that of an untreated control group. Examples of these early models include the Gardner lymphosarcoma, Ridgeway osteogenic sarcoma, Walker 256 carcinoma, P388 lymphocytic leukemia, and L1210 leukemia. Solid tumor models have also been employed in evaluating the antitumor activity of these compounds. These models are usually conducted by implanting a small section of tumor in the flank of a syngeneic host and measuring the effect of the compound under test on either the survival of the treated animals or the inhibition of tumor growth. Inhibition of tumor growth in the animals is measured as the weight or volume of the tumor at the end of the treatment period. Murine solid tumors that are typically responsive to treatment with bisindole alkaloids include the 6C3HED lymphosarcoma and the C3H mammary adenocarcinoma.

The most profound clinical impact that the bisindole alkaloids have made is in the treatment of leukemias and lymphomas (*25*). In many respects, the goals of the medicinal chemist working in the area of vinca alkaloids has been to develop compounds with improved efficacy or broader therapeutic applications. The activities of vinblastine and vincristine appear to be quite divergent in their clinical application. Vinblastine, in combination with other agents, afforded striking efficacy in Hodgkin's disease, testicular cancer, and ovarian cancer as well as considerable efficacy in non-Hodgkin's lymphoma (*26*). In contrast, vincristine offers curative effects in childhood lymphocytic leukemia and in Hodgkin's disease (*26,27*), and vindesine, a semisynthetic analog of vinblastine, has demonstrated considerable activity in lung cancer (*28*).

The clinical use of vinca alkaloids is not without toxic liabilities. Vinblastine often produces dose-limiting myelosuppressive toxicity (*29*), whereas vincristine usually produces dose-limiting neurotoxicity (*30*). Interestingly, vindesine has been shown to be both myelosupressive and neurotoxic, depending on the clinical protocol (*31*). Several structural modification programs have been undertaken with the goal of reducing these major toxicities. A more detailed review of the clinical applications of these compounds is presented elsewhere in this volume (Chapter 6).

III. Physicochemical Considerations

A. Acid–Base Properties

The bisindole alkaloids of *Catharanthus* exhibit ambiphilic behavior in solution, owing, in large part, to their gross lipophilic character combined with the presence of four nitrogen atoms of varying basicity. The "upper

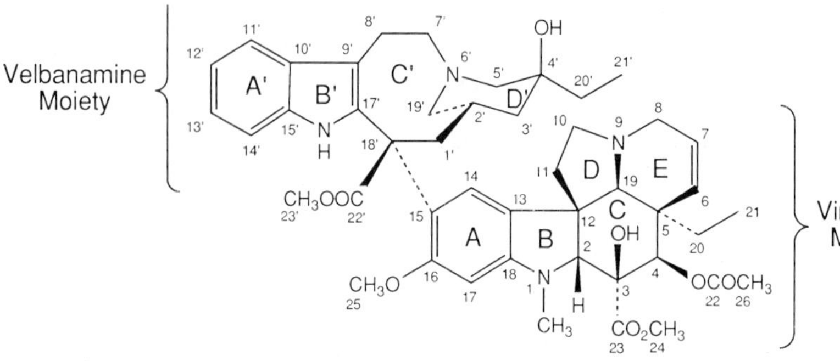

FIG. 1. Atomic numbering and designation of rings in vinblastine.

half" or velbanamine portion of these compounds is a polycylic structure having an indole nucleus fused to a conformationally fluxional nine-membered ring (C′) that contains a basic nitrogen atom and the bridging carbon of the piperidine ring (D′, Fig. 1). The velbanamine fragment is connected by a single carbon–carbon bond to an intact vindoline molecule between C-18′ of the velbanamine moiety and C-15 of the vindoline piece. The "lower half" of the bisindole is considerably less fluxional, and N-1 and N-9 are much less basic than N-6′ of the upper half (Table I). Nuclear magnetic resonance (NMR) studies suggest that the lack of basicity of N-9 is due to a boat conformation of the C-ring wherein N-9 is protonated by an internal hydrogen bond from the β-hydroxyl group attached to C-3, thus decreasing the availability of the electrons on nitrogen (*32–35*).

B. MOLECULAR SIZE AND SHAPE ANALYSES

The molecular weight of a typical bisindole alkaloid is in excess of 800 g/mol, yet the high degree of unsaturation (9 rings) and shared bonds (6

TABLE I
DISSOCIATION CONSTANTS OF SELECTED BISINDOLE ALKALOIDS

Alkaloid	pK_a Values	Ref.
Vinblastine	5.4, 7.5	*33*
Vincristine	5.0, 7.4	*34*
Leurosine	5.5, 7.5	*33*
Leurosidine	5.0, 8.8	*34*

ring fusion bonds and three bridging atoms) present in these compounds contribute to a relatively compact molecular structure. The molecular volume of these compounds expressed as a calculated molar refractivity (CMR) is on the order 230 $Å^3$ (*36*). The central, connecting carbon–carbon bond is essentially hidden by the two "monomer" domains, each of which expresses a single planar aromatic domain at the surface of the molecule. The nonbonding electron pair at N-6′ is also accessible from the outer surface of the molecule (Fig. 2).

C. Conformations

A great deal of insight as to the preferred conformation of the bisindole nucleus has been deduced from the solution chemistry of the compounds and from NMR studies of numerous derivatives (*37,38*). The relatively low basicity of N-6′ in compounds with a β-hydroxyl group at C-4′ suggests a chair conformation for ring D′ of these compounds, and this conclusion is supported by ^{13}C-NMR studies (*37*). Epimerization at C-4′ leads to a series of compounds in which the ^{13}C chemical shifts of the ring D′ carbons do not correspond with those having a β-hydroxyl group at C-4′, implying that the conformation of the piperidine ring is inverted and exists in the alternate chair conformation.

X-Ray diffraction analyses of vincristine methiodide (*39*) and vinzolidine 1-naphthalene sulfonate (*40*) provide atomic coordinates of compounds that are either modified in the velbanamine portion (alkylation at N-6′, Fig. 3) or in the vindoline portion (a spiro-fused oxazolidinedione at C-3 and C-23, Fig. 4). These structures show a chair–boat conformation for ring C′ with C-8′ exhibiting an endo pucker. Ring D′ is clearly in a chair conformation, but, in contrast to the conclusions drawn from ^{13}C-NMR studies, the N-6′–C-7′ bond is axial relative to the piperidine ring,

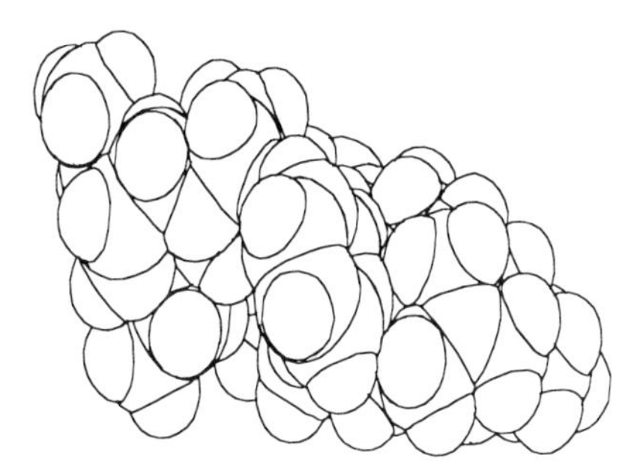

FIG. 2. Computer-generated space-filling representation of vinblastine.

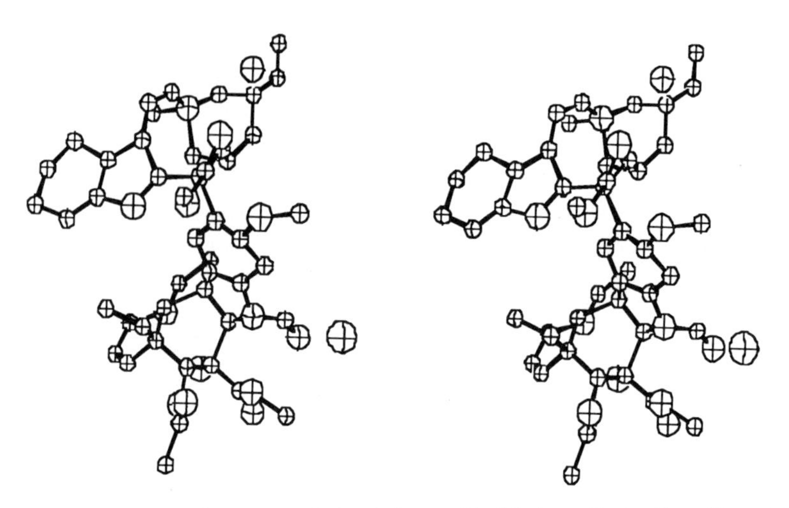

FIG. 3. X-Ray crystal structure of vincristine methiodide [coordinates from Moncrief and Lipscomb (*39*)].

thus requiring the nonbonded lone pair at N-6′ to be exo when either protonated or alkylated. Unfortunately, there has been no similarly definitive report for any of the naturally occurring bisindole alkaloids.

D. PARTITION PROPERTIES

As previously noted, the partition properties of the bisindole alkaloids tend to be lipophilic. The octanol–water partition coefficients of several bisindole derivatives have been measured and compared to various bio-

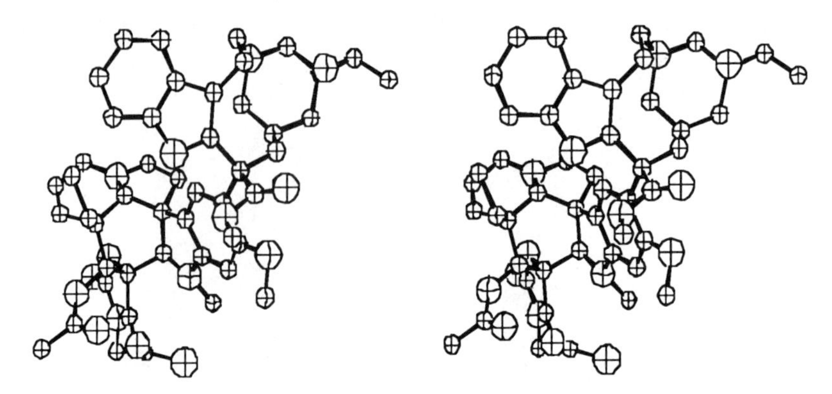

FIG. 4. X-Ray crystal structure of vinzolidine 1-naphthalene sulfonate [coordinates from Jones *et al.* (*40*)].

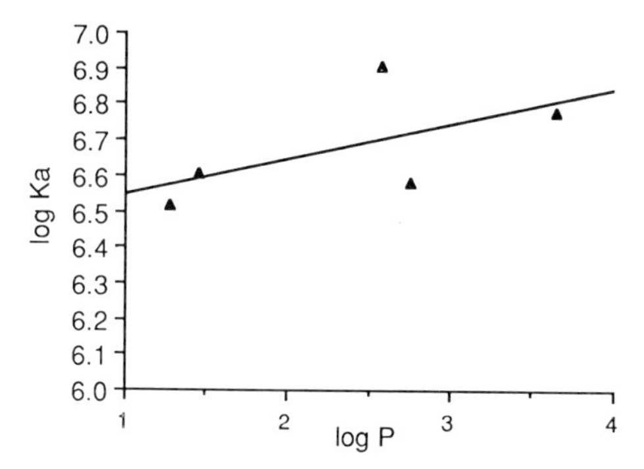

FIG. 5. Correlation of bisindole alkaloid tubulin binding affinity (K_a) with partition coefficients at pH 7.4 [data from Owellen *et al.* (*41*)].

chemical parameters including tubulin binding affinity and antitumor activity. Owellen *et al.* (*41*) described an ambitious study of these correlations using data obtained for compounds in different structural series, and they proposed a quadratic relationship between the log of the partion coefficient and the tubulin association constant as well as the log of the antitumor activity against P388 leukemia. By restricting the range of structures under consideration to a single series, a simple correlation between the log of the partition coefficient and the tubulin binding affinity is apparent (Fig. 5).

IV. Naturally Occurring Bisindole Alkaloids from *Catharanthus*

The medicinal chemistry of the bisindole alkaloids from *Catharanthus* focuses mainly on functional group modifications carried out on naturally occurring substrates. The choice of starting material for these transformations is usually based on an intent to modify the biological activity of the parent; however, the availability of the starting bisindole has often determined this decision. It is not surprising that both active and inactive naturally occurring alkaloids have been used to initiate these programs.

The relatively high yield of vinblastine from *Catharanthus* extracts and the remarkable biological activity of this alkaloid render this compound a "natural" choice for structural modification. Vinblastine is subject to attack by electrophilic reagents at a number of sites (Fig. 6), but by careful selection of reagents and conditions it is possible to operate in this structurally complex theater with considerable selectivity. Potentially electrophilic sites are, in some cases, immune from electrophilic attack.

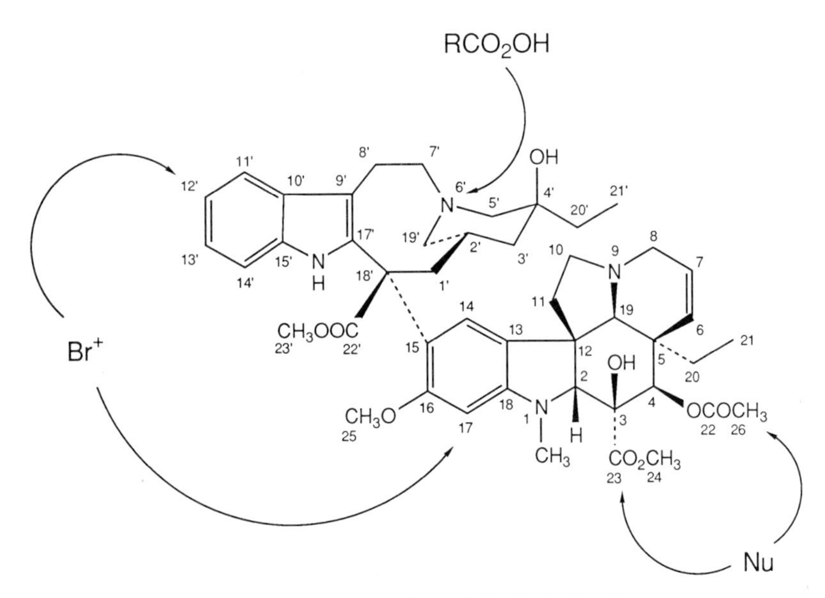

FIG. 6. Sites of chemical reactivity in vinblastine.

For example, treatment of **1** with bromine in aprotic media results in electrophilic aromatic substitution rather than attack of the C-6,7 double bond, and reaction with peracids gives rise to selective oxidation of N-6′ rather than N-9. Protonation of N-6′ is an effective method of protecting this site from oxidizing conditions, rendering other sites accessible to oxidizing reagents. Thus, it is possible to convert vinblastine (**1**) to vincristine (**2**) under a variety of acidic oxidizing conditions. Reactions with nucleophilic reagents rarely modify functionality other than the C-4 acetate and the C-3 carbomethoxy function. Hydrolysis of the ester at C-22′ is not observed except under forcing conditions.

As a substrate for chemical modification, vincristine (**2**) represents much more challenge. The extracted yield of **2** from *Catharanthus* is approximately 10% that of **1**, making this compound a much scarcer substrate (*42*). Furthermore, the chemical lability of the N-1 formyl bond is extreme under a variety of conditions. The preparation of compounds in this series is usually accomplished by performing the requisite functional group interconversions with compounds bearing an N-1 methyl and reserving the oxidation of this methyl for the last step.

Leurosine (**3**) represents a useful entry for the synthesis of compounds in which the stereochemistry at C-4′ is inverted (*43,44*). The reactivity patterns exemplified by vinblastine (**1**) are also present in the chemistry

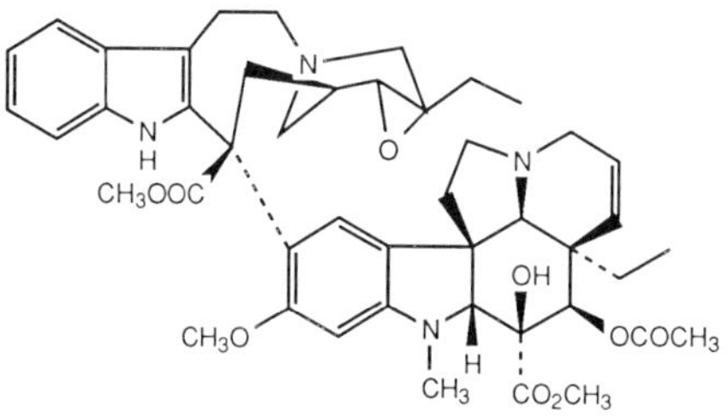

3 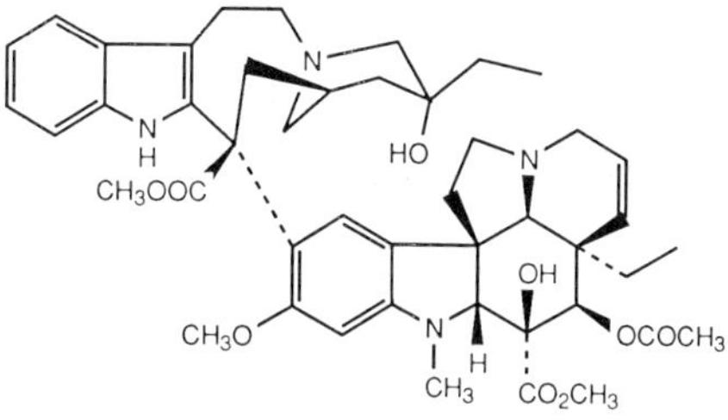 **4**

of **3**; however, by reductive opening of the C-3′,4′ epoxide, compounds with reduced and inverted stereochemistry are available. These products are closely related to those available from leurosidine (**4**) in which the stereochemical configuration at C-4′ is inverted relative to vinblastine (**1**).

V. Modifications of the Upper Half (Velbanamine Portion) of Bisindole Alkaloids

A. C-12′ and Other Aromatic Substitutions

Relatively few bisindole derivatives having unnatural aromatic substituents have been prepared. The most reactive aromatic center to electrophilic substitution is certainly at C-12′, where steric hindrance is minimized and electron density favors stabilization of positive charge. Treatment of vinblastine (**1**) with less than 1.0 equiv of bromine in dichloromethane results in selective bromination at C-12′ to give 12′-bromovinblastine (**5**) (*45,46*). If excess bromine is employed, then bromination in the dihydroindole ring is also observed, and mixtures of (**5**) and 12′,17-dibromovinblastine (**6**) are obtained (*46*).

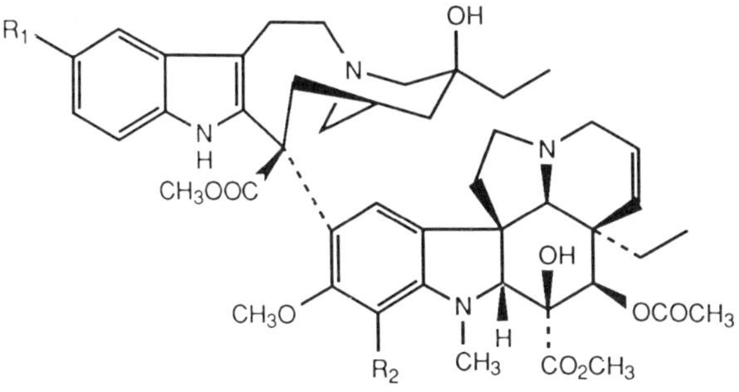

5, R_1=Br; R_2=H
6, R_1=Br, R_2=Br

Electrophilic aromatic iodination affords considerably more regioselectivity, owing in large part to the more congested steric environments at C-17, but efficient iodination at C-12′ requires unusual conditions. Reaction of **1** in acetonitrile with ferrous perchlorate in the presence of tetra(*n*-propyl)ammonium periodate and catalytic ruthenium dioxide gives 12′-iodovinblastine (**7**) as the sole product in 43% isolated yield (*47*). This unusual reaction probably proceeds through a reactive iron(III) iodide resulting from the *in situ* generation of I·. If any of the reactants are omitted from the reaction mixture, no product is formed.

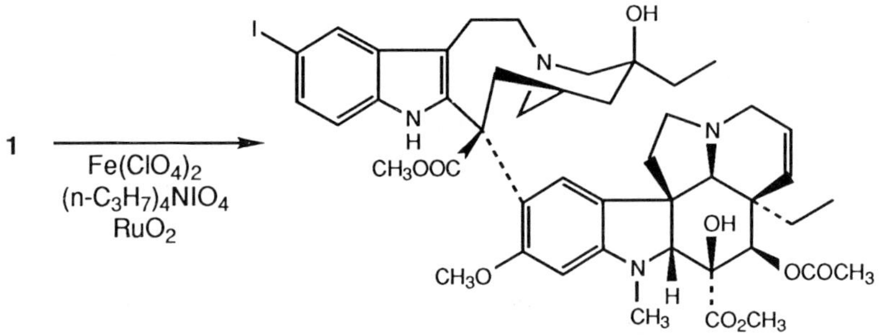

1 $\xrightarrow[\text{RuO}_2]{Fe(ClO_4)_2,\ (n\text{-}C_3H_7)_4NIO_4}$ **7**

7

Reaction of either **1** or **2** with nitrous acid at low temperature results in mixtures of nitration products (*48*). Nitration occurs at positions 12′, 9′, and 17, with the 12′-nitro product (**8**) predominating.

In general, aromatic substitution of the bisindole nucleus results in reduced antitumor activity. 12′,17-Dibromovinblastine (**6**) shows approximately one-tenth the potency of vinblastine in cell culture, and the monohalogenated products show only about 25% of the parent's potency against L5178Y cells in culture (*46*). 12′-Iodovinblastine (**7**) is an active antitumor compound that inhibits the growth of B16 (melanoma), P1534(J) (solid leukemia), and CA755 (mammary adenocarcinoma) at doses ranging

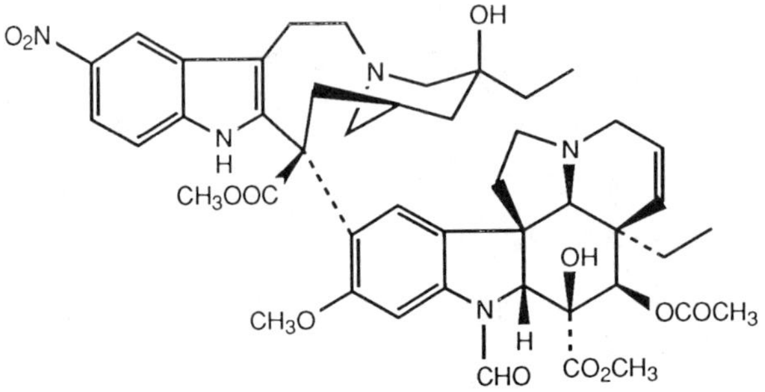

8

from 6.25 to 12.5 mg/kg/day × 10 (*46*). 12′-Nitrovincristine (**8**) is active against P388 leukemia, giving a 217% increase in life span (ILS) at 8.0 mg/kg/day × 8 (*48*).

B. C-4′ SUBSTITUTIONS

Functional group addition or substitution reactions at C-4′ have produced a variety of bisindole derivatives, and structure modifications at this position are often accompanied by unexpected shifts in biological activity. As a result these studies have produced several compounds with intriguing biological properties.

Reaction of vinblastine (**1**) with cold, concentrated sulfuric acid gives a mixture of the C-4′ elimination products $\Delta^{3'}$-dehydro-4-deacetylvinblastine (**9**), $\Delta^{4a'}$-dehydro-4-deacetylvinblastine (**10**), and $\Delta^{4b'}$-dehydro-4-deacetylvinblastine (**11**), isolated in 14, 12, and 15% yield, respectively,

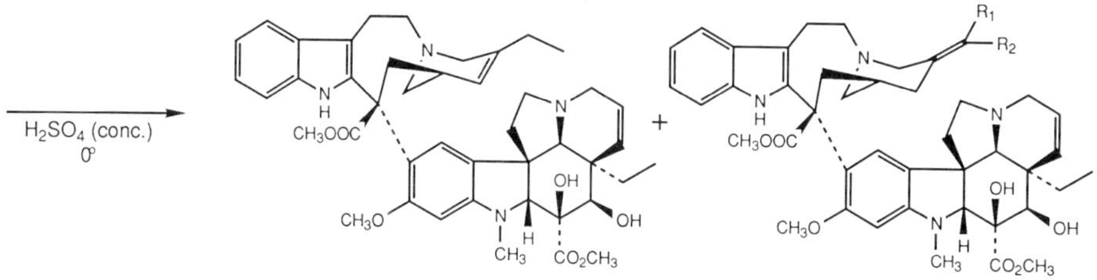

H_2SO_4 (conc.)
0°

9

10, $R_1=CH_3$, $R_2=H$
11, $R_1=H$, $R_2=CH_3$

after preparative thin-layer chromatography (*49*). The dehydro derivatives were shown to be unstable unless they were isolated and stored as their acid salts. If the foregoing reaction is carried out in the presence of acetonitrile at low temperature, a mixture of iminium addition products is obtained in which the 4′-acetylamino derivative (**12**) predominates.

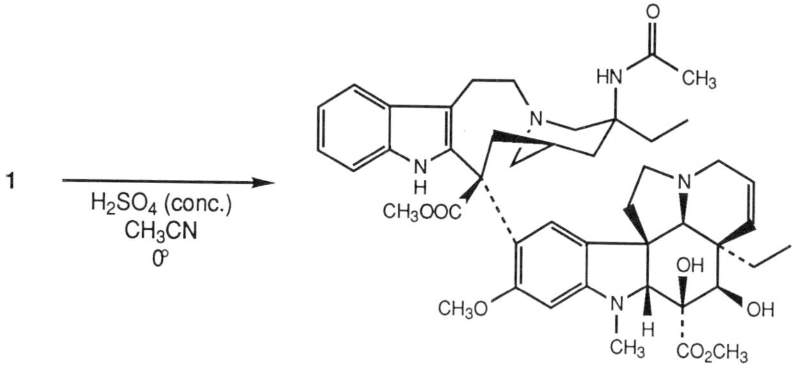

1 → H_2SO_4 (conc.), CH_3CN, 0°

12

The dehydro derivatives **9**, **10**, and **11** were less toxic and considerably less active against Gardner lymphosarcoma *in vivo* when compared to 4-deacetylvinblastine (*49*). This observation is consistent with their lack of antimitotic activity in Chinese hamster ovary (CHO) cells and suggests that changes in the D′ ring conformation (as induced by rehybridization at C-4′) can adversely effect biological activity. The 4′-acylamino derivatives, in which the gross conformation of ring D′ is unchanged, retain their activity. Compound **12** is active against the P1534(J) leukemia model at 7.5 mg/kg/day, inhibiting tumor growth by 71%.

C. N-6′ Derivatives

The chemical reactivity of N-6′ (or $N^{b'}$), directed entirely by the basicity of this atom, is controlled by the nature and stereochemistry of the substituents at C-4′ *(vide supra)*. Oxidation of N-6′ occurs under mild conditions in several naturally occurring bisindole alkaloids. Thus, treatment of a dichloromethane solution of leurosine (**4**) with *m*-chloroperbenzoic acid at −20°C for 4 hr gives the $N^{6'}$-oxide (**15**) in greater than 90% yield after preparative reversed-phase chromatography (*46*). Leurosine $N^{6'}$-oxide has also been isolated from *Catharanthus roseus* and should therefore be considered a naturally occurring bisindole (*50*). The analogous conversion of vinblastine (**1**) to its $N^{6'}$-oxide (**16**) proceeds under similar conditions but requires longer exposure to the peracid (24 hr) (*51*); 3′,4′-anhydrovinblastine is converted to its $N^{6'}$-oxide (**17**) in 10 min at 0°C

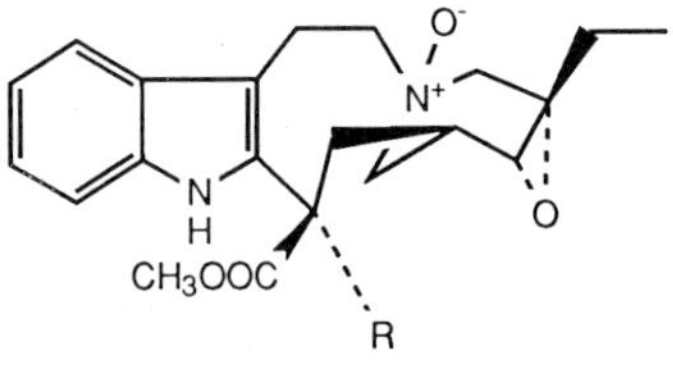

15

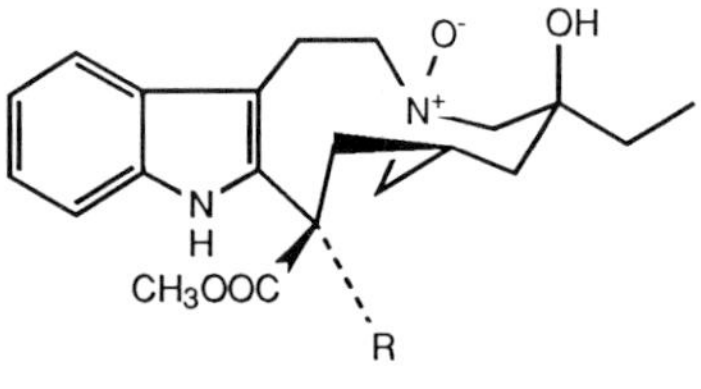

16

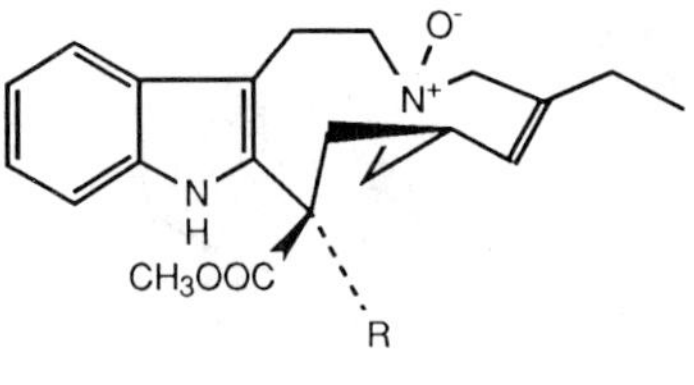

17

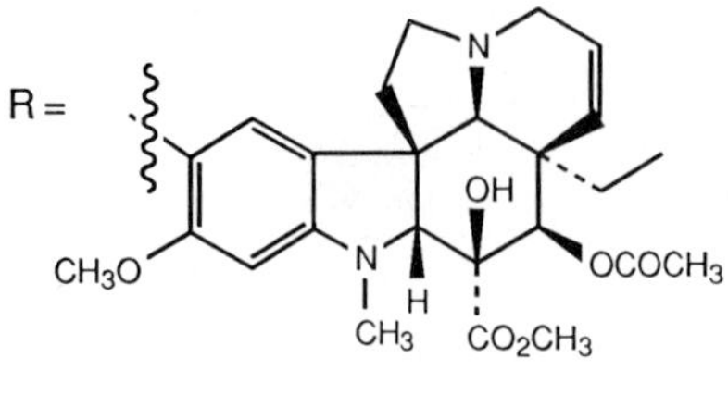

R =

(*52*). This later derivative is a useful intermediate for the preparation of ring C′ contraction products (Section V,G).

There is little mention of N-6′ quaternerization reactions in the literature. $N^{6'}$-Methylvincristine, a methiodide of **2**, was prepared by reaction of vincristine (**2**) with iodomethane, but few experimental details are available (*39*).

The $N^{6'}$-oxides show substantially reduced biological activity. Tubulin binding affinity is reduced by 50-fold, and biological potency *in vitro* and *in vivo* is reduced by a comparable amount (*46*).

D. C-18′ Substitutions (Decarboxy Derivatives)

Hydrolysis of the methyl ester and decarboxylation at C-18′ occur only under forcing conditions. Alkaline hydrolysis of the C-18′ ester of vinblastine requires refluxing in 5 *N* sodium hydroxide for several hours to give the diacid (**18**), and ammonialysis of this position in anhydrous methanol is accomplished in a sealed vessel at 100°C for 60 hr to yield the 18′-decarbomethoxy-4-deacetylvinblastine amide (**19**) (*53*). Bisindole derivatives lacking the C-22′ carboxyl have also been prepared by coupling the vindoline portion with an appropriately chosen ibogane precursor (Section V,G) (*54*).

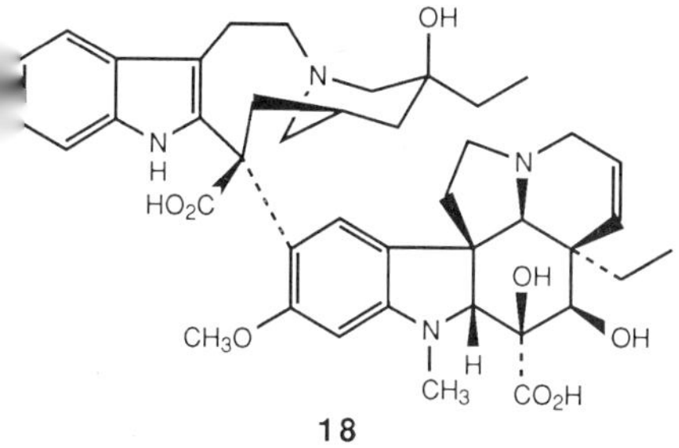

18 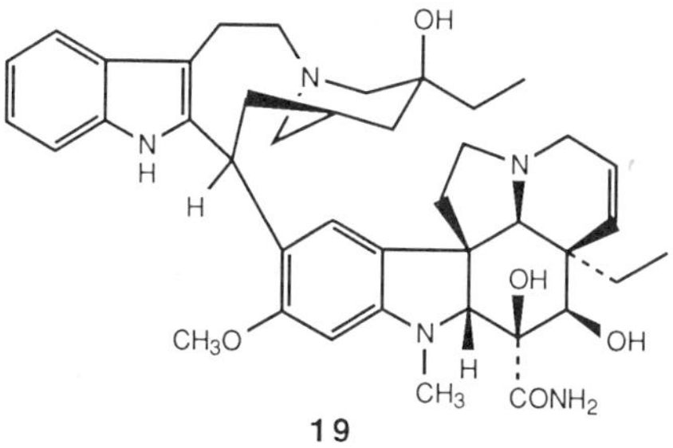 **19**

18′-Decarbomethoxy bisindole derivatives show minimal biological activity in *in vitro* or *in vivo* systems. This lack of activity may be rationalized on the basis of an altered conformation of the bisindole since removal of the C-18′ group increases the number of low energy conformers available by rotation about the C-18′–C-16 bond.

E. Ring D′ Cleavage Products

Reaction of bisindoles with certain oxidants leads to ring D′ cleavage products. Treatment of 4′-deoxyleurosidine (**22**) with potassium permanganate gives a 22% yield of vinamidine (**23**, catharinine) plus trace amounts of 19′-oxo-4′-deoxyleurosidine (**24**) (*55,56*). Reactions of this type also occur in the C-4′-oxy series. Reaction of leurosine (**3**) with *tert*-butyl hydroperoxide gives catharine (**25**) together with 19′-oxoleurosine in 48% yield (*57*). Air oxidation of 3′,4′-anhydrovinblastine (**26**) gives **25** in 34% yield. The C-19′-oxo by-products of these reactions can be prepared under mild conditions by reaction of the starting bisindole with iodine and sodium bicarbonate (*58*). Ring D′ cleavage results in total loss of biological activity.

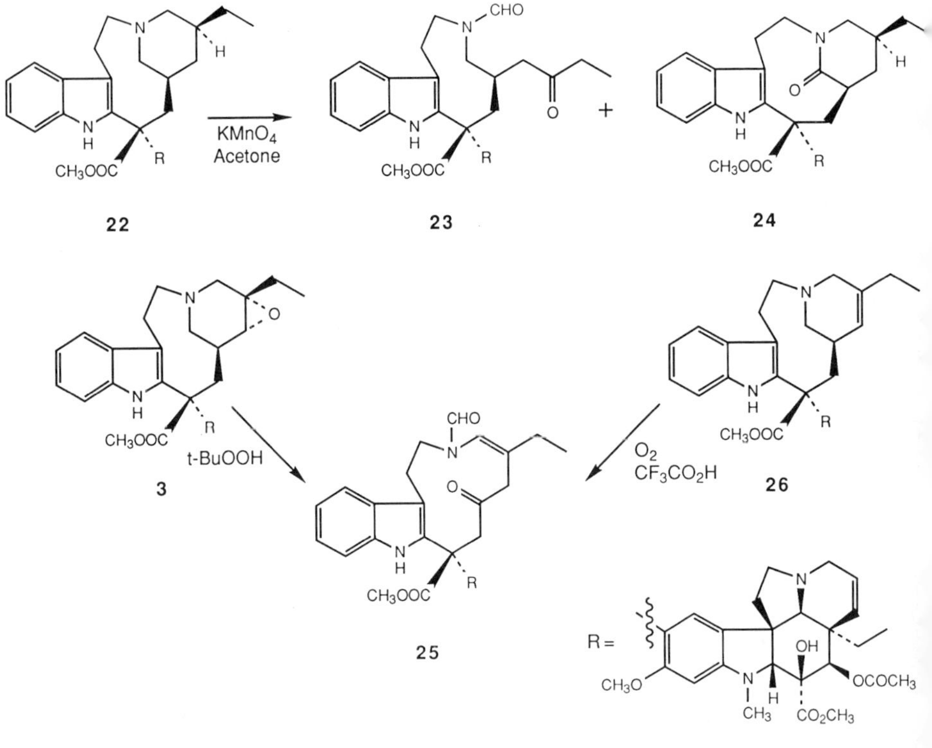

$KMnO_4$ Acetone

22 **23** **24**

t-BuOOH

O_2 CF_3CO_2H

3 **25** **26**

F. Ring D′ Deletion Derivatives

Partial structures (lacking ring D′) of the bisindole alkaloids have been prepared by reaction of a substituted chloroimine with vindoline (**21**) under conditions that promote ionization to yield the coupled products as a mixture of diastereomers (*59*). Reaction of **27** with silver fluoroborate followed by addition of **21** gives **29** as a mixture of diastereomers in 49% yield. Treatment of this mixture with potassium borohydride under acid conditions gives the D′ ring deletion analog (**31**) along with its diastereomer (**32**). Kuehne *et al.* employed a similar strategy in the preparation of an intermediate (**34**) that was used in the synthesis of 4′-desethyl-4-deoxyvinblastine (*60*).

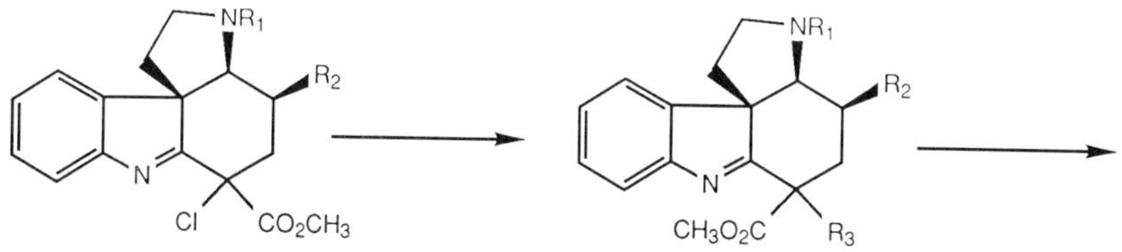

27 $R_1=CH_3$, $R_2=H$
28 $R_1=CH_2Ph$, $R_2=CH_2CH_2CH_2OTs$

29 $R_1=CH_3$, $R_2=H$
30 $R_1=CH_2Ph$, $R_2=CH_2CH_2CH_2OTs$

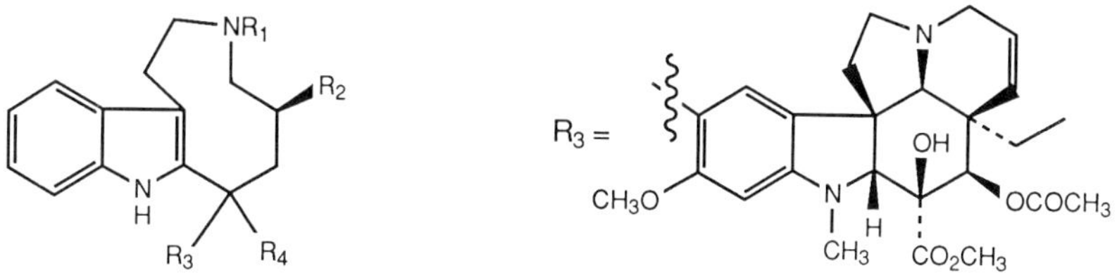

31 $R_1=CH_3$, $R_2=H$, $R_4=\alpha$-CO_2CH_3
32 $R_1=CH_3$, $R_2=H$, $R_4=\beta$-CO_2CH_3
33 $R_1=CH_2Ph$, $R_2=CH_2CH_2CH_2OTs$, $R_4=\alpha$-CO_2CH_3
34 $R_1=CH_2Ph$, $R_2=CH_2CH_2CH_2OTs$, $R_4=\beta$-CO_2CH_3

In an approach to the synthesis of bisindole model systems developed by Magnus and co-workers, the inherent reactivity of precursor (**35**), activated by reaction with phenylchloroformate, provides bisindole (**37**) as a mixture of diastereomers (*61*). In this reaction the indolenium intermediate (**36**) is probably the reacting species.

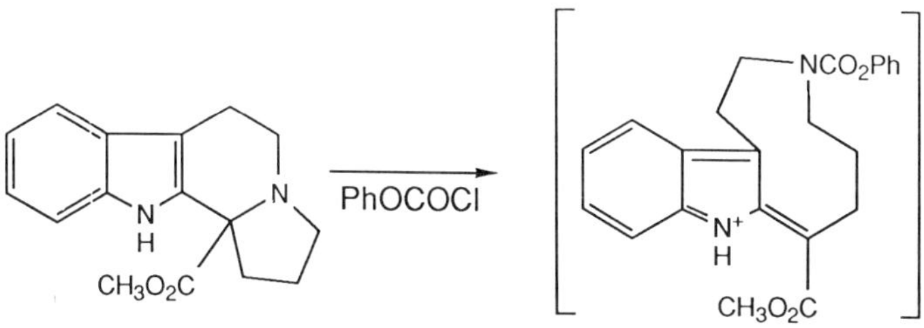

PhOCOCl

35

36

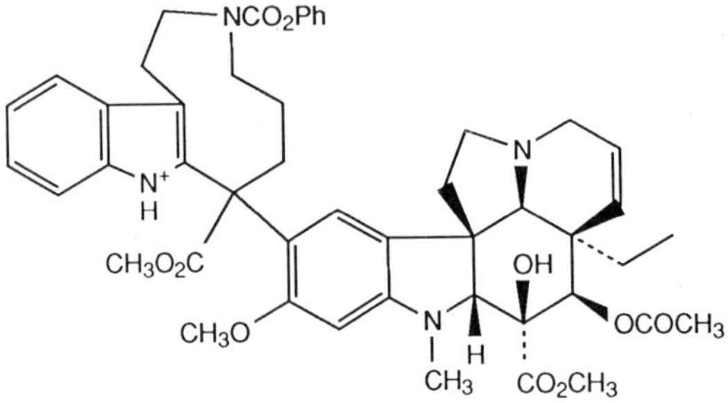

37

Although the D′ ring deletion analogs are interesting models for bisindole synthesis, there are no reports of their biological activity. If the gross conformation of these partial structures is important in conferring antitubulin activity, then it is reasonable to assume that these compounds are inactive. The observed differences in the activities of compounds modified in ring D′ support this assumption.

G. Ring C′ Contraction Derivatives: Noranhydrovinblastine (Navelbine)

Although there are many examples of bisindole derivatives that are prepared by reaction of a naturally occurring alkaloid with various reagents, there are few examples of compounds that result from chemistry that alters the skeletal features of these compounds. Much interest has been directed at the synthesis of the "dimeric" compounds by the coupling of the "monomer" units vindoline (**21**) and catharanthine (**38**), in part because of the necessity of using this strategy to prepare bisindoles by total synthesis.

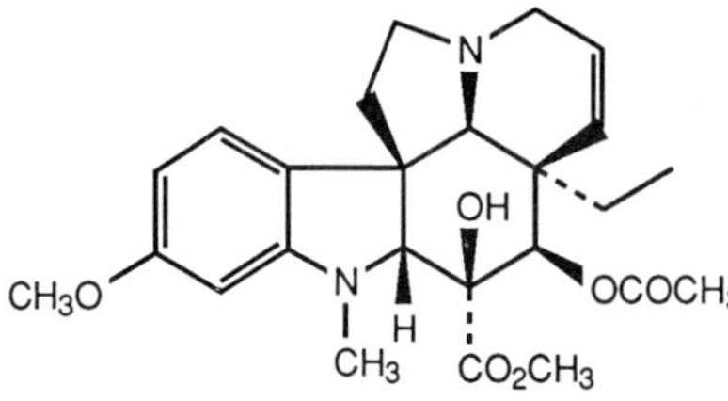

21 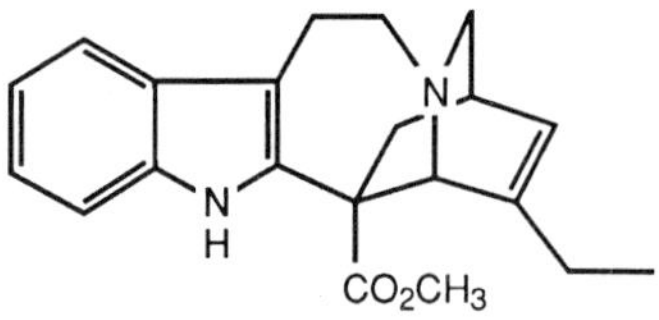 **38**

Reaction of 16-hydroxydihydrocleavamine (**20**) with vindoline (**21**) under conditions originally described by Büchi produces a dimeric product (**39**) in which the stereochemistry at C-18′ is inverted (*62*). A later, biomi-

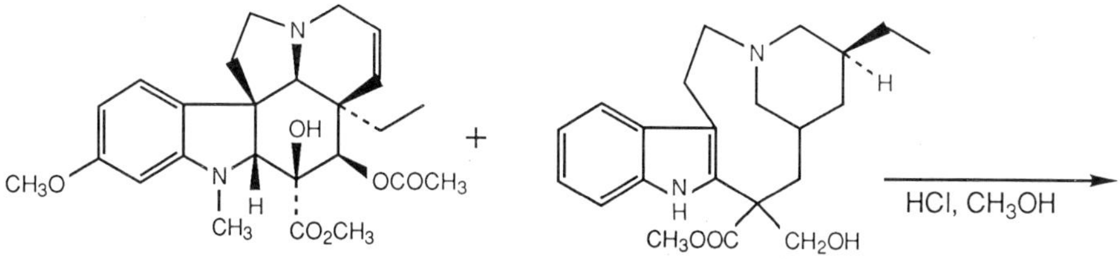

21 + 20 —HCl, CH_3OH→ 39

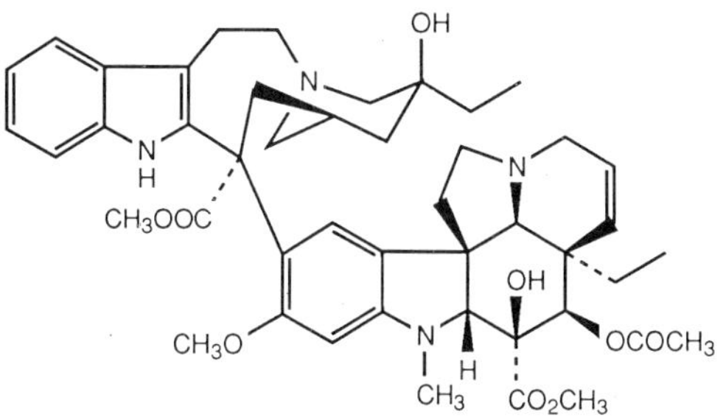

metic synthesis employed a Polonovski fragmentation strategy (*63,64*). Catharanthine *N*-oxide (**40**) is converted to its *O*-trifluoroacetate and treated with vindoline (**21**) to produce a mixture of isomeric iminium salts (**41**). Reduction of this mixture with sodium borohydride affords 3′,4′-anhydrovinblastine (**42**) in 50% yield together with the 18′-epimer (**43**) in

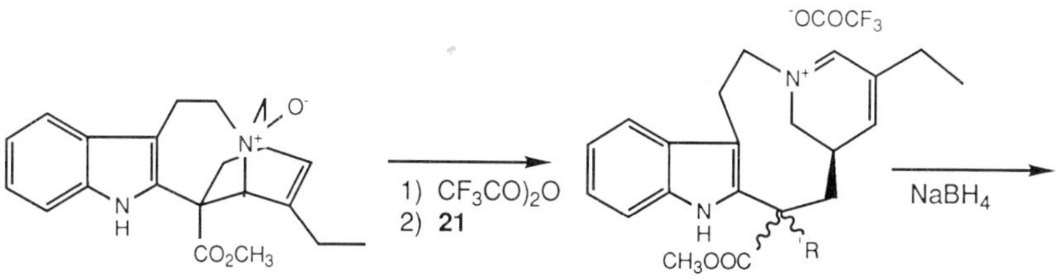

40 —1) $(CF_3CO)_2O$, 2) **21**→ 41 —$NaBH_4$→

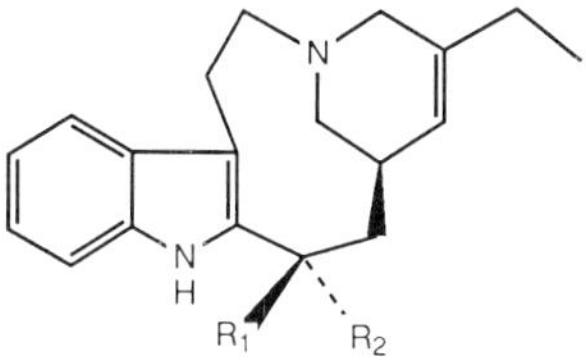

42 $R_1=CO_2CH_3$, R_2=Vind.
43 R_1= Vind., $R_2=CO_2CH_3$

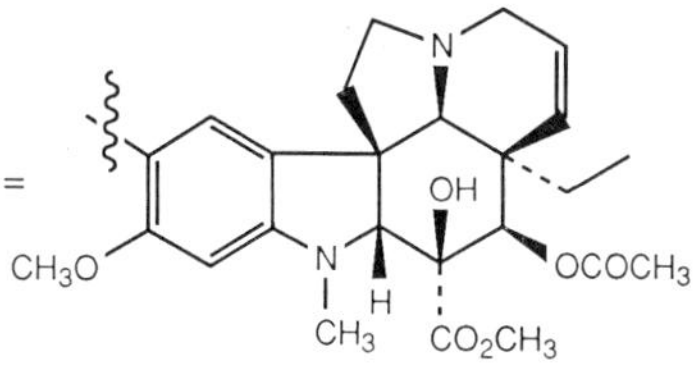

12% yield. This modified Polonovski coupling has been employed in the synthesis of a wide variety of naturally occurring bisindoles.

The successful application of a bond fragmentation process in the synthesis of bisindoles suggested the possibility of inducing the reverse process in the intact alkaloid (*65,66*). Treatment of 3′,4′-anhydrovinblastine N^6-oxide (**44**) with trifluoroacetic anhydride gives an unstable salt (**45**) which undergoes elimination and fragmentation to produce a bisiminium salt (**46**). Reduction of **46** with sodium cyanoborohydride in methanol gives a mixture of products consisting of ring C′ cleavage products **47** and **48** together with the iminium salt reduction products **42** and **53**.

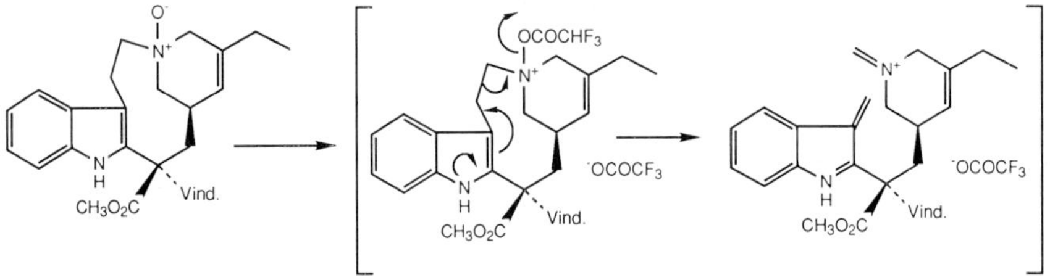

44 45 46

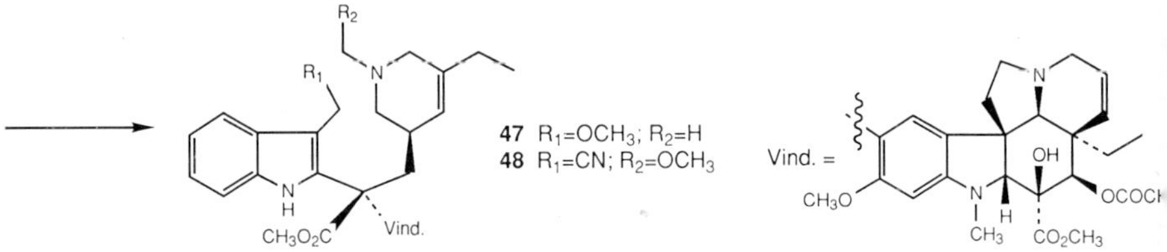

47 R_1=OCH_3; R_2=H
48 R_1=CN; R_2=OCH_3

Vind. =

Treatment of a tetrahydrofuran solution of **46** with water led to formation of a new product, noranhydrovinblastine (**49**, navelbine, NVB), in 27% yield. The formation of this new product can be explained as follows: Addition of 2 mol of water to the bisiminium salt (**46**) gives an unstable diol (**50**) which then undergoes deformylation at N-6′, rendering N-6′ nucleophilic. Bond formation between N-6′ and C-8′ occurs after elimination of water and addition to the exocyclic carbon of **51**.

Noranhydrovinblastine (**49**) represents a new class of semisynthetic bisindole alkaloids. The application of the Polonovski fragmentation strategy to produce a ring contraction product is relevant to understanding the biogenetic origins of the bisindole alkaloids. The biosynthetic formation of ring contraction derivatives of this type was first suggested by Potier and Janot in 1973 (*67*). This hypothesis was subsequently confirmed by two observations. Stemmadenine (**52**) was shown to be converted to apparicine (**53**) in the plant (*68*) and to vallesamine (**54**) in plant tissue culture (*69*).

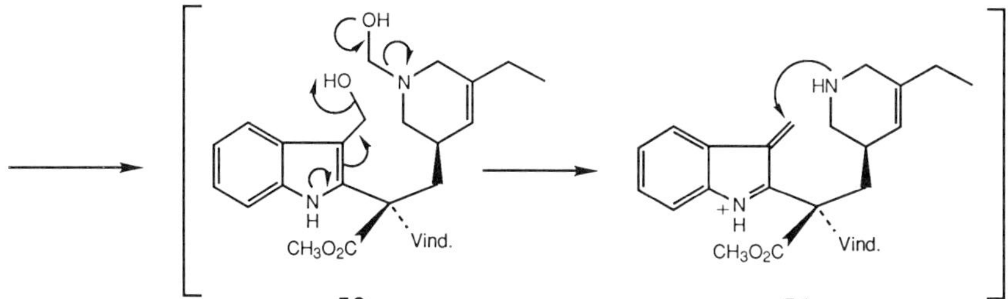

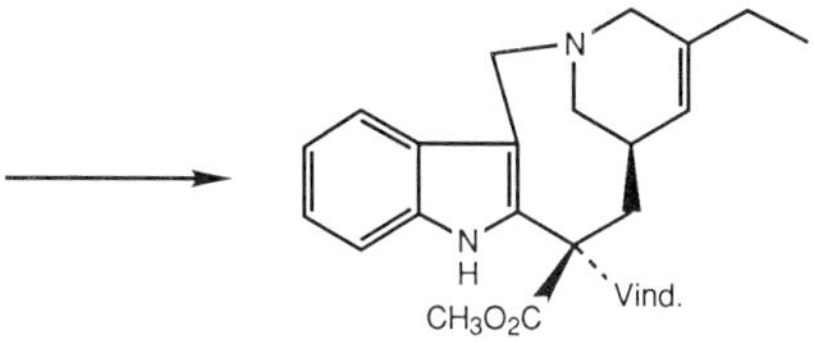

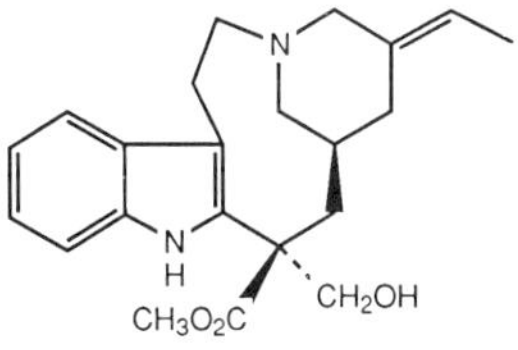

The experimental antitumor activity of navelbine has been evaluated in several murine tumor models including P388 leukemia, L1210 leukemia, Lewis lung carcinoma, B16 melanoma, TM2 mammary adenocarcinoma, colon 26 carcinoma, ICIG-C4 fibrosarcoma, and glioma 26 (*70*). Navelbine was inactive in all carcinoma and fibrosarcoma models (as measured by increases in median life span) but showed significant activity against the leukemia models (Table II). When compared to vinblastine and vincristine, navelbine appears to be approximately 10-fold less potent at its maximally effective dose. Vinblastine and vincristine are more effective than navelbine against L1210 leukemia, and vincristine is more effective that either navelbine or vinblastine against P388 leukemia. Interestingly,

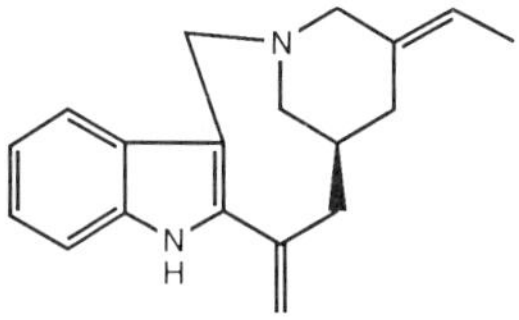

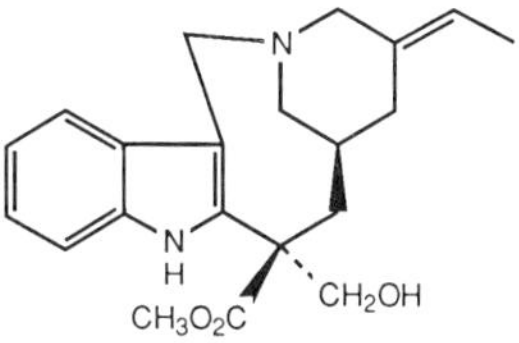

TABLE II
EXPERIMENTAL ANTITUMOR ACTIVITY OF NAVELBINE, VINBLASTINE, AND VINCRISTINE AGAINST P388 AND L1210 MURINE LEUKEMIAS

Tumor[a]	Compound[b]	Dose (mg/kg)[c]	*T/C*[d]	Survivors[e]
L1210	Navelbine	10	233	2/10
	Vinblastine	1.0	>500	5/9
	Vincristine	1.0	333	2/9
P388	Navelbine	10	196	4/8
	Vinblastine	1.0	196	3/8
	Vincristine	1.0	283	7/8
P388/VCR	Navelbine	10	123	0/8
	Vinblastine	1.0	130	0/8
	Vincristine	1.0	103	0/8

[a] L1210, 10^5 cells inoculated i.p. on day 0; P388, 10^6 cells inoculated i.p. on day 0; P388/VCR, 10^6 cells inoculated i.p. on day 0.

[b] Navelbine administered as its tartrate salt; vinblastine and vincristine administered as their sulfate salts.

[c] Maximum effective dose administered on days 1, 5, and 9, except for the P388 tumor in which drug was administered on days 3, 5, and 9.

[d] (Median survival time of the treated animals)/(median survival time of the control animals) × 100.

[e] L1210, fraction of surviving animals on day 60; P388 and P388/VCR, fraction of surviving animals on day 30.

navelbine and vinblastine are somewhat more effective against the vincristine-resistant subline of P388, P388/VCR, but the magnitude of these differences is minimal (<30%), suggesting that a degree of cross-resistance is evident in this group of compounds.

VI. Modifications of the Lower Half (Vindoline Portion) of Bisindole Alkaloids

A. N-1 SUBSTITUTIONS

The functionality associated with N-1 of the vindoline portion of the bisindole alkaloids confers much biological diversity to these compounds. This property is exemplified by the differences in the cellular pharmacology, antitumor efficacy, and toxicity of vinblastine (**1**) and vincristine (**2**). Since **1** is isolated from extracts of *Catharanthus* in approximately 10-fold greater yield, it is not surprising that several oxidative methods have been developed for the conversion of **1** to **2**. These methods have led to a number of compounds bearing "unnatural" functionality at N-1.

Treatment of **1** with chromium trioxide or chromic acid in a mixture of acetone, acetic acid, and acetic anhydride at −60°C produces a mixture of **2** and N^1-demethylvinblastine (**55**) (*71*). When this crude mixture is treated with formic acid in acetic anhydride, **2** is obtained in 50–70% yield based on starting vinblastine. A modification of this procedure involves the use of preformed chromic acid with either tetrahydrofuran or ethyl acetate as solvent. Using these methods, yields of **2** as high as 90% have been obtained (*72*).

An interesting alternative to the use of chromium(VI) oxidants for the conversion of **1** to **2** involves the use of a low-valent iron reagent prepared *in situ* by the action of hydrogen peroxide on an iron(II) complex of **1** (*73*). Vinblastine (as the free base) is treated with 2 equiv of perchloric acid in acetonitrile at −20°C. Ferrous perchlorate is then added, followed by the addition of excess 30% hydrogen peroxide. Work-up of the reaction mixture with a saturated solution of ammonium hydroxide gives **2** in yields of 35–50% after chromatography.

The oxidative conversions described above probably proceed via formation of an N-1-oxymethyl intermediate which undergoes further oxidation to produce **2**. If **1** is treated as before with chromium trioxide in the presence of excess methanol, N^1-methoxymethylvinblastine (**56**) is isolated in 64% yield (*74*). Additional N-1-alkyloxymethyl or cycloalkyloxymethyl compounds have been prepared by substituting other alcohols for methanol in this reaction.

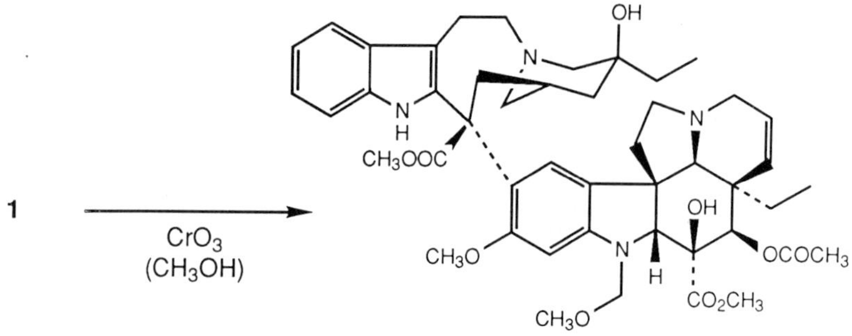

1 → CrO_3 (CH_3OH) → 56

Reaction of **56** with thiols in dichloromethane gives the analogous alkylthiomethyl and cycloalkylthiomethyl derivatives [e.g., treatment with ethanethiol gives N^1-(2-hydroxyethylthiomethyl)vinblastine (**57**)] (*75*). Compound **56** is also a useful intermediate for the preparation of N-1-furanyl derivatives. For example, treatment of **56** with furfuryl alcohol in methylene chloride in the presence of trace acid gives N^1-[(5-hydroxymethyl)furfuryl]vinblastine (**58**).

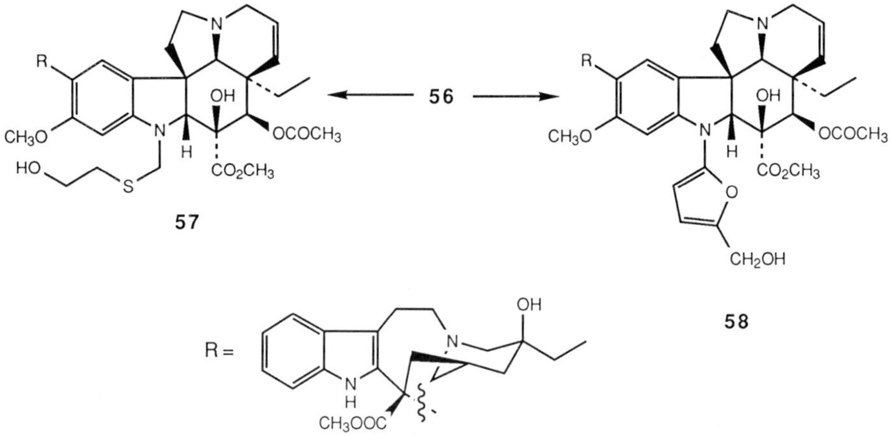

57 ← 56 → 58

Many of the N-1 derivatives described above show antitumor activity. Compound **56** is active against P388 murine lymphocytic leukemia at a dose of 6.0 mg/kg/day × 8 (i.p.), giving an ILS of 82%, and compound **57** shows an improvement in the therapeutic index of vinblastine. Compound **58** was also active against P388 at very low doses (0.1 mg/kg/day × 8).

B. C-3 Substitutions

1. Deacetylvinblastine Amide Derivatives

An extensive structure modification program was initiated in the early 1970s with the goal of producing a more potent bisindole alkaloid with improved antitumor characteristics. A drug with the combined spectrum of clinical activity of vinblastine and vincristine but lacking the severe, dose-limiting neurotoxicity of vincristine would fulfill this objective. There was substantial prejudice that the clinical toxicity of vincristine was in some way related to its N-1-formyl group, and this notion propelled the search in the direction of compounds having a vinblastinelike N-1-methyl group (*76*). Furthermore, derivatives that were in some way modified in the velbanamine portion such as leurosine (**3**) or leurosidine (**4**) were not expected to demonstrate useful efficacy since these naturally occurring congeners were substantially less active in experimental antitumor models when compared to vinblastine or vincristine. Initial experiments were, therefore, directed at modifying the lower-half of the vindoline moiety of vinblastine.

Reaction of vinblastine (**1**) with excess ammonia in anhydrous methanol in a sealed pressure vessel at 100°C for 60 hr produces a mixture of products which gives, after chromatography on silica gel, 4-deacetylvinblastine amide (vindesine, VDS, **59**) and 4-deacetyl-18′-decarbomethoxyvinblastine amide (**60**) in yields of 42 and 15%, respectively (*53*). A much improved preparation of vindesine (**59**) was realized by hydrazinolysis of

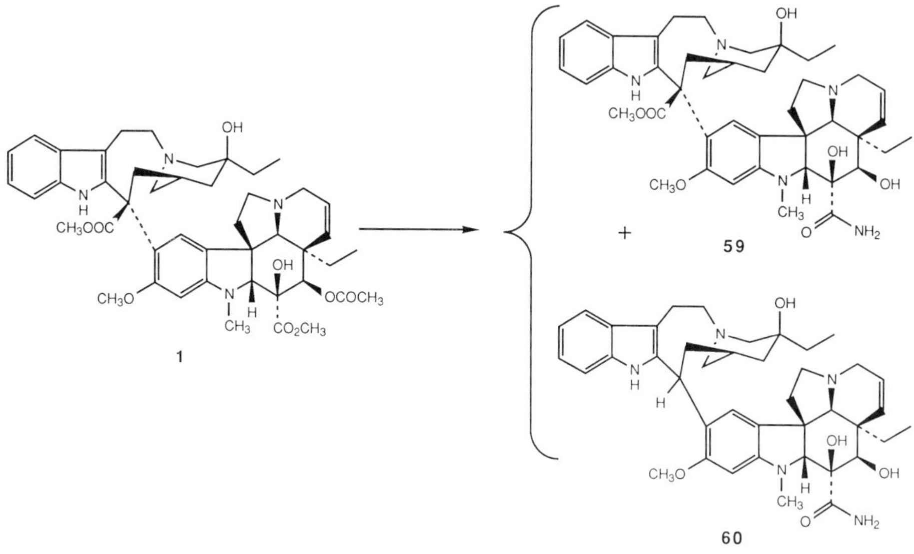

1, followed by either oxidation of the acylhydrazide and displacement of the corresponding acylazide with ammonia or direct hydrogenolysis of the N—N bond (*77*). Thus, treatment of **1** with excess anhydrous hydrazine in methanol at 60°C gives 4-deacetylvinblastine hydrazide (**61**) in 83% yield. Only trace amounts of the decarbomethoxy product (**60**) could be detected. Reduction of the acylhydrazide (**61**) with Raney nickel in methanol at reflux gives vindesine (**59**) in 57% yield after crystallization. In the alternate procedure, reaction of **61** with nitrous acid provides the acylazide (**62**), which is used in subsequent reactions without purification. Treatment of a dichloromethane solution of the azide (**62**) with anydrous ammonia at 0°C gives vindesine (**59**).

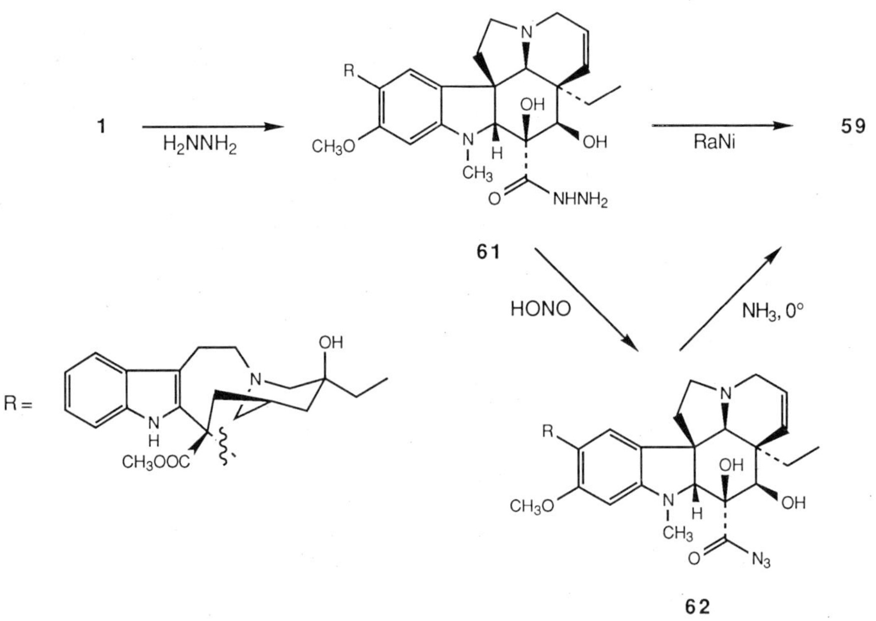

Deacetylvinblastine acylazide (**62**) was later shown to be an exceptionally versatile intermediate for the preparation of C-3 amides. Since nucleophilic displacement of azide occurs at relatively low temperatures under mild conditions, a wide variety of C-3 derivatives have been prepared (Scheme 1, Table III). This observation is in contrast to the direct aminolysis of vinblastine which usually fails when the amine employed is substituted (e.g., β-hydroxyethylamine) or secondary (dimethylamine). The reactions can be conveniently followed by the disappearance of the $CO—N_3$ infrared band at 2135 cm^{-1} with the concomitant appearance of the $CO—NH_2$ band in the region 1665–1675 cm^{-1}. Acetylation of the

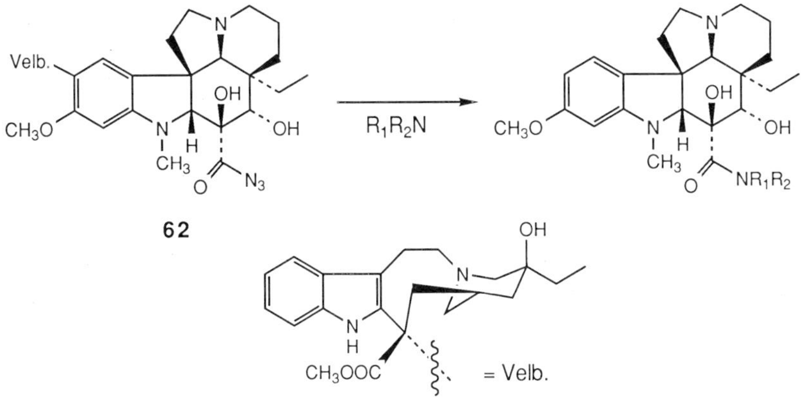

SCHEME 1.

C-4 hydroxyl group is accomplished under standard conditions (acetic anhydride, pyridine). If an additional reactive functionality has been introduced at C-3 (e.g., β-aminoethyl), then acetylation also occurs at the second reactive site. It is not unreasonable to conclude that **62** has been employed in the preparation of more vinblastine derivatives than any other chemical intermediate.

As a group, the 4-deacetylvinblastine amides are quite active antitumor agents (*78*). The tubulin binding activity of several of these agents has been described and correlates in a qualitative fashion with the ability of the compounds to arrest the replication of CHO cells during mitosis (Table IV). These data suggest the importance of a hydrogen bond donor within the C-3 substituent if *in vitro* activity superior to vinblastine (**1**) is to be realized. Since relative tubulin binding affinity is measured in a cell-free environment and mitotic arrest can only be measured in intact cells, it is possible to infer that increasing polarity of the C-3 substituent increases activity up to the point when passive cellular uptake is no longer operative.

The comparative antitumor activity of these compounds has been extensively studied in a variety of tumor systems including Gardner lymphosarcoma, Ridgeway osteogenic sarcoma, P1534(J) leukemia, B16 melanoma, and P388 leukemia (sensitive and resistant sublines) (Table V) (*78*). The most active compounds in the series are vindesine (**59**), deacetylvinblastine methylamide (**59a**), β-hydroxyethylamide (**75**), and 3-hydroxypropylamide (**76**). The comparative antitumor activity of these compounds indicated that vindesine was somewhat better than its hydroxyalkylamide congeners, and an exhaustive study of the antitumor

TABLE III

ACTIVITY OF 4-DEACETYLVINBLASTINE AMIDES AGAINST GARDNER LYMPHOSARCOMA IN C3H MICE

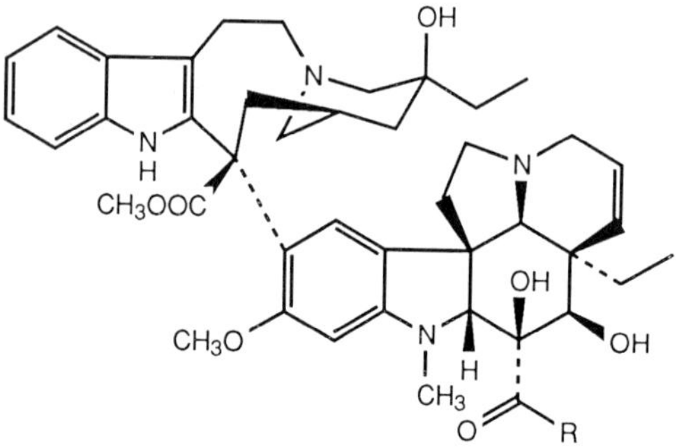

Compound[a]	R	Dose[b] 0.05	0.10	0.20	0.30	0.40
59	NH_2	−	++	+++	+++	+++
59a	$NHCH_3$	−	−	+++	+++	+++
61	$NHCH_2CH_3$		−	+++		T
62	$NHCH(CH_3)_2$		−	T		T
63	NH-*n*-hexyl			+		−
64	NH-*n*-cyclohexyl			−		+
65	$NHCH_2Ph$			+	−	−
66	$NHCH_2CH{=}CH_2$		−	+++		T
67	$NHCH_2C{\equiv}CH$		++	+++		T
68	$N(CH_3)_2$			+	T	T
69	—Pyrrolidine			−		T
70	$NHCH_2CH_2NH_2$			−		−
71	$NHCH_2CH_2NHCOCH_3$		+++	+++		+++
72	$NHCH_2CH_2N(CH_2)_2$			−		−
73	$NHCH_2CN$	−	+++	T		
74	$NHCH_2CH_2CN$	−	−	+++		+++
75	$NHCH_2CH_2OH$	+++	+++	+++		+++
76	$NHCH_2CH_2CH_2OH$			++		+++
77	$NHCH_2CH(OH)CH_2OH$			+		++
78	$NHCH_2CH_2OCOCH_3$	+	++	+++		T
79	$NHCH_2CH_2OCOCH_2CH_2CH_3$	−	++	+++		T
80	$NHCH_2CH_2OCOC_{17}H_{25}$	−	−	++		+++
81	$NHCH_2CH_2OCOCH{=}CH_2$	++	+++	T		T
82	$NHCH_2CH_2OCH_3$		−	−		−
83	$NHCH_2CH_2SH$		−	+	++	+++
84	$NHCH_2CH_2SCH_3$		−	−		T
85	$NHCH_2CH_2CH_2SCH_3$		−	−		T
86	$NHCH_2CH_2SCPh_3$		−	−		−
87	$NHCH_2CH_2SC(p\text{-}CH_3OPh)_3$		−	−		−
88	$NHCH_2CH_2SSCH_2CH_2NHCOCH_3$		−	−	+	++
89	$NHCH_2CHO$	−	+	+++		+++

TABLE III (*Continued*)

Compound[a]	R	Dose[b]				
		0.05	0.10	0.20	0.30	0.40
90	$NHCH_2CH(OCH_3)_2$	−	+ + +	+ + +		T
91	$NHCH_2CHC({=}NH)OCH_3$		−	−	−	−
92	$NHCH_2CH_2$(Ph-4-OH)		−	−		+
93	$NHCH_2CH_2[Ph\text{-}3,4\text{-}(OH)_2]$		−	−		+
94	$NHCHCOSCH_2CH_2$		−	−	T	T
95	$NH(CH_2S)_2$	−	−	+ +	+ + +	+ + +

[a] Compounds tested as sulfate salts except for **81, 94** (free base), and **91** (hydrochloride salt).
[b] Dose in mg/kg × 8–10 days (i.p.). Key to activity: −, 0–25% tumor growth inhibition; +, 26–50% tumor growth inhibition; + +, 51–75% tumor growth inhibition; + + +, 76–100%, tumor growth inhibition; T, toxic in 50% of treated animals.

TABLE IV
ANTIMITOTIC AND TUBULIN BINDING ACTIVITY OF 4-DEACETYLVINBLASTINE AMIDES AND RELATED COMPOUNDS

Compound[a]	Relative binding affinity[b]	Mitotic arrest (μg/ml)[c]			
		0.002	0.02	0.20	2.0
59	1.8	+	+ +		
60		−	+ +		
66	1.1				
71	1.0				
73	1.4				
84		−	+ +	+ +	
93		−	+ +	+ +	
95	1.3				
101	1.0				
104		−	+	+ +	
1	1.0	−	+ +	+ + +	
2	3.9	+	+ + +		
98	1.8	−	+ + +		

[a] Compounds tested as sulfate salts.
[b] Affinity determined by displacement of pig brain tubulin-bound [^{3}H]vinblastine. Data are relative to vinblastine binding (1.00, K_b = 106).
[c] Key: −, 3–7% of CHO cells in mitotic arrest over control; +, 10–15%; + +, 15–40%; + + +, 40–50%.

TABLE V
Comparative Antitumor Activity of Selected 4-Deacetylvinblastine Amides against Several Murine Tumor Models

Tumor	Compound[a]					
	59	**60**	**75**	**76**	**1**	**2**
Ridgeway osteogenic sarcoma[b]	+ + +	+ + +	+ + +	+ + +	−	+ + +
P1534(J) leukemia[c]	+ + +	+ + +	+ +	+ + +	+ +	+ + +
B16 melanoma (i.p.)[d]	+ + +		+		+	+
P388 leukemia[e]	+ +				+	+ +
P388/VCR leukemia[f]	−				−	−

[a] Compounds tested as sulfate salts administered i.p. at the maximum tolerated dose for each combination of tumor and host animal.

[b] In AKR mice treated (daily × 8) 8 days posttumor inoculation. Key: −, 0–25% tumor growth inhibition; +, 26–50% tumor growth inhibition; + +, 51–75% tumor growth inhibition; + + +, 76–100%, tumor growth inhibition; T, toxic in 50% of treated animals.

[c] In DBA2 mice treated daily × 8. Key: same as in note *b*.

[d] In C57BL/6 mice treated on days 1, 5, and 9 following i.p. inoculation of 10^6 tumor cells. Key: −, <50% increase in life span of treated animals compared to untreated controls; +, 51–100%; + +, 101–150%; + + +, >151%.

[e] In CD2F mice treated on days 1, 5, and 9 following i.p. inoculation of 10^6 tumor cells. Key: same as in note *d*.

[f] Vincristine-resistant subline. All other parameters same as in note *e*.

properties of the compound in several murine solid tumors indicated that vindesine provided similar or superior efficacy when compared to vinblastine (**1**) or vincristine (**2**). This observation was particularly evident in the Ridgeway osteogenic sarcoma and B16 melanoma models. Furthermore, *in vitro* studies on the effect of vindesine on axoplasmic transport in cat sciatic nerve suggested the compound had less neurotoxic potential than did vincristine. As a result of these observations, vindesine was advanced to clinical trial.

The Phase I studies of vindesine determined the maximum tolerated dose of the compound to be 3–4 mg/m^2 (body surface area) when given once weekly (i.v.) (*78*). Mild leukopenia ensues, reaching a nadir after about 4 days, and is accompanied by neurotoxicity. Several patients responded to drug treatment during this initial trial, including several complete responders with acute leukemias and partial responders with malignant melanoma and non-Hodgkin's lymphoma. Subsequent studies indicated that vindesine was active in vincristine-resistant childhood lymphocytic leukemia and, when combined with cisplatin, was effective in treating non-small cell lung cancer (*28*). The drug is now approved for the treatment of leukemia in several countries.

2. Oxazolidinedione Derivatives

The vast majority of compounds having the bisindole nucleus characteristic of the vinca structure with antitumor activity in experimental tumor systems bear a carboxyl function and a free tertiary hydroxyl group at C-3 of the vindoline moiety. The group of semisynthetic derivatives distinguished by the presence of a C-3 spiro-fused oxazolidine-1,3-dione are important exceptions to this generalization.

Treatment of a toluene solution of vinblastine (**1**) with a large excess of methyl isocyanate followed by warming for 8 hr gives a 64% yield of *N*-methyloxazolidinedione (**96**) (*79,80*). If this reaction is repeated under more forcing conditions (higher concentration and temperature) reaction at C-4′ also occurs, and the mixed carbamate (**97**) is isolated.

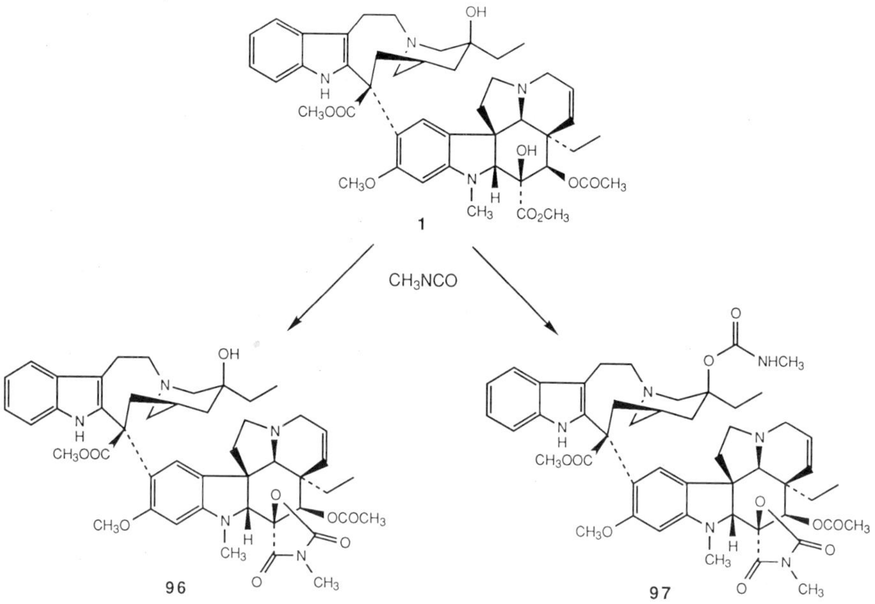

Reaction with isocyanates is also observed in derivatives having a free hydroxyl at C-4. Reaction of 4-deacetylvinblastine (**98**) under conditions identical to those employed in the preparation of **96** affords the mixed carbamate (**99**). The preparation of oxazolidinedione derivatives with a free hydroxyl at C-4 can be accomplished by acid-catalyzed hydrolysis of the C-4 acetate **96** by refluxing in 0.5 *N* HCl to give **100**.

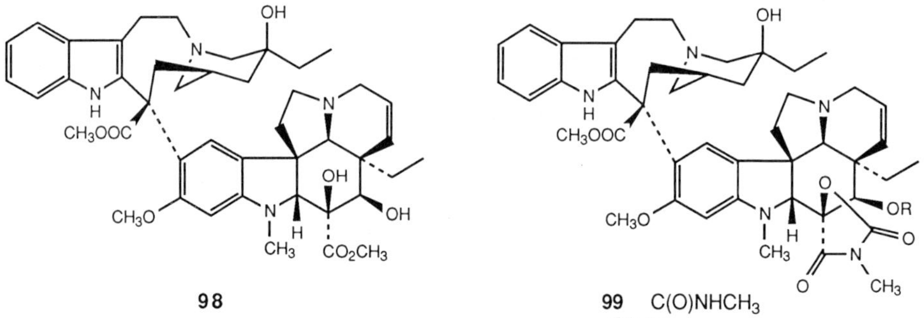

98

99 C(O)NHCH$_3$
100 H

Unsubstituted oxazolidinediones are prepared directly from the unsubstituted C-3 amides. Treatment of 4-deacetylvinblastine amide (**59**) with sodium hydride in tetrahydrofuran followed by the addition of dimethyl carbonate gives 4-deacetyl-3-oxazolidinedione (**101**) in 40% yield.

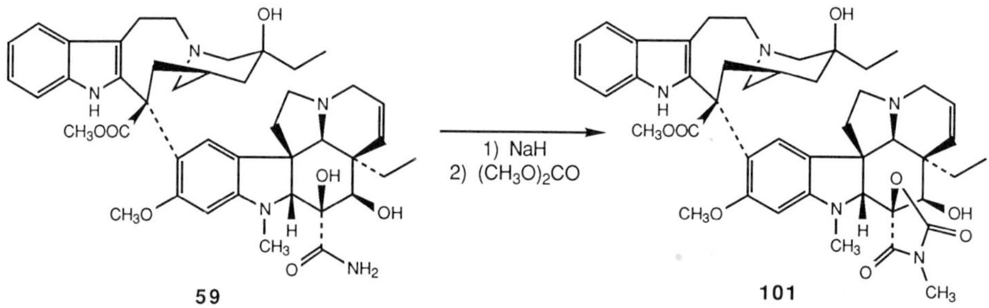

59 → 1) NaH 2) $(CH_3O)_2CO$ → **101**

The oxazolidinedione-forming reaction appears to be a general method for the preparation of spiro-fused derivatives, and a number of compounds having substituted oxazolidinediones have been prepared by this approach (**81**). An interesting variation of this procedure involves the use of bisisocyanoethyl disulfide (**103**) for the preparation of mercaptans and disulfide-bridged dimers of this class. The reagent is prepared by the following series of reactions (Scheme 2). Cystamine hydrochloride is treated with 4-nitrophenyl chloroformate to prepare carbamate **102**. Reaction of **102** with trimethylsilyl chloride in the presence of triethylamine followed by heating affords the disulfide (**103**), which is used without further purification after removal of the triethylamine hydrochloride by filtration.

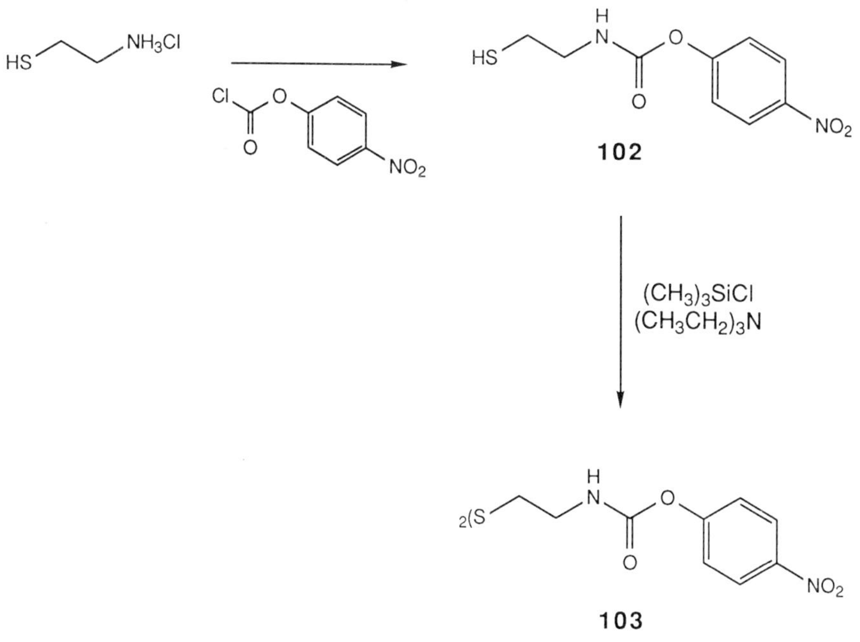

HS NH3Cl → 102 → (CH3)3SiCl, (CH3CH2)3N → 103

SCHEME 2.

Reaction of **1** with excess **103** in toluene gives the mixed carbamate (**104**) in 21% yield after chromatography on silica gel. Reduction of the S—S bond present in this carbamate is accomplished with zinc dust in acetic acid to give the *N*-(2′-mercaptoethyl) oxazolidinedione (**105**) in high yield. Dimerization of **105** is then achieved by reaction with excess aqueous potassium ferricyanide to give the disulfide (**106**).

Oxazolidinedione derivatives of **1** were found to be exceptionally stable to acidic hydrolysis (**81**). Selective hydrolysis of the C-4 acetate in **96** without affecting the integrity of either the oxazolidinedione ring or any

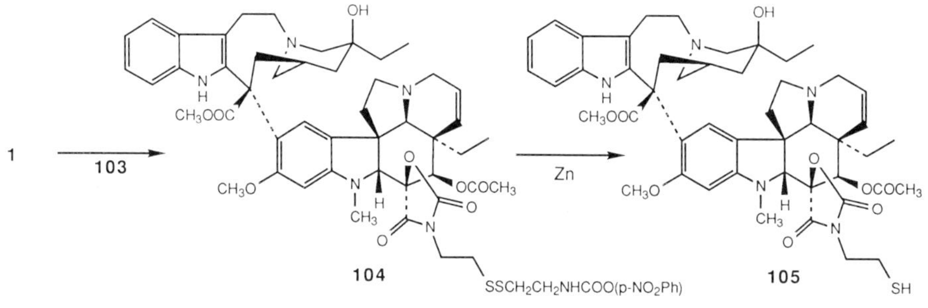

1 → 103 → 104 → Zn → 105

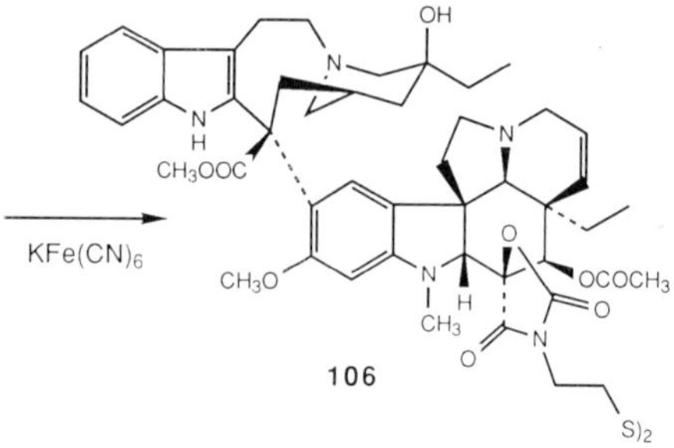

KFe(CN)6

106

functionality in the remainder of the molecule is an example of this unusual stability. The spiro-fused ring system is considerably more labile to base-catalyzed reactions. Treatment of derivatives of this type with mild base usually gives the N-substituted vinblastine amides as the sole reaction products, indicating that nucleophilic attack at C-3″ is preferred over attack at C-1″ of the heterocycle (Scheme 3).

A number of substituted bisindole oxazolidinedione derivatives were prepared using the methods described above (Table VI) (*79–81*). When examined for their ability to induce a mitotic block in CHO cells, these compounds were active from 2 to 0.002 μg/ml (Table VII), and compounds having the greatest degree of potency in this *in vitro* system were evaluated in several experimental tumor systems (Table VIII). The *in vitro* potency of these compounds appears to correlate with their *in vivo* potency as reflected by the minimum effective dose that could be administered without toxicity. Several of these compounds were evaluated for their antitumor efficacy when administered by the oral route (Table IX). Compounds **96** and **107** were shown to be exceptionally active in this regard.

Prior studies had established that bisindole alkaloid antitumor drugs are not well absorbed when orally administered (**82**). This fact, taken with

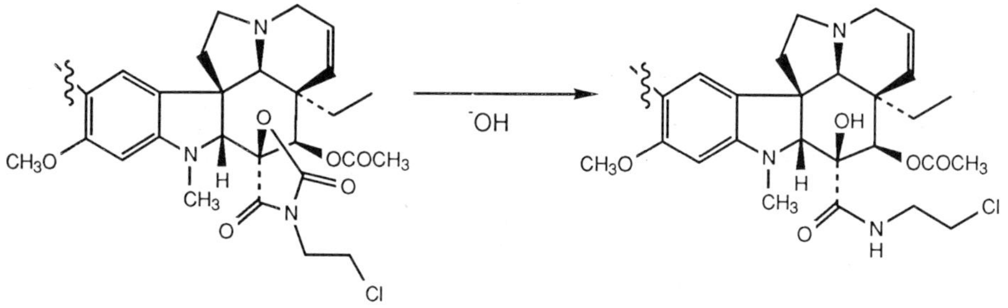

SCHEME 3. Regioselective hydrolysis of oxazolidinedione derivatives.

TABLE VI
C-3 OXAZOLIDINEDIONE DERIVATIVES

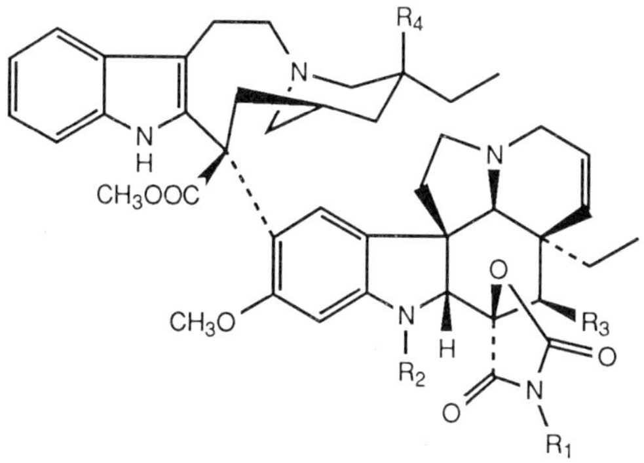

Compound	R_1	R_2	R_3	R_4
96	CH_3	CH_3	$COCH_3$	β-OH
107	CH_2CH_2Cl	CH_3	$COCH_3$	β-OH
108	$CH_2CH_2CH_3$	CH_3	$COCH_3$	β-OH
109	$CH_2CH_2CH_2CH_3$	CH_3	$COCH_3$	β-OH
110	$CH_2CH{=}CH_2$	CH_3	$COCH_3$	β-OH
111	CH_3	CHO	$COCH_3$	β-OH
112	CH_3	CH_3	$COCH_3$	3′,4′-α-epoxide
113	CH_3	CH_3	$COCH_3$	α-H
114	CH_3	CHO	$COCH_3$	3′,4′-α-epoxide
115	CH_2CH_2Cl	CH_3	$COCH_3$	α-H
116	CH_3	CH_3	$COCH_3$	3′,4′-C=C
117	CH_3	CH_3	$COCH_3$	β-$OCONHCH_3$
118	CH_3	H	$CONHCH_3$	β-$OCONHCH_3$
119	CH_3	CH_3	$CONHCH_3$	β-OH
120	CH_3	H	$CONHCH_3$	β-OH
121	$CH_2CH_2SCH_3$	CH_3	$COCH_3$	β-OH

the experimental antitumor spectrum of **107** (LY104208, vinzolidine), was sufficient to support the initiation of clinical studies with this drug. Phase I trials revealed that the drug was generally well absorbed when given by this route, but extreme patient to patient variability was observed. Nevertheless, an objective response to drug treatment was noted, but, owing to the unpredictable nature of the drug's absorption and disposition, clinical studies with orally administered vinzolidine were discontinued, and a follow-up study with drug administration via the i.v. route was initiated. These studies indicated that the clinical activity of vinzolidine was not significantly different from vinblastine, and the clinical trial was discontinued.

TABLE VII
ANTIMITOTIC ACTIVITY OF C-3 OXAZOLIDINEDIONE DERIVATIVES

Compound[a]	Mitotic index[b]
96	+ + +
107	+ + +
108	+ +
109	+ +
110	+ +
111	+ + +
112	±
113	+
115	+
116	±
118	−
119	+ + +
120	±
121	+ +

[a] Compounds tested as sulfate salts.
[b] Key: Minimum effective dose to produce a mitotic accumulation in CHO cells at + + +, 0.002 μg/ml; + +, 0.02; +, 0.2; ±, 2.0; −, >2.0.

TABLE VIII
ANTITUMOR ACTIVITY OF C-3 OXAZOLIDINEDIONE DERIVATIVES

	Compound[a]							
Tumor	**96**	**107**	**108**	**109**	**110**	**111**	**119**	**121**
Gardner lymphosarcoma[b]	+ + +	+ +		−	+	+ +	+ + +	
6C3HED lymphosarcoma[b]	+ + +	+ + +						+
P1534(J) leukemia[c]	+ + +	+ + +	+					+
B16 melanoma[d]	+ + +	+ + +	+					

[a] Compounds tested as their sulfate salts at the maximum tolerated dose.
[b] In C3H mice treated daily × 8. Key: −, 0–25% tumor growth inhibition; +, 26–50% tumor growth inhibition; + +, 51–75% tumor growth inhibition; + + +, 76–100%, tumor growth inhibition.
[c] In DBA2 mice treated daily × 8. Key: same as in note *b*.
[d] In C57BL/6 mice treated on days 1, 5, and 9 following i.p. inoculation of 10^6 tumor cells. Key: −, <50% increase in life span of treated animals compared to untreated controls; +, 51–100%; + +, 101–150%; + + +, >151%.

TABLE IX
ANTITUMOR ACTIVITY OF C-3 OXAZOLIDINEDIONE DERIVATIVES WHEN ADMINISTERED ORALLY

Tumor	Compound[a]			
	96	**107**	**108**	**1**
Gardner lymphosarcoma[b]	+ + +	+ + +		+ +
P1534(J)[c]	+ + +	+ + +	+	+ + +
B16 melanoma[d]	−	+	−	−

[a] Compounds administered by oral gavage as their sulfate salts.

[b] In C3H mice treated daily × 8. Key: −, 0–25% tumor growth inhibition; +, 26–50% tumor growth inhibition; + +, 51–75% tumor growth inhibition; + + +, 76–100%, tumor growth inhibition.

[c] In DBA2 mice treated daily × 8. Key: same as in note *b*.

[d] In C57BL/6 mice treated on days 1, 5, and 9 following i.p. inoculation of 10^6 tumor cells. Key: −, <50% increase in life span of treated animals compared to untreated controls; +, 51–100%; + +, 101–150%; + + +, >151%.

3. Additional C-3 Derivatives

Peptide derivatives of the bisindole alkaloids have been prepared by appending amino acids at C-3. Reaction of acylazide **62** with an α-amino acid ester affords amide derivatives of this type (**122**) (*46*). Conversely, the attachment of the amino acid can be inverted by reacting a C-3 amide derivative with an activated amino acid ester. Thus, treatment of 3-(β-aminoethyl)-4-deacetylvinblastine amide (**70**) with an N-protected α-aminoacylazide gives the alternative amide derivative (**123**). These techniques have been used to prepare di-, tri-, and tetrapeptide conjugates.

An application of the peptide conjugate chemistry was used to prepare vinca prodrugs as substrates for cleavage by plasmin or other fibrinolytic enzymes. Several tumors had been noted to produce high levels of plasminogen activator, a proteolytic enzyme that converts the zymogen plasmin to the serine protease plasmin, resulting in high local levels of plasmin within (or in the vicinity of) tumor (*83–87*). Antitumor drugs coupled to peptides containing plasmin-specific sequences would be inactive with respect to their antitumor activity. Plasmin cleavage of the peptide would "activate" the prodrug with the release of the active antitumor drug. This approach was applied to the antitumor drugs acivicin, melphalan, phenylenediamine mustard, and doxorubucin with demonstrable *in vitro* activity (*88,89*), and a similar approach was investigated with 4-deacetylvinblastine amide (*46*). These derivatives lacked *in vivo* activity, presumably owing to plasmin activation in the extracellular space, leading to nonspecific distribution of the drug throughout the circulatory system. A similar series of peptide conjugates was effective in treating L1210 and P388 leukemia when drug and tumor were confined to the peritoneal space (*90*).

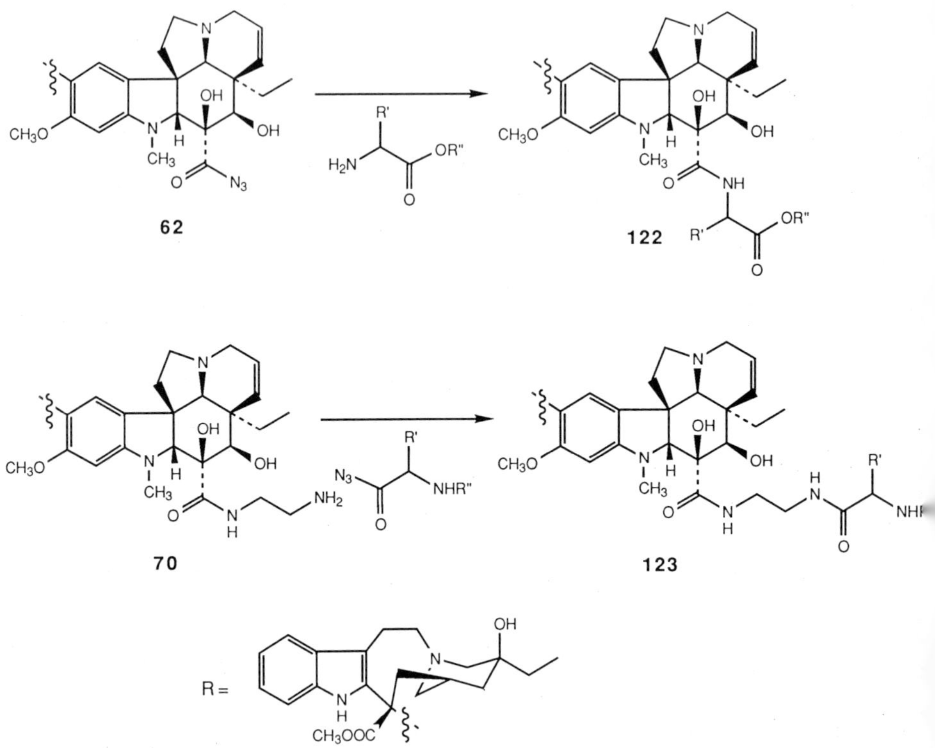

C. C-4 Substitutions: 4-Deacetyl-4-acyl Vinblastine Derivatives

The 4-acyl derivatives of 4-deacetylvinblastine (**98**) are among the earliest semi-synthetic compounds prepared in exploring the medicinal chemistry of the bisindole alkaloids from *Catharanthus*. By 1965, the clinical utility of vinblastine (**1**) and vincristine (**2**) was firmly established, and the naturally occurring congeners leurosine (**3**) and leurosidine (**4**) had been well characterized with respect to their experimental antitumor activity. Since these compounds were substantially less active in animal tumor models than **1** or **2**, structural modification programs were initiated in order to find new derivatives with enhanced antitumor properties (*91*).

The medicinal chemistry of this period relied on simple modifications of the complex alkaloids. Deacetylation of **1** produced 4-deacetylvinblastine (**98**), a perfect substrate for these first experiments. Treatment if **1** with hydrogen chloride in anhydrous methanol gives **98** directly (*91*). Cleavage of the 4-acetyl function can also be effected by methanolysis, but during this period the acid-catalyzed methanolysis method was regarded as the preferred method.

Reaction of **98** with a variety of acid anhydrides in the presence of pyridine produced a series of 4-acyl derivatives (**A**) in which only the group at C-4 varies (Scheme 4, Table X). By reacting the chloroacetyl derivative (**A**, R = CH_2Cl) with various amines, a series of 4- (α-aminoacyl) derivatives (**B**) were prepared.

The P1534(J) leukemia model was employed in the early experimental antitumor studies with the naturally occurring bisindoles. Therefore, this tumor model was chosen for the first antitumor evaluations of the 4-acyl and 4- (α-aminoacyl) derivatives (*91*). The results of these experiments indicate that increasing the size of the acyl substitutent at C-4 has a

Velb. CH_3O OH $OCOCH_3$ N CH_3 H CO_2CH_3 **1** → HCl CH_3OH → **98** (OH) → $(RCO)_2O$ Py →

A (OCOR) → R = CH_2Cl, $R_2'NH$ → **B** ($OCOCH_2NR_2'$)

Velb. = [indole unit, N, H, CH_3OOC, OH]

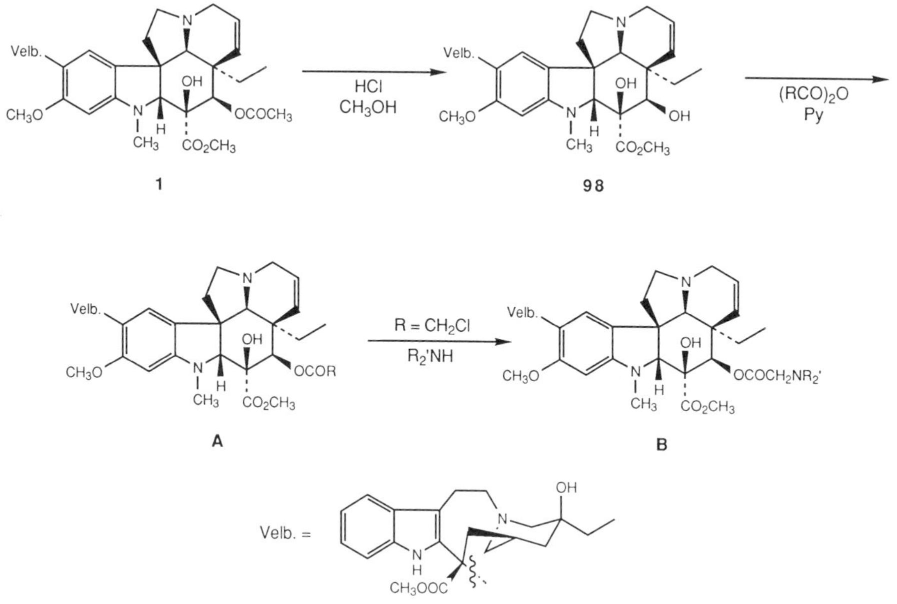

SCHEME 4.

TABLE X
4-DEACETYL-4-ACYLVINBLASTINE AND 4-DEACETYL-4-(α-AMINOACYL)VINBLASTINE DERIVATIVES

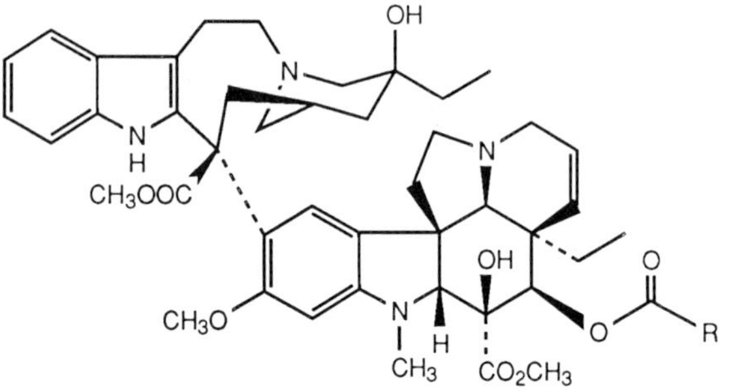

Compound	R
1	CH_3
124	CH_2CH_3
125	$CH_2CH_2CH_3$
126	$CH(CH_3)_2$
127	CH_2Ph
128	Ph
129	CH_2Cl
130	$CHCl_2$
131	CH_2CN
132	CH_2NHCH_3
133	$CH_2N(CH_3)_2$
134	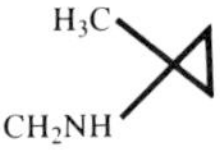 H_3C / CH_2NH
135	$CH_2N(CH_2CH_3)_2$
136	CH_2N 
137	CH_2N 
138	CH_2N  O
139	CH_2N NH
140	CH_2N NCH_3
141	CH_2N  NPh
142	CH_2N  NCH_2CH_2OH

deleterious effect on antitumor activity (Table XI). This observation is especially consistent in the 4-acyl series. Interestingly, the chloroacetyl derivative (**129**) shows substantially improved activity over vinblastine (**1**). The cause of this effect has not been determined, but it nevertheless presents some intriguing possibilities, most notably the introduction of an electrophilic carbon atom at a position where S_N2 reaction with a receptor-associated nucleophile is possible. The observation that the three derivatives presenting displaceable functionality at this position (**129**, **130**, and **131**) are also the most potent compounds in this series supports this notion.

TABLE XI
ACTIVITY OF 4-DEACETYL-4-ACYLVINBLASTINE AND 4-DEACETYL-4-(α-AMINOACYL)VINBLASTINE DERIVATIVES AGAINST P1534(J) LEUKEMIA

Compound[a]	Dose[b]	Activity[c]
124	0.55	+[d]
125	0.75	+[d]
126	0.9	−
127	2.2	−
128	0.75	−
129	0.35	+ +
130	0.3	−
131	0.3	−
132	3.0	+ + +[e]
133	5.0	+ + + +[f]
134	3.75	+ + + +[g]
135	1.5	−
136	7.5	+ +
137	3.0	−
138	15.0	+ +
139	1.5	−
140	15.0	−
141	1.5	−
142	30.0	+ + +[h]

[a] Compounds administered as their sulfate salts.
[b] Maximum tolerated dose (determined by 5-day toxicity tests in non-tumor-bearing mice).
[c] In DBA/2 mice. Key: + + + +, >200% increase in life span; + + +, 151–200%; + +, 101–150%; +, 51–100%; −, <50%.
[d] 20% of treated animals surviving on day 45.
[e] 60% of treated animals surviving on day 45.
[f] 100% of treated animals surviving on day 45.
[g] 80% of treated animals surviving on day 45.
[h] 40% of treated animals surviving on day 45.

The α-aminoacyl analogs in this series show striking differences in toxicity and efficacy. Although these compounds are generally less toxic than their 4-acyl counterparts, it is interesting to note that the most toxic compounds in this group are also the least active. The compounds bearing simple, substituted amines exhibit striking efficacy against the P1534(J) tumor, and 4-deacetyl-4-(α-dimethylaminoacyl) vinblastine (**133**), vinglycinate) is curative in this model. This observation marked **133** for extended study.

In antitumor spectrum testing, vinglycinate demonstrated measurable activity against several leukemia models but was inactive in solid tumor models with the exception of Gardner lymphosarcoma. The compound was also orally absorbed and demonstrated activity superior to vinblastine against the P1534(J)leukemia when administered by this route at a dose of 10 mg/kg/day × 10. In summary, the experimental antitumor spectrum of vinglycinate was found to very similar to that of vinblastine, although its activity in P1534(J) leukemia and Gardner lymphosarcoma was superior to that of the parent alkaloid. Furthermore, the oral activity of this compound distinguished it from other related bisindole derivatives. As a consequence of these findings, clinical studies with vinglycinate were initiated; thus, the utility of chemical modifications of the bisindole alkaloids as a viable method for the discovery of new antitumor agents was established.

In its initial clinical experience, vinglycinate was given to 31 patients (*92*). The maximum tolerated dose was found to be 75–100 mg/kg given once weekly. Dose-limiting toxicity was leukopenia with a nadir in leukocyte count being observed 5 days posttreatment. This toxicity was noted to be similar to that observed with vinblastine. No neurotoxicity was observed or associated with vinglycinate treatment. Nineteen evaluable patients yielded partial responders in Hodgkin's disease, bronchogenic carcinoma, and chondrosarcoma. These responses were not durable, but a lack of cross-resistance to vinblastine and vincristine was observed in 3 patients with disease refractory to these agents. As a consequence of this study, the perceived benefits of vinglycinate were not judged to be superior to the established benefits of vinblastine, and the continued clinical development of vinglycinate was not pursued.

VII. Multiple Modifications: 4′-Epideoxyvincristine

4′-Epideoxyvinblastine (**143**) shows unremarkable biological properties; however, when **143** is converted to its *N*-1-formyl derivative, 4′-epi-4′-deoxyvincristine (**144**), a compound with striking biological properties is obtained (*93,94*). Reaction of **143** with chromium trioxide in acetic anhydride at low temperature produces **144** (vinepidine) in 48% yield (*95*). While these two, tandem functional group substitutions appear minimal with respect to the overall complexity of the bisindole alkaloid, they trigger gross conformational changes. These effects are manifested in chemical shift differences observed in the ^{1}H- and ^{13}C-NMR spectra of vinepidine when compared to that of vinblastine (*96*). These differences are particularly evident in the resonances of the velbanamine portion of 144 and are suggested by the tubulin binding and antimitotic characteristics of this compound.

CrO_3, $(CH_3CO)_2O$, -60°

143 → **144**

The relative tubulin binding properties of a collection of bisindole alkaloids reflect the importance of combined modifications at C-4′ and N-1 (Table XII) (*94*). Replacement of the vinblastine C-4′ hydroxyl with hydrogen results in reduction in tubulin affinity by over 50%. However, when this substitution is combined with inversion of configuration at C-4′, the tubulin binding affinity is increased by 84%. Replacement of the N-1 methyl with formyl increases tubulin affinity an additional 37%. The

TABLE XII
EFFECT OF CONFIGURATION AND SUBSTITUTION AT C-4′ AND N-1 ON RELATIVE TUBULIN BINDING AFFINITY

Compound[a]	Relative affinity[b]
Vincristine	3.87
4′-Epideoxyvincristine	2.21
Vindesine	1.92
4′-Epideoxyvinblastine	1.84
Vinblastine	1.00
4′-Deoxyvinblastine	0.424
4′-Epivinblastine	0.021

[a] Compounds tested as their sulfate salts.

[b] Extent of displacement of [^{3}H]vinblastine from purified pig brain tubulin. Values are relative to vinblastine (= 1.00).

structural effects of these changes are manifested in an altered conformation of the piperidine (D′) ring and an increase in the local dipole of the lower portion of the vindoline moiety. This effect is evidenced in the pK_a of the piperidine nitrogen (pK_a of vinblastine, 7.4; pK_a of 4′-epivinblastine, 8.8) and the observation that substituting the C-3 carbomethoxy and C-4 hydroxyl groups with aminocarbonyl and hydrogen, respectively (vindesine), increases the polarity of the molecule. These observations indicate that these sites are important in interacting with the tubulin binding site.

An increase in tubulin binding affinity relative to vinblastine prompts the prediction that vinepidine should be more potent than vinblastine in causing a mitotic block in proliferating cells. Surprisingly, this prediction is not borne out. Vinepidine was found to be much less potent than vinblastine, vincristine, or vindesine in arresting cycling CHO cells during mitosis (Table XIII), and close inspection of the DNA histograms produced by FACS analysis of treated cells suggested that a slight accumulation of cells in S phase preceded arrest during M phase (*95*). Do these results indicate a subtle change in mechanism? This question was amplified by studying the effect of vinepidine on cultured rat midbrain cells.

In an attempt to find an *in vitro* assay to predict differences in the neurotoxic potential of bisindole alkaloids, an assay using cultured rat midbrain cells was developed. This system provided a qualitative measure of the effect of compounds on neuronal tissue, and when several compounds (for which clinical toxicity data were available) were evaluated using this method the results were consistent in rank order with the compounds' clinical manifestation of neurotoxicity. When vinepidine was studied in this system, it was found to produce a minimal effect (Fig. 7).

TABLE XIII
MITOTIC ACCUMULATION OF CHO CELLS AFTER TREATMENT WITH 4′-EPIDEOXYVINCRISTINE AND RELATED COMPOUNDS

	Mitotic index[b] (μgml)			
Compound[a]	2.0	0.2	0.02	0.002
Vinepidine	+ +	+	−	−
Vindesine	+ +	+ +	+	−
Vinblastine	+ +	+ +	+ +	−
Vincristine	+ +	+ +	+ +	−

[a] Compounds tested as their sulfate salts.

[b] Measured after 5 hr of exposure to drug using a fluorescence activated cell sorter. Key: + +, >75% of cells in mitosis; +; 50–75%; −, <50%.

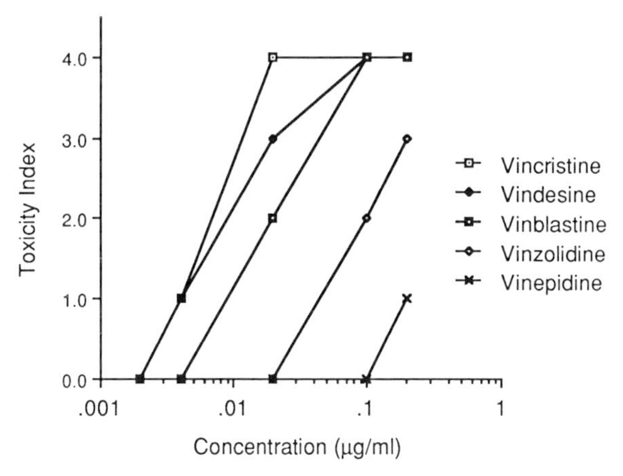

FIG. 7. Effects of vinblastine, vindesine, vincristine, vinzolidine, and vinepidine on neuronal processes in cultured rat midbrain cells [data from Boder *et al.* (*93*)].

The experimental antitumor activity of vinepidine was found to be very similar to that of vincristine (Table XIV) (*94*). The compound was somewhat more effective than vincristine in the vinca-sensitive tumors 6C3HED lymphosarcoma and P1534(J) leukemia, and also demonstrated improved activity against the colon 26 carcinoma, a tumor considered to be vinca resistant. In all cases the maximum tolerated dose of vinepidine was significantly lower that that of vincristine when the drugs were administered on a staggered regimen. These trends are reflected in data obtained against P388 leukemia (Fig. 8).

TABLE XIV
EXPERIMETAL ANTITUMOR ACTIVITY OF VINEPIDINE AND VINCRISTINE

Tumor	Vinepidine[a]		Vincristine[a]	
	Dose (mg/kg)	Activity[b]	Dose (mg/kg)	Activity[b]
6C3HED lymphosarcoma	0.42	+ + + +	1.2	+ + + +
P1534(J) leukemia	0.6	+ + + +	1.5	+ + +
Ca755 adenocarcinoma	0.7	+ +	1.5	+ +
Colon 26 carcinoma	0.42	+ +	1.0	−
M5076 ovarian carcinoma	0.4	+	1.0	+
Lewis lung carcinoma	0.6	+	1.5	+

[a] Compounds tested as sulfate salts administered i.p. on days 1, 5, and 9 at the maximum tolerated dose for each combination of tumor and host animal.

[b] Key: −, 0–25% tumor growth inhibition; +, 26–50%; + +, 51–75%; + + +, 76–99%; + + + +, 100%.

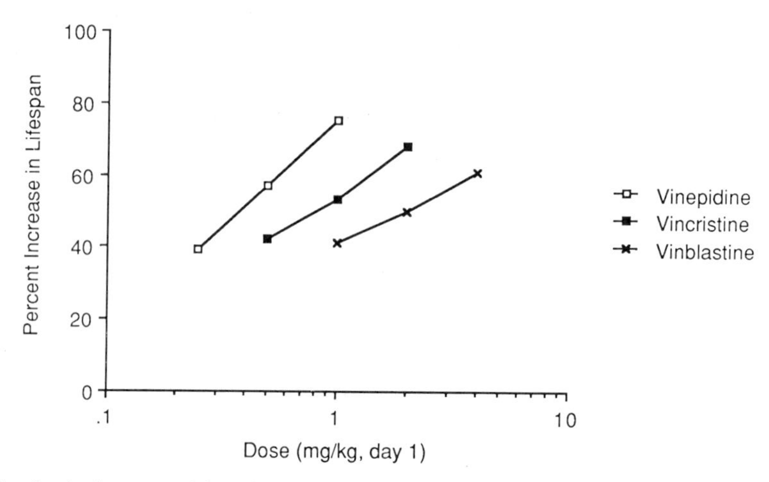

FIG. 8. Antitumor activity of vinepidine, vincristine, and vinblastine against P388 leukemia. Compounds were administered i.p. as their sulfate salts on day 1 following i.p. inoculation of 1×10^6 tumor cells.

Because of the unusual nature of the biochemical pharmacology of vinepidine, particularly the tubulin binding properties of the compound and its minimal effect on neuronal cell processes, as well as its antitumor efficacy and potency, a clinical study was initiated in early 1983. The initial dose-ranging studies were conducted by intravenous administration of vinepidine once weekly, and extreme neuromuscular toxicity was observed at moderate dose levels. Other forms of toxicity were minor (alopecia, gastrointestinal) or absent (leukopenia). The observed neuromuscular toxicity was similar to that of vincristine and was clearly the dose-limiting toxicity. The clinical trail of vinepidine was discontinued as a result of these findings. In a later study of the efficacy and pharmacokinetics of vinepidine in a rhabdomyosarcoma xenograft model, Houghton noted an unusually long half-life of drug clearance from muscle tissue relative to plasma clearance (*120*). The results indicate that the plasma pharmacokinetics of vinepidine may not be predictive in the association of a neurotoxic dose and suggest a need for additional studies with this intriguing substance.

VIII. New Concepts in Medicinal Chemistry of Bisindole Alkaloids

A. DRUG TARGETING: MONOCLONAL ANTIBODIES

In so far as the goal of structure modification programs is the discovery of new antineoplastic agents with improved therapeutic selectivity and an expanded spectrum of antitumor activity, the studies directed at targeting a cytotoxic agent to specific tissues are exemplary. Monoclonal antibodies (MoAbs) that are reactive with tumor-associated antigens have been evaluated as vehicles for delivering the toxic bisindole alkaloids to tissues that present these antigens (*97*). By appropriate selection of the antigen–antibody pair it should be possible to direct the cytotoxic alkaloid to tumor tissue. The high molecular weight complex should not bind to cells that do not bear the appropriate antigen, thereby eliminating "innocent bystander" tissues. The affinity of a MoAb–drug conjugate with a tumor-specific antigen should deliver drug to the vicinity of the tumor or, if internalization of the antigen–antibody complex occurs, directly to intracellular targets (Fig. 9).

The task of attaching a complex bisindole alkaloid to a MoAb with a molecular weight in the vicinity of 150,000 without degrading the antigen combining capacity of the protein or reducing the cytotoxic effectiveness of the alkaloid is a complex problem. Several covalent attachment strategies have been applied to the vinca alkaloids. Reaction of 4-deacetylvinblastine azide (**62**) with antibody leads to conjugates (**145**) in which the MoAb–drug bond is between C-23 and the ε-amino group of lysine residues in the protein. Conjugates of this type have been prepared with antibodies directed at carcinoembryonic antigen (CEA), an antigen that is often associated with colon carcinoma (*98*).

The biological activity of these conjugates was evaluated in athymic nude mice bearing human tumor xenografts and was shown to be sensitive to the stoichiometry of conjugation. When the molar ratio of drug to antibody exceeded 8 : 1, the conjugates lost biological activity, owing at least in part to the effect of high conjugation ratios on the physicochemi-

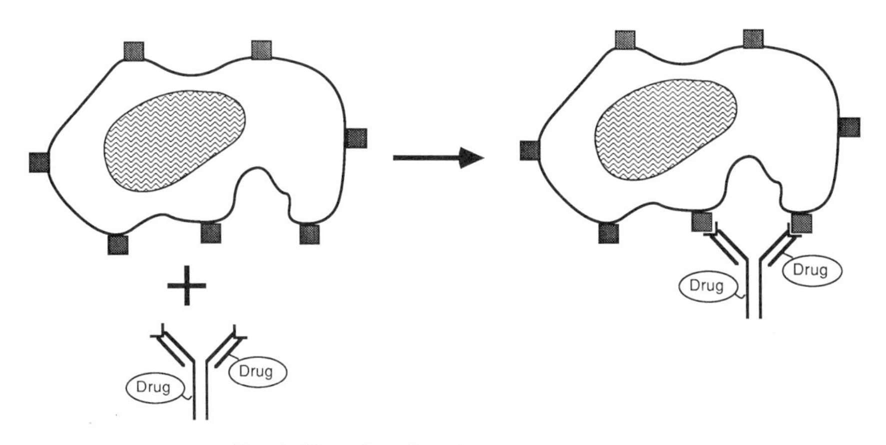

FIG. 9. Targeting of antibody–drug complexes.

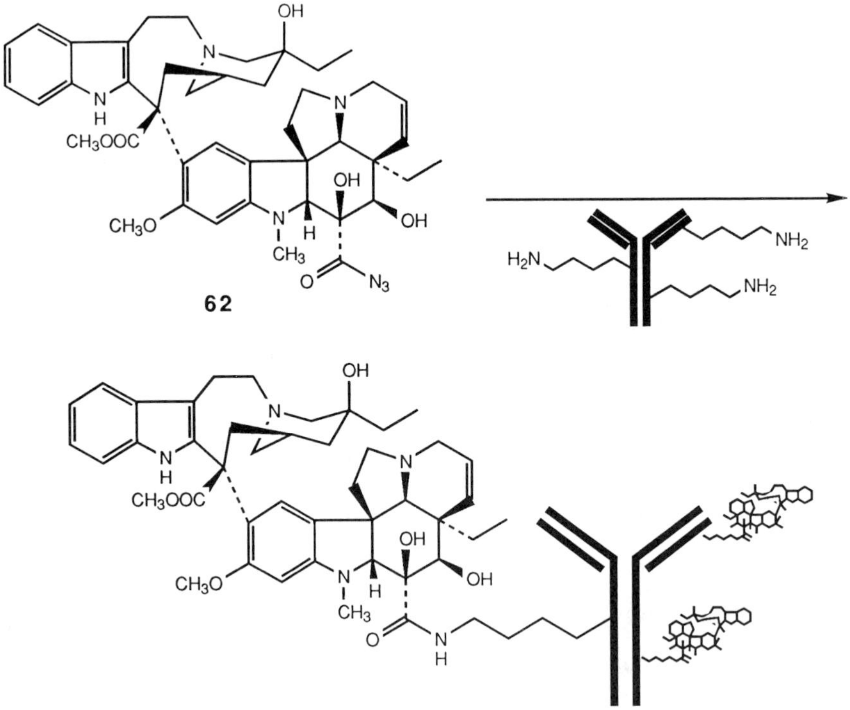

62

145

cal characteristics of the conjugates and possibly by blocking the antigen combining sites with drug. Antibodies with conjugation ratios below 5 : 1 were less effective since these constructions were unable to deliver a critical dose of drug to the tumor. The chemistry of this conjugation was extremely sensitive to reaction conditions since the starting azide undergoes hydrolysis in weakly basic aqueous solution.

An alternative strategy "tethers" the alkaloid via a succinyl bridge (*99*). Treatment of deacetylvinblastine (**98**) with succinic anhydride and pyridine gives a succinate half-ester (**146**) with is converted to the active ester (**147**) by reaction with *N*-hydroxysuccinimide. The active ester (**147**) proved to be a more useful reagent under a variety of conditions than acylazide **62** and has been successfully coupled with several MoAbs as well as the specific receptor ligand transferrin. Again, protein lysine residues are the presumed target of this compound, and the biological activity of its conjugates was also found to be sensitive to stoichiometry. Reaction of **147** with the KS1/4 antibody gave a conjugate with notable activity in the P3UCLA human lung xenograft.

A third conjugation strategy was developed in which carbohydrate resi-

98

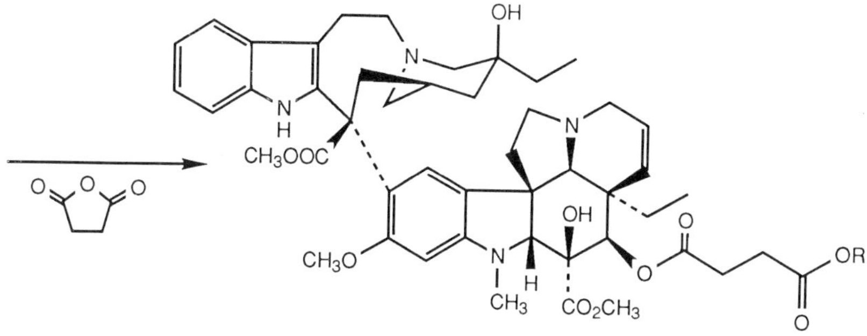

146 R = H

147 R = 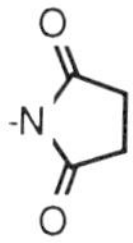

dues located within the "hinge" region of the antibody are modified by reaction with periodate (*100*). This method effectively cleaves vicinal diol functionalities in these groups to produce protein-linked aldehyde residues. Since the carbohydrate residues are located primarily within the hinge region of the antibody, far removed from the antigen combining

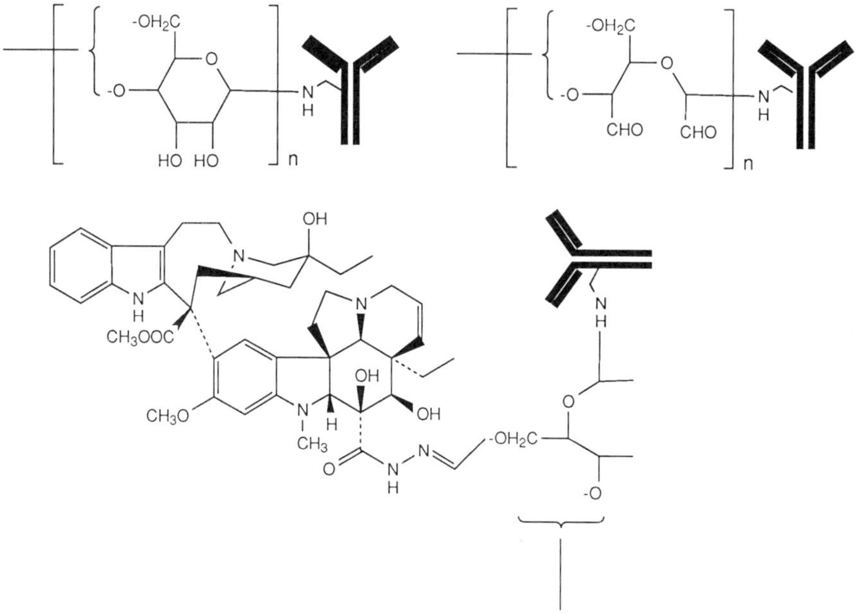

SCHEME 5.

TABLE XV
RELEASE OF 4-DEACETYLVINBLASTINE HYDRAZIDE FROM 4-DEACETYLVINBLASTINE HYDRAZINE–KS1/4S2 CONJUGATE AS A FUNCTION OF TEMPERATURE AND pH

Temperature (°C)	pH	Free 4-desacetylvinblastine hydrazide (%)[a]
4	5.3	17
37	5.3	28
4	7.4	3
37	7.4	10

[a] Determined by HPLC.

sites, it is much less likely that the attached drug molecules will interfere with tumor binding. Reaction of 4-deacetylvinblastine hydrazide (**61**) with antibody modified in this way gives a conjugate in which drug attachment is effected through formation of a Schiff base (Scheme 5). The Schiff base linkage is reversible, and may release drug on reaction with water (Table XV).

The rate of hydrolysis is slow relative to the rate of attachment to tumor, however, and effectively delivers drug to tumor tissue as evidenced by the extreme biological potency of conjugates of this type (Table XVI).

TABLE XVI
ACTIVITY OF 4-DEACETYLVINBLASTINE HYDRAZIDE–MoAb CONJUGATES AGAINST HUMAN TUMOR XENOGRAFTS

MoAb[a]	Human tumor xenograft, ED_{50} (mg/kg)[b]				
	P3UCLA[c]	HT29[d]	LS174[e]	T222[f]	M14[g]
KS1/4 S2	<0.0625	<0.25	—	—	>2.0
L1KS	0.05	0.0625	0.15	—	—
9.2.27	—	—	—	—	<0.0625
Desacetylvinblastine	1.0	1.0	1.0	2.0	0.5

[a] KS1/4 S2 isotype, IgG_{2a} recognizes KS1/4 antigen; L1KS isotype, IgG_{2b} recognizes KS1/4 antigen; 9.2.27 isotype, IgG_{2a} recognizes melanoma-associated antigen.

[b] Dose (in vinca equivalents), administered on days 2, 5, 8, required to produce a 50% suppression of tumor growth relative to controls. The tumors were measured 28 days after tumor inoculation (s.c., 1×10^7 cells). T222: Animals pretreated with 350 R γ-radiation 24 hr prior to tumor inoculation and dosed on days 3, 6, and 9.

[c] Lung adenocarcinoma, KS1/4 positive.

[d] Colon adenocarcinoma, KS1/4 positive.

[e] Colon adenocarcinoma, carcinoembryonic antigen positive.

[f] Lung squamous carcinoma.

[g] Melanoma.

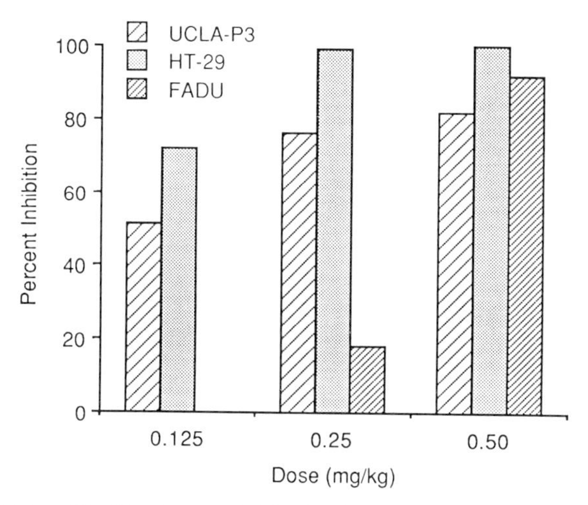

FIG. 10. Activity of L1KS–deacetylvinblastine hydrazide conjugate against three human tumor xenografts. Conjugate was administered i.v. after s.c. inoculation of 1×10^7 tumor cells in the flank of *Nu*/*Nu* mice. For P3UCLA lung adenocarcinoma, conjugate was administered on days 2, 4, and 7; for HT-29 colon carcinoma, on days 2, 5, and 8; and for FADU pharyngeal squamous carcinoma, on days 3, 5, and 7. Tumors were measured on day 22. Conjugate dose is given as 4-deacetylvinblastine equivalents (mg/kg). [Data from Laguzza *et al.* (*100*).]

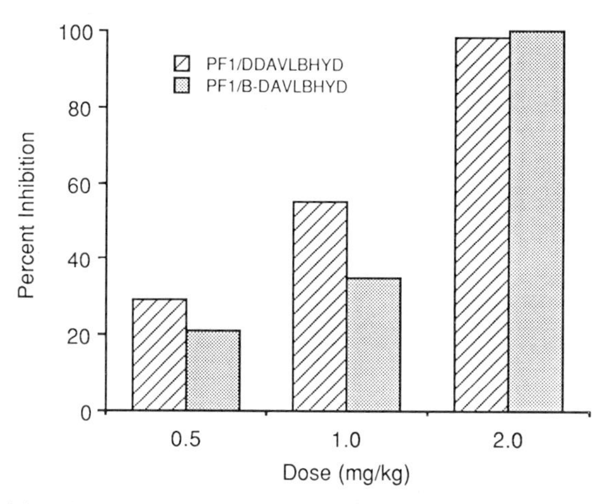

FIG. 11. Activity of PF1/D– and PF1/B–deacetylvinblastine hydrazide conjugates against the T222 squamous cell carcinoma xenograft. Conjugates were administered i.v. on days 3, 6, and 9 after s.c. inoculation of 1×10^7 tumor cells in the flank of *Nu*/*Nu* mice. Tumors were measured on day 22. Dose is given as 4-deacetylvinblastine equivalents (mg/kg).

The 4-deacetylvinblastine hydrazide–L1KS conjugate was found to be extremely effective against several human tumor xenografts bearing the KS1/4 antigen (Fig. 10) (*102*). Monoclonal antibodies PF1/D and PF1/B, reactive to human squamous cell carcinoma, were also linked to 4-deacetylvinblastine hydrazide (**61**) by the oxidizied carbohydrate–Schiff base method and demonstrated excellent efficacy against T222 human squamous carcinoma xenograft (Fig. 11) (*101, 103*).

B. Drug Resistance

1. Multidrug Resistance

The treatment of several human malignacies with chemotherapeutic agents often results in the initial regression of the tumor burden, producing a remissive state that may be of temporary duration. In cases in which the tumor regrows, the success of continued therapy with the initially effective regimen is often limited, and the disease is characterized as chemoresistant. This phenomenon is easily mimicked in cell culture by culturing tumor cells in the continual presence of increasing concentrations of drug; a resistant phenotype is produced that shows significantly reduced sensitivity to the drug with which this phenotype was initially derived. Interestingly, there are many examples in which the resistant cells are also resistant to other chemotherapeutic agents that differ in structure and mechanism of action from the agent to which the resistant phenotype was derived. Cells that demonstrate these properties exhibit the property of multidrug resistance (MDR). This topic has been the subject of several recent reviews (*104–107*).

The vinca alkaloids vinblastine and vincristine are capable of producing the MDR phenotype in a wide variety of cell types. Furthermore, cells that are made resistant to antitumor drugs such as doxorubicin, actinomycin D, or the epipodophyllotoxins etoposide (VP-16) and teniposide (VM-26) are often resistant to the effects of the bisindole alkaloids. The structural and mechanistic diversity of these compounds is even more striking against the backdrop of collateral resistance.

By examining the retention of drug in resistant cells, is was soon demonstrated that resistance was associated with the intracellular concentration of drug (*108*). Thus, resistant cells were incapable of maintaining effective intracellular concentrations of drug. Examination of membrane preparations derived from resistant cells revealed the high-level expression of a phosphorylated glycoprotein, P-gp, present in these cells that was not evident in the sensitive parental cells (*109*). The murine and human genes that code for P-gp have been identified and cloned (*110*). Transfection of normal cells with either of these genes results in a clonogenic population of cells that express P-gp and demonstrate multidrug resistance (*111*).

2. Agents That Reverse (Modulate) MDR

In 1983 Tsuruo demonstrated an enhancement in the cytotoxicity of vinblastine in a resistant variant of P388 leukemia by simultaneous treatment of these cells with verapamil (**148**) (*112*). It was later shown that the enhancement of drug cytotoxicity by verapamil was not limited to vinblastine but was operable with other drugs such as doxorubicin and that other noncytotoxic agents such as chloroquine (**149**) and nifedipine (**150**) were capable of reversing drug resistance in multidrug-resistant cells (*113,114*). These experiments revealed an additional dimension of complexity in understanding the molecular mechanism of MDR since drug resistance as well as its reversal spanned several classes of seemingly unrelated substances.

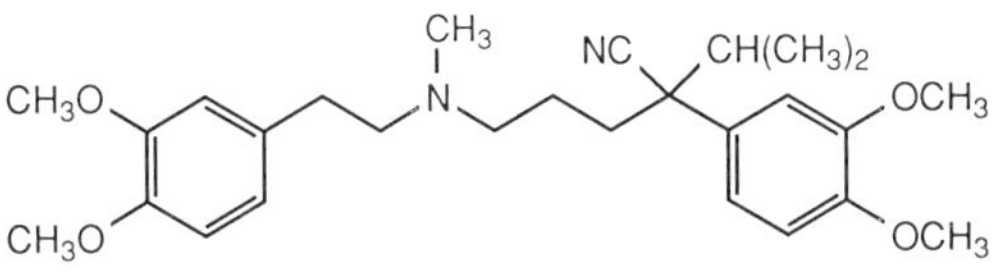

148

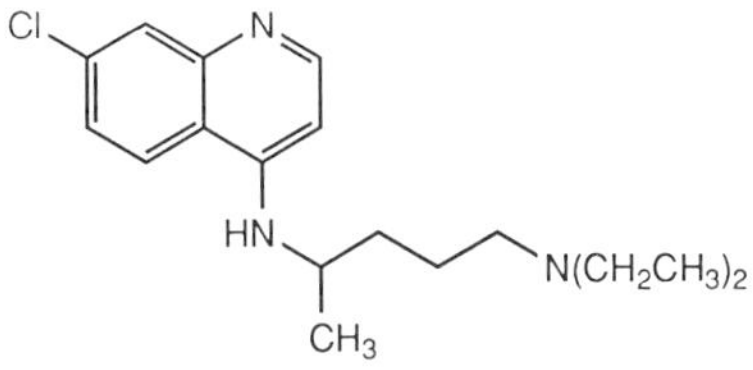

149

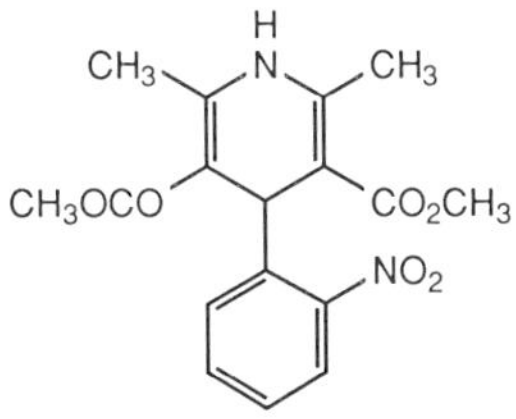

150

3. Structural Comparisons

The apparent partition coefficient (log P) and the calculated molar refractivity (CMR) are physicochemical parameters that correlate with the ability of a compound to be involved in MDR, and nonobvious structural homologies between compounds as dissimilar as verapamil (**148**) and vinblastine (**1**) exist as revealed by computer modeling studies (*115*). A far more quantitative method of studying the molecular mechanism of P-gp-associated MDR was introduced by Safa *et al.* with the preparation of two photoactivatible bisindole derivatives, **151** and **152** (*116,117*). These substances effectively bind tubulin and mimic the cellular pharmacology of vinblastine.

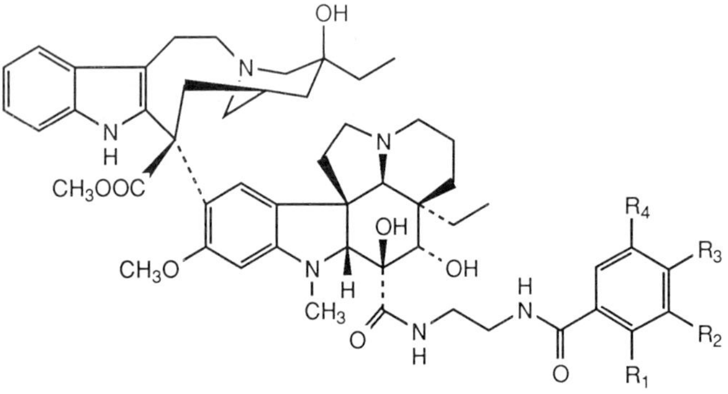

	R_1	R_2	R_3	R_4
151	OH	^{125}I	N_3	H
152	H	^{3}H	N_3	^{3}H

Treatment of a membrane vesicle preparation derived from multidrug-resistant cells with **151** or **152** followed by irradiation results in specific labeling of membrane-associated P-gp. Furthermore, it is possible to compete for the photoactivatible binding of **151** or **152** with agents such as verapamil (**148**). These studies establish the role of P-gp as an active efflux system and have made possible the discovery of more potent modulators such as reserpine (*118,119*). A more detailed conformational analysis of the structural domains required for drug binding to P-gp indicates that the MDR "pharmacophore" consists of two aromatic domains and a basic nitrogen atom (*119*). These domains are represented in the bisindole alkaloids by rings A′ and A and either N-6′ or N-9.

4. Implications

The bisindole alkaloids are clearly closely associated with MDR from both historical and structural perspectives. The clinical relevance of this mechanism of drug resistance has not been unambiguously established; however, the tissue-specific expression of P-gp has been noted for several normal as well as malignant tissues (*105*). Therefore, the natural resistance of many human solid tumors to bisindole treatment may be due, at least in part, to the presence of this resistance mechanism, and argues for the development of new strategies to circumvent resistance by blocking the action of P-gp or by site-specific targeting of drug.

IX. Conclusions

The medicinal chemistry of the bisindole alkaloids from *Catharanthus* spans a vast chemical territory as well as a major chapter of cancer pharmacology. It is illustrative of the growth of medicinal chemistry and its integration into several allied disciplines. The development of an analog series based on single group substitution is represented by the early studies that led to the selection of vinglycinate (**133**) as a new derivative with enhanced antitumor properties as measured by the pharmacologic and clinical standards of that era.

Increasing sophistication in chemical synthesis and pharmacology led eventually to vindesine (**59**), a substance that resulted from a combination of two modifications to the parent structure. The vindesine story is exemplified by increasing mechanistic understanding regarding the mechanism of action of these compounds, namely, tubulin binding, and a realization that experimental antitumor models were useful indicators of an underlying pharmacology. Chemistry that was developed during this campaign set the stage for more sophisticated studies and provided chemical intermediates that have demonstrated their importance in the development of monoclonal antibody conjugates.

The importance of the C-3 functionality and its attendant effect on conformation is illustrated by the discovery of the oxazolidinedione vinzolidine (**107**). This group of compounds represents a serendipitous observation of an orally active vinca alkaloid, but unfortunately the clinical application of this compound was not fully realized.

Multiple single-point modification of both halves of the bisindole framework provided vinepidine (**144**), a bisindole alkaloid that exhibited experimental antitumor activity similar to vincristine without demonstrating the acute neuronal toxicity of this alkaloid. Ultimately, the vinepidine experience was both educational and an exercise in humility, since the compound showed promise in preclinical neurotoxicology models yet demonstrated considerable neurotoxicity in clinical trials. Nevertheless, vinepidine brings a rich example of the combined effects of ring D′ conformation (with its associated effect on N-6′ basicity) and the presence of an N-1 formyl group.

The preparation of navelbine (**49**) by rearrangement of the bisindole carbon framework produced a new class of antitumor compounds. The success of this chemistry is indicative of the potential for new discovery in the bisindole series and again highlights the importance of ring D′ conformation in determining new pharmacology. Finally, the horizon for discovery of new bisindole alkaloid antitumor agents may be defined by emerging concepts in drug targeting via monoclonal antibody–drug conjugates and an understanding of the molecular mechanisms of drug resistance.

Acknowledgments

The author wishes to acknowledge several colleagues, all of Lilly Research Laboratories, who have contributed to the preparation of this chapter. Dr. Gerald Thompson made available a large collection of literature and data which established the basis for much of the data presented herein. Drs. Thomas Bumol, Bennet Laguzza, David Johnson, James Starling, and William Scott provided data in the form of several new publications concerning monoclonal antibody–drug conjugates and provided valuable criticism of the manuscript.

References

1. D. V. Jackson, Jr., and R. A. Bender, *Appl. Methods Oncol.* **1**, 277. (1978).
2. M. Tin-Wa and N. R. Farnsworth, *in* "*Catharanthus* Alkaloids," (W. I. Taylor and N. R. Farnsworth, eds.), p. 85. Dekker, New York, 1975.
3. W. T. Beck, *Handb. Exp. Pharmacol.* **72**, 569 (1984).
4. R. A. Bender, *in* "Cancer Chemotherapy," (T. M. Becker, ed.), Vol. 3, p. 273. Little, Brown, Boston, Massachusetts, 1981.
5. G. A. Cordell, *Recent Adv. Nat. Prod. Res., Proc. Int. Symp.*, 65. (1980).
6. G. H. Svoboda and D. A. Blake, *in* "*Catharanthus* Alkaloids," (W. I. Taylor and N. R. Farnsworth, eds.), p. 45. Dekker, New York, 1975.
7. W. A. Creasey, *Biochem. Pharmacol.* **23** (Suppl. 2), 217 (1973).
8. N. Bruchovsky, *Cancer Res.* **25**, 1232 (1965).
9. P. Ernst, *Biomedicine* **18**, 484 (1973).
10. J. A. Budman, *J. Natl. Cancer Inst.* **37**, 331 (1966). J. A. Donoso, L. S. Green, I. E. Heller-Bettinger, and F. E. Samson, *Cancer Res.* **37**, 1401 (1977).
11. F. J. Seil and P. W. Lampert, *Exp. Neurol.* **21**, 219 (1968).
12. I. S. Johnson, W. W. Hargrove, P. N. Harris, H. F. Wright, and G. B. Boder, *Cancer Res.* **26**, 2431 (1966).
13. H. Madoc-Jones and F. Mauro, *J. Cell. Physiol.* **72**, 185 (1968).
14. R. S. Camplejohn, B. Schultze, and W. Maurer, *Cell Tissue Kinet.* **13**, 239 (1980).
15. A. M. Lengsfeld, B. Schultze, and W. Maurer, *Eur. J. Cancer* **17**, 307 (1980).
16. R. J. Owellen, C. A. Hartke, R. M. Dickerson, and F. O. Hains, *Cancer Res.* **36**, 1499 (1976).
17. F. O. Hains, R. M. Dickerson, L. Wilson, and R. J. Owellen, *Biochem. Pharmacol.* **27**, 71 (1978).
18. L. Wilson, K. Anderson, L. Grisham, and D. Chin, *Microtubules Microtubule Inhibitors, Proc. Int. Symp.*, p. 103 (1975).

19. R. H. Adamson, S. M. Sieber, J. Wang-Peng, and H. B. Wood, *Proc. Am. Assoc. Cancer Res.* **17**, 42 (1976).
20. A. C. Sartorelli and W. A. Creasey, *Annu. Rev. Pharmacol.* **9**, 51 (1969).
21. W. A. Creasey and M. E. Markiw, *Biochim. Biophys. Acta* **87**, 601 (1964).
22. W. A. Creasey and M. E. Markiw, *Biochim. Biophys. Acta* **103**, 635 (1965).
23. D. Schmäl and H. Osswald, *Arzneim.-Forsch.* **20**, 1461 (1970).
24. Y. Seino, M. Nagao, T. Yahagi, A. Hoshi, T. Kawachi, and T. Sugimura, *Cancer Res.* **38**, 2148 (1978).
25. C. G. Zubrod *in* "Antineoplastic and Immunosupressive Agents, Part I" (A. C. Sartorelli and D. G. Johns, eds.), p. 1. Springer-Verlag, New York, 1974.
26. V. T. DeVita, Jr., and S. Hellman *in* "Cancer: Principles and Practice of Oncology" (V. T. DeVita, Jr., S. Hellman, and S. A. Rosenberg, eds.), 2nd Ed., p. 1331. Lippincott, Philadelphia, Pennsylvania, 1985.
27. J. V. Simone, J. R. Cassady, and R. M. Filler *in* "Cancer: Principles and Practice of Oncology" (V. T. DeVita, Jr., S. Hellman, and S. A. Rosenberg, eds.), 2nd Ed., p. 1254. Lippincott, Philadelphia, Pennsylvania, 1985.
28. J. D. Minna, G. A. Higgins, and E. J. Glatstein, *in* "Cancer: Principles and Practice of Oncology" (V. T. DeVita, Jr., S. Hellman, and S. A. Rosenberg, eds.), 2nd Ed., p. 396. Lippincott, Philadelphia, Pennsylvania, 1985.
29. L. Einhorn and J. P. Donohue, *Ann. Int. Med.* **87**, 293 (1977).
30. H. D. Weiss, M. D. Walker, and P. H. Wiernik, *N. Engl. J. Med.* **291**, 127 (1974).
31. R. W. Dyke and R. L. Nelson, *Cancer Treat, Rep.* **4**, 135 (1977).
32. B. K. Hunter, L. D. Hall, J. K. M. Sanders, *J. Chem. Soc., Perkin Trans. I*, 657 (1983).
33. N. Neuss, M. Gorman, G. H. Svoboda, G. Maciak, and C. T. Beer, *J. Am. Chem. Soc.* **81**, 4754 (1959).
34. G. H. Svoboda, *Lloydia* **24**, 17 (1961).
35. E. Wenkert, D. W. Cochran, E. W. Hagaman, F. M. Schell, N. Neuss, and A. S. Katner, *J. Am. Chem. Soc.* **95**, 4990 (1973).
36. J. M. Zamora, H. L. Pearce, and W. T. Beck, *Mol. Pharmacol.* **33**, 454 (1988).
37. E. Wenkert, E. W. Hagaman, B. Lai, G. E. Gutowski, A. S. Katner, J. C. Miller, and N. Neuss, *Helv. Chem. Acta* **58**, 1560 (1975).
38. M. R. Yagudaev, *Khim. Prir. Soedin.* **3** (1986); *Chem. Abstr.* **105**, 153369 (1986).
39. J. W. Moncrief and W. N. Lipscomb, *J. Am. Chem. Soc.* **87**, 4963 (1965).
40. N. D. Jones, S. Rolski, and J. K. Swartzendruber, *Acta Crystallogr., Sect. A. Cryst. Phys. Diffr. Theor. Gen. Crystallogr.* **40** (Suppl. 3), 3 (1984).
41. R. J. Owellen, D. W. Donigian, C. A. Hartke, and F. O. Hains, *Biochem. Pharmacol.* **26**, 1213 (1977).
42. G. H. Svoboda, *Lloydia* **24**, 173 (1961).
43. D. J. Abraham and N. R. Farnsworth, *J. Pharm. Sci.* **58**, 694 (1969).
44. N. Neuss, L. L. Huckstep, and N. J. Cone, *Tetrahedron Lett.*, 811 (1967).
45. A. B. L. Trouet, J. A. A. J. Hannart, and K. S. B. Rao, U.S. Patent No. 4,388,305, 4,639,456 A 870,127; *Chem. Abstr.* **82**, 41804 (1975).
46. H. L. Pearce, Lilly Research Laboratories, personal communication,
47. H. L. Pearce, U.S. Patent No. 4,430,269 A 840,207; *Chem. Abstr.* **100**, 192142 (1984).
48. C. Santay, L. Szabo, K. Honty, T. Keve, T. Acs, S. Eckhardt, J. Sugar, Z. Somafi, E. Ivan, and Z. Kneffel, Eur. Patent No. 207,336 A2 870,107; *Chem. Abstr.* **106**, 113543 (1987).
49. J. C. Miller, G. E. Gutowski, G. A. Poore, and G. B. Boder, *J. Med. Chem.* **20**, 409. (1977).

50. A. El-Sayed, G. A. Hardy, and G. A. Cordell, *J. Nat. Prod.* **46**, 517 (1983).
51. M. D. Jablonski, personal communication.
52. P. Mangeney. R. Z. Andriamialisoa, J.-Y. Lallemand, N. Langlois, Y. Langlois, and P. Potier, *Tetrahedron* **35**, 2175 (1979).
53. C. J. Barnett, C. J. Cullinan, K. Gerzon, R. C. Hoying, W. E. Jones, W. M. Newlon, G. A. Poore, R. L. Robison, M. J. Sweeney, and G. C. Todd, *J. Med. Chem.* **21**, 88 (1978).
54. J. Harley-Mason and Atta-ur-Rahman, *Tetrahedron* **36**, 1057 (1980).
55. J. P. Kutney, J. Balsevich, and B. R. Worth, *Can. J. Chem.* **57**, 1682 (1979).
56. J. P. Kutney, J. Balsevich, and B. R. Worth, *Heterocycles* **11**, 69 (1978).
57. J. P. Kutney, J. Balsevich, and B. R. Worth, *Heterocycles* **9**, 493 (1978).
58. J. P. Kutney, J. Balsevich, G. H. Bokelman, T. Hibino, T. Honda, I. Itoh, A. H. Ratcliffe, and B. R. Worth, *Can. J. Chem.* **56**, 62 (1978).
59. M. Doe de Maindreville and J. Levy, *Bull. Soc. Chim. Fr.* **5–6**, Pt. II, 179 (1981).
60. M. E. Kuehne, T. C. Zebovitz, W. G. Bornmann, and I. Marko, *J. Org. Chem.* **52**, 4340 (1987).
61. P. Magnus, B. Hewitt, K. Cardwell, P. Cairns, and P. Boniface, *Stud. Surf. Sci. Catal.* **25**, 203 (1986).
62. J. Harley-Mason and Atta-ur-Rahman, *Tetrahedron* **36**, 1057 (1980).
63. P. Potier, N. Langlois, Y. Langlois, and F. Guéritte, *J. Chem. Soc., Chem. Commun.*, 670 (1975).
64. N. Langlois, F. Guéritte, Y. Langlois, and P. Potier, *J. Am. Chem. Soc.* **98**, 7017 (1976).
65. P. Mangeney, R. Z. Andriamialisoa, N. Langlois, Y. Langlois, and P. Potier, *J. Org. Chem.* **44**, 3765 (1979).
66. P. Mangeney, R. Z. Andriamialisoa, J.-T. Lallemand, N. Langlois, Y. Langlois, and P. Potier, *Tetrahedron* **35**, 2175 (1979).
67. P. Potier and M.-M. Janot, *C. R. Acad. Sci. Paris* **276C**, 1727 (1973).
68. J. P. Kutney, *Heterocycles* **4**, 429 (1976).
69. A. I. Scott, C. L. Yeh, and D. Greenslade, *J. Chem. Soc., Chem. Commun.*, 947 (1978).
70. R. Maral, C. Bourut, E. Chenu, and G. Mathé, *Cancer Lett.* **22**, 49 (1984).
71. K. Jovanovics, K. Szasz, G. Fekete, E. Bittner, E. Dezseri, and J. Eles, U.S. Patent No. 3,899,493-750,812; *Chem. Abstr.* **83**, 179360 (1975).
72. R. A. Conrad, Eur. Patent No. 37,290 A1 811,007; *Chem. Abstr.* **96**, 104583 (1982).
73. H. L. Pearce, Eur. Patent No. 37,289 A1 811,007; *Chem. Abstr.* **96**, 85834 (1982).
74. Gedeon Richter Vegyeszeti Gyar Rt., Belg. Patent No. 889,990 A1 820,217; *Chem. Abstr.* **97**, 216542 (1982).
75. C. Szantay, L. Szabo, K. Honty, T. Keve, T. Acs, S. Eckhardt, J. Sugar, Z. Somafai, E. Ivan, and Z. Kneffel, Eur. Patent No. 205,169 A2 861,217; *Chem. Abstr.* **107**, 023569 (1987).
76. K. Gerzon, *in* "Anticancer Agents Based on Natural Product Models" (J. M. Cassady and J. D. Douros, eds.). Academic Press, New York, 1980.
77. R. A. Conrad, G. J. Cullinan, K. Gerzon, and G. A. Poore, *J. Med. Chem.* **22**, 391 (1979).
78. M. J. Sweeney, G. B. Boder, G. J. Cullinan, H. W. Culp, W. D. Daniels, R. W. Dyke, K. Gerzon, R. E. McMahon, R. L. Nelson, G. A. Poore, and G. C. Todd, *Cancer Res.* **38**, 2886 (1978).
79. J. C. Miller and G. E. Gutowski, W. Ger. Patent No. 2,753,791-780,608; *Chem. Abstr.* **89**, 129778 (1978).

80. K. Gerzon and J. C. Miller, Eur. Patent No. 55,602 A2 820,707; *Chem. Abstr.* **97**, 163310 (1982).
81. J. C. Miller, personal communication.
82. I. S. Johnson, *in* "Cancer Medicine" (J. F. Holland and E. Frei, eds.), p. 844. Lee & Febiger, Philadelphia, Pennsylvania, 1973.
83. D. B. Rifkin, Z. M. Loeb, G. Moore, and E. Reich, *J. Exp. Med.* **139**, 1317 (1974).
84. W. E. Laug, P. A. Jones, and W. F. Benedict, *J. Natl. Cancer Inst.* **54**, 173 (1975).
85. W. S. Tucker, W. M. Kirsch, M. Martinez-Hernandez, and L. M. Fink, *Cancer Res.* **38**, 297 (1978).
86. B. Nagy, J. Ban, and B. Brdar, *Int. J. Cancer* **19**, 614 (1977).
87. S. T. Rohrlich and D. B. Rifkin, *in* "Annual Reports in Medicinal Chemistry" (H. Hess, ed.), p. 229. Academic Press, New York, 1979.
88. P. L. Carl, P. K. Chakravarty, J. A. Katzenellenbogen, and M. J. Weber, *Proc. Natl. Acad. Sci. U.S.A.* **77**, 2224 (1980).
89. P. L. Carl, *in* "Target-Oriented Anticancer Drugs" (Y.-C. Chenh, ed.), p. 143. Raven, New York, 1983.
90. K. S. Bhushana Rao, M. P. Collard, and A. Trouet, *Anticancer Res.* **5**, 379 (1985).
91. I. S. Johnson, W. W. Hargrove, P. N. Harris, H. F. Wright, and G. B. Boder, *Cancer Res.* **26**, Part 1, 2431 (1966).
92. J. G. Armstrong, R. W. Dyke, P. J. Fouts, J. J. Hawthorne, C. J. Jansen, and A. M. Peabody, *Cancer Res.* **27**, Part 1, 221 (1967).
93. G. B. Boder, W. W. Bromer, G. A. Poore, G. L. Thompson, and D. C. Williams, *Proc. Am. Assoc. Cancer Res.* **73**, 793 (1982).
94. G. L. Thompson, G. B. Boder, W. W. Bromer, G. B. Grindey, and G. A. Poore, *Proc. Am. Assoc. Cancer Res.* **73**, 792 (1982).
95. G. L. Thompson, U.S. Patent No. 4,143,041.
96. G. L. Thompson, Lilly Research Laboratories, personal communication.
97. I. S. Johnson, M. E. Spearman, G. C. Todd, J. L. Zimmerman, and T. F. Bumol, *Cancer Treatment Rev.* **14**, 193 (1987).
98. G. F. Rowland, C. A. Axton, R. W. Baldwin, J. P. Brown, J. R. F. Corvalon, M. J. Embleton, V. A. Gore, I. Hellstrom, K. E. Hellstrom, E. Jacobs, C. H. Marsden, M. V. Pimm, R. G. Simmons, and W. Smith, *Can. Immunol. Immunother.* **19**, 1 (1985).
99. T. F. Bumol, A. L. Baker, E. L. Andrews, S. V. DeHerdt, S. L. Briggs, M. E. Spearman, and L. D. Apelgren, *in* "Antibody-Mediated Delivery Systems" (J. D. Rodwell, ed.), p. 55. Dekker, New York, 1988.
100. B. C. Laguzza, C. L. Nichols, S. L. Briggs, G. J. Cullinan, D. A. Johnson, J. J. Starling, A. L. Baker, and T. F. Bumol, J. R. F. Corvalan *J. Med. Chem.* **32,** 548 (1989).
101. D. A. Johnson, J. L. Zimmerman, B. C. Laguzza, and J. N. Eble, *Can. Immunol. Immunother.* **27,** 241 (1988).
102. J. J. Starling, R. S. Maciak, N. A. Hinson, C. L. Nichols, S. L. Briggs, and B. C. Laguzza, *Can. Immunol. Immunother.* **28,** 171 (1989).
103. D. A. Johnson and B. C. Laguzza, *Cancer Res.* **47**, 3118 (1987).
104. W. T. Beck, *Biochem. Pharmacol.* **36**, 2879 (1987).
105. I. Pastan and M. Gottesman, *N. Engl. J. Med.* **316**, 1388 (1987).
106. J. R. Riordan and V. Ling, *Pharmacol. Ther.* **28**, 51 (1985).
107. M. M. Gottesman and I. Pastan, *J. Biol. Chem.* **263**, 12163 (1988).
108. A. Fojo, S.-I. Akiyama, and M. M. Gottesman, *Cancer Res.* **45**, 3002 (1985).
109. W. T. Beck, T. J. Mueller, and L. R. Tanzer, *Cancer Res.* **39**, 2070 (1979).
110. I. B. Robinson, J. E. Chin, and K. G, Choi, *Proc. Natl. Acad. Sci. U.S.A.* **83**, 4538, (1986).

111. P. Gros, Y. Ben Neriah, and J. M. Croop, *Nature (London)* **323**, 728 (1986).
112. T. Tsuruo, H. Iida, K. Haganuma, S. Tsukagoshi, and Y. Sakurai, *Cancer. Res.* **43**, 808 (1983).
113. W. T. Beck and J. M. Zamora, *Biochem. Pharamcol.* **35**, 4303 (1986).
114. A. R. Safa, C. J. Glover, J. L. Sewell, M. B. Meyers, J. L. Bieddler, and R. L. Felsted, *J. Biol. Chem.* **262**, 7884 (1987).
115. J. M. Zamora, H. L. Pearce, and W. T. Beck, *Mol. Pharmacol.* **33**, 454 (1988).
116. A. R. Safa, C. J. Glover, M. B. Meyers, J. L. Biedler, and R. L. Felsted, *J. Biol. Chem.* **261**, 6137 (1986).
117. M. M. Cornwell, A. R. Safa, R. L. Felsted, M. M. Gottesman, and I. Pastan, *Proc. Natl. Acad. Sci. U.S.A.* **83**, 3847 (1986).
118. S.-I. Akiyama, M. M. Cornwell, M. Kuwano, I. Pastan, and M. M. Gottesman, *Mol. Pharmacol.* **33**, 454 (1988).
119. H. L. Pearce, N. J. Bach, M. C. Cirtain, A. R. Safa, M. A. Winter, J. M. Zamora, and W. T. Beck, *Proc. Am. Assoc. Cancer Res.* **29**, 299 (1988).
120. K. Mullin, P. J. Houghton, J. A. Houghton, and M. E. Horowitz, *Biochem. Pharmacol.* **34,** 1975 (1985).

—Chapter 5—

PHARMACOLOGY OF ANTITUMOR BISINDOLE ALKALOIDS FROM *CATHARANTHUS*

John J. McCormack

Department of Pharmacology, and
Vermont Regional Cancer Center
University of Vermont
Burlington, Vermont 05405

I. Introduction

Noble and colleagues, in 1958, found that extracts of *Catharanthus roseus* (*Vinca rosea* L.) G. Don were able to produce pronounced bone marrow depression in animals. These investigators were able to purify the alkaloid vinblastine (Fig. 1) from such extracts and to demonstrate that this alkaloid was responsible for the observed biological effect. It is of interest to note that these investigators were evaluating possible hypoglycemic actions of the plant extracts because of reports that such extracts possessed antidiabetic activity (*1*). At essentially the same time, Johnson and colleagues observed that extracts of *Catharanthus roseus* were active against a murine leukemia and that the activity of such extracts could be duplicated by administration of a pure alkaloid obtained from the extracts (*2*).

The alkaloids vinblastine and vincristine (Fig. 1) have attained a significant role in cancer chemotherapy. A relatively large number of analogs of these alkaloids have been evaluated for biological activity, but very few have advanced beyond initial clinical trials. Vindesine and vinzolidine (Fig. 2) are examples of compounds that have been clinically evaluated. These compounds represent structural modifications of the lower portion (vindoline) of the "dimeric" molecule at position 3. Thus, vinde-

Copyright © 1990 by Academic Press, Inc.
All rights of reproduction in any form reserved.

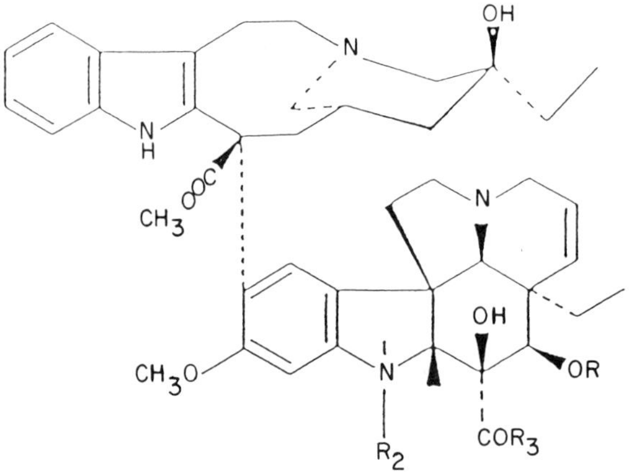

FIG. 1. Structures of vinblastine (R_1 = $COCH_3$, R_2 = CH_3, R_3 = OCH_3) and vincristine (R_1 = $COCH_3$, R_2 = CHO, R_3 = OCH_3).

sine is obtained, formally, by replacing the carbomethoxy function of vinblastine by a carboxamide. Despite the close structural resemblance of vinblastine, vincristine, and vindesine, the profiles of biological activities of the compounds are not identical. Vinblastine treatment is much less likely to result in neurotoxicity than is treatment with vincristine; on the other hand, vinblastine produces much more perturbation of bone marrow function than does vinblastine. Vincristine is more effective than vinblastine in the treatment of acute leukemia in children.

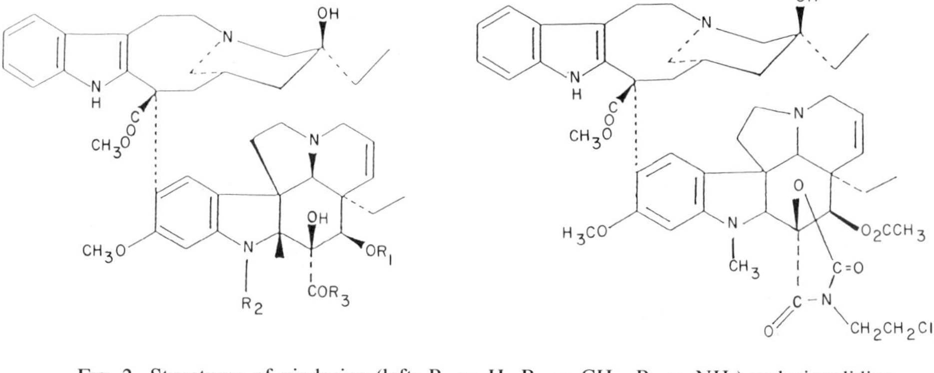

FIG. 2. Structures of vindesine (left; R_1 = H, R_2 = CH_3, R_3 = NH_2) and vinzolidine (right).

II. Biochemical Pharmacology

Vincristine and vinblastine interfere with the mitotic process by producing arrest of cell division in metaphase. These drugs interact with tubulins, dimeric cellular proteins that play several critical roles in cell structure and function, including organization of chromosomes on a matrix (spindle), which is an essential feature of normal cell division. The aggregation (assembly) of tubulin units into microtubules is required not only for formation of the mitotic spindle but also for other processes such as the transport of materials in nerves (axonal transport). Inhibition of microtubule formation can be demonstrated with low concentrations of vinblastine and vincristine *in vitro*. Recent developments in synthetic chemistry have made possible extraordinarily interesting structure–activity studies that may shed light on the detailed mechanisms of efficacy and toxicity of vinblastine and analogs and also lead to design of more selective congeners for use in cancer chemotherapy (see Chapters 3 and 4, this volume).

Vinblastine, vincristine, and structurally related analogs inhibit microtubule polymerization by 50% at concentrations in the range 0.1–1 μM, and the process of tubulin addition to preformed microtubules, at steady state, is comparably sensitive to inhibition by these agents (*3*). As shown in Table I, the differences in K_i values for inhibition of steady-state tubulin addition by vinblastine, vincristine, vindesine, and vinepidine were relatively small, but the pattern of activity in the tubulin addition system did not parallel that observed when the compounds were evaluated for effects on the proliferation of B16 melanoma cells *in vitro*. Vinepidine was more than twice as potent as vinblastine as an inhibitor of steady-state tubulin addition but nearly 10-fold less potent than vinblastine as an inhibitor of cell growth (*3*).

It is important to emphasize that the binding of vinblastine and vincristine to tubulin occurs at sites on the protein different from that involved in interactions of tubulin with other drugs such as colchicine (Fig. 3). In addition to tubulins, microtubules are composed of other proteins, the so-called microtubule-associated proteins, that may facilitate the formation of microtubules by serving to cross-link tubulin units (*4*). Effects of vinblastine and vincristine on microtubules *in vitro* are concentration dependent. At low concentrations (submicromolar) the characteristic effects on microtubule assembly, described above, are observed, while at higher concentrations (10 μM and above) formation of spiral aggregates is induced by the drugs. Borman and colleagues have found that inhibition of microtubule assembly and induction of spiral aggregate formation are not

TABLE I

INHIBITION OF TUBULIN ADDITION AND CELL PROLIFERATION BY VINBLASTINE AND RELATED COMPOUNDS

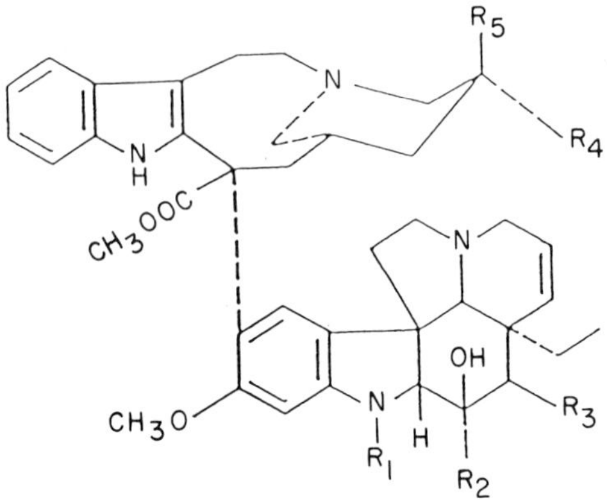

Compound	R_1	R_2	R_3	R_4	R_5	Relative Potency[a] Tubulin	Relative Potency[a] Cell growth
Vinblastine	CH_3	$COOCH_3$	$OCOCH_3$	CH_2CH_3	OH	0.4	1.0
Vindesine	CH_3	$CONH_2$	OH	CH_2CH_3	OH	0.7	0.8
Vincristine	CHO	$COOCH_3$	$OCOCH_3$	CH_2CH_3	OH	1.0	0.5
Vinepidine	CHO	$COOCH_3$	$OCOCH_3$	H	CH_2CH_3	1.0	0.1

[a] The K_i for vinblastine estimated for inhibition of tubulin addition was 0.18 μM, and vinblastine inhibited the growth of B16 melanoma cells completely at a concentration of 4 nM (*3*).

uniformly expressed in vinblastine analogs, indicating subtle differences in the underlying molecular events involved in vinblastine–microtubule interactions at low and high vinblastine concentrations (*5*; see also Chapter 3, this Volume).

Earlier studies indicated that vinblastine binding to tubulin dimers involves interaction of two molecules of vinblastine with each dimeric pro-

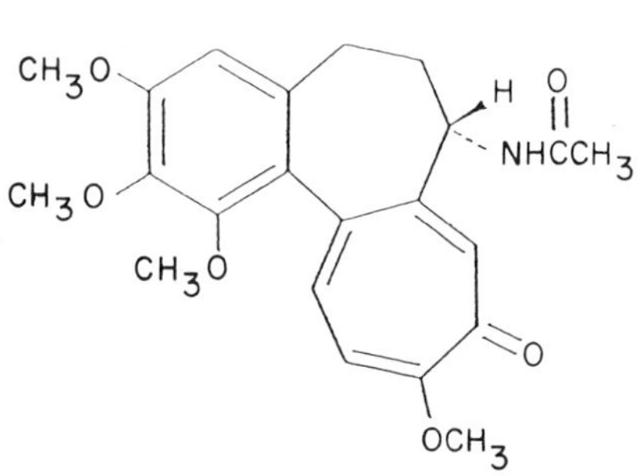

FIG. 3. Structure of colchicine.

tein molecule. Na and Timasheff have provided evidence that the interaction of vinblastine with calf brain tubulin is complicated, in that the protein dimer actually contains only one specific binding site (binding constant 40,000/*M*) and that binding of vinblastine to the dimer results in self-association of the protein. The specific binding of vinblastine to the polymerized tubulin is estimated to be at least 100-fold greater than binding to the dimer. Nonspecific interactions of vinblastine with tubulin also occur, but these are characterized by binding constants at least 10-fold lower than those calculated for the simple interaction with the tubulin dimer (*6*). The interaction of vinblastine with tubulin is promoted by magnesium ions, and discrepancies in the various characterizations of the interaction by different investigators can be related to differences in the concentrations of magnesium ions used in different laboratories (*7*).

A wide variety of other biochemical effects has been reported to be associated with treatment of cells with vinblastine, vincristine, and related compounds (*8*). These effects include inhibition of the biosynthesis of proteins and nucleic acids and of aspects of lipid metabolism; it is not clear whether such effects contribute to the therapeutic or toxic actions of vincristine and vinblastine. Vinblastine and vincristine inhibit protein kinase C, an enzyme system that modulates cell growth and differentiation (*9*). The pharmacological significance of such inhibition has not been established, however, and it must be emphasized that the concentrations of the drugs required to inhibit protein kinase C are several orders of magnitude higher than those required to alter tubulin polymerization phenomena (*10*).

III. Cellular Pharmacology

Vincristine and vinblastine are generally considered to act specifically on the metaphase portion of the mitotic (M) stage of the cell cycle as a consequence of perturbations of the structure and function of tubulin. A characteristic action of the drugs is production of mitotic arrest in which the percentage of cells in mitosis in a given population of cells will rise from a few percent to 50% and more after treatment with a drug such as vinblastine. There are reports, however, that these drugs can interfere with other phases of the cell cycle in ways not clearly related to interference with tubulin function (*8*).

Treatment of cells with concentrations of vincristine or vindesine that produce relatively little effect on cell viability results in an accumulation of cells in the M and G_2 (gap after DNA synthesis) phases of the cell cycle. The cellular effects, as measured by techniques such as flow cytometry, are concentration dependent, and cell killing can be demonstrated for cells in the S (DNA synthesis) phase of the cell cycle (*11*; Fig. 4). The abilities of vincristine and vinblastine to produce lethality during the S phase of the cell cycle was observed in early studies by Madoc-Jones and Mauro (*12*). These investigators also reported that vinblastine, but not vincristine, is lethal for HeLa cells in the G_1 phase (gap before DNA synthesis) of the cell cycle.

In view of the substantial current interest in drug actions at the cell membrane and the relationships between such effects and anticancer drug action, it is interesting to note that both vincristine and vinblastine produce crenellation (wrinkling) of cell membranes as a characteristic morphological effect (*12*). That cytotoxicity produced by treatment with vinblastine may be linked to perturbations of membrane function, suscep-

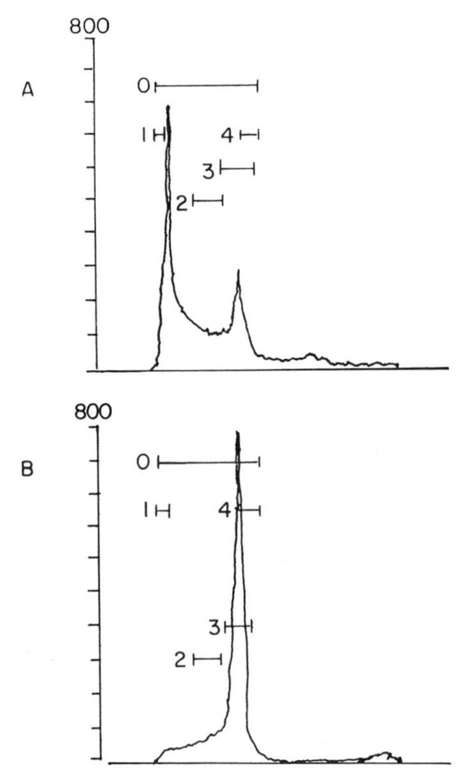

FIG. 4. Flow cytometric patterns for L1210 mouse leukemia cells in the absence (A) and presence (B) of vinblastine (9 hr exposure to 8 n*M* vinblastine). (Data courtesy of Dr. Linda Borman, Vermont Regional Cancer Center.)

tible to modification by agents such as verapamil, has been suggested by Beck and colleagues (*13*), and these investigators also described morphological alterations in cells treated with vinblastine that included apparent perturbations of membrane structure as well as effects on intracellular structures. Early observations of morphological changes in neoplastic cells treated with nontoxic concentrations of vinblastine might be interpreted as a "differentiative" effect (*1b*).

Granting that precise mechanisms responsible for the characteristic anticancer activity and mammalian toxicity of vinblastine, vincristine, and related compounds have not been rigorously established, it nevertheless is important to describe representative biological and biochemical actions of the drugs that may have mechanistic pertinence. The susceptibility to mitotic spindle dissolution of cell lines with 100-fold differences in sensitivity to vinblastine has been investigated (*14*). There was an excellent correlation between drug concentrations required to produce inhibition of cell colony formation and those required to dissolve mitotic spindles. It is noteworthy that effects on the mitotic spindle of vinblastine occur very rapidly and can be detected within 30 sec.

Safa and colleagues (*15*) have briefly reviewed mechanisms other than interaction with tubulin that may play a role in the biological actions of vinblastine and related compounds, and they state, "the mechanism(s) responsible for the cytotoxic and antitumor activities of *Vinca* [*sic*] alkaloids is not clear." These investigators used a synthetic photoaffinity-labeled derivative of vinblastine to identify alkaloid binding sites in P388 murine leukemia cells. The photoactive derivative *N*-(*p*-azidobenzoyl)-*N'*-β-aminoethylvindesine (NABV) inhibited the growth of P388 cells, *in vitro*, at concentrations (1 n*M* range) comparable to those required for equivalent growth inhibition by vinblastine. Photolabeling of cells with NABV resulted in the identification of three major cellular proteins, one of which appears to be a tubulin subunit on the basis of immunochemical analysis. It remains to be established whether nontubulin proteins, which appear to be specific for vinblastine and related alkaloids, contribute in a pharmacologically important way to the anticancer activity or toxicity of such compounds.

Beck has contributed an incisive review of investigations from his laboratory and those of other scientists on the cellular pharmacological aspects of resistance to vinblastine and related compounds (*16*). Important cellular characteristics that are frequently associated with resistance to vinblastine include cross-resistance, not only to closely related compounds such as vincristine and vindesine but also to other basic naturally occurring compounds. Concomitant resistance to vinblastine and representative anthracyclines (e.g., doxorubicin) may involve reduced abilities of resistant cells to accumulate these cytotoxic drugs and the expression of a moderately sized (150,000 dalton range) surface glycoprotein. Expression of this glycoprotein, the so-called P-glycoprotein, is a hallmark of the multiple drug resistance (MDR) phenotype. The phenomenon of multiple drug resistance has also been termed "pleiotropic drug resistance," but most investigators in this field of research prefer the use of the former descriptive term.

Cells that exhibit high (several hundredfold) levels of resistance to vinblastine, vincristine, and vindesine have an extremely limited capacity to accumulate radiolabeled vinblastine; for example, essentially no increase in radioactivity associated with human leukemic lymphoblastic cells resistant to vinblastine could be detected over a 60-min incubation period in the presence of concentrations of tritiated vinblastine that were cytotoxic to parent cells. The parent cells, highly sensitive to vinblastine, were observed to accumulate vinblastine to levels seven-fold higher than those observed for the resistant cells (*16*).

Several investigators have postulated, but not proved, that the P-glycoprotein functions as an efflux pump, acting at the cell membrane to accelerate the transfer of drugs from the cell interior to the extracellular milieu. It has been observed that treatment of cells resistant to vincristine and Adriamycin (doxorubicin) with the calcium entry blocker verapamil increases the susceptibility of the cells to the antitumor drugs (*17*). Beck has shown that the susceptibility of vinblastine-resistant cells can be increased nearly 100-fold in the presence of moderately high (10 μM) concentrations of verapamil (*16*). It is generally agreed that the enhancement of the cytotoxicity of vinblastine and other drugs by verapamil is not a consequence of simple calcium channel block but reflects alteration of cell membrane processes essential to the accumulation/retention of the drugs within target cells. The suggestion has been made that multiple drug resistance may, at least in part, be a consequence of membrane depolarization, and the action of verapamil and other compounds that modulate this resistance could be due to a restoration of membrane potential (*18*).

Human KB carcinoma cells resistant to vinblastine and other drugs have been shown to exhibit increased membrane vesicular binding of tritiated vinblastine, and this binding is correlated with photoaffinity labeling of a 150,000- to 170,000-dalton protein in the vesicles. Labeling of this protein is inhibited by vinblastine, vincristine, and verapamil but not by colchicine (*19*). The failure of colchicine to inhibit the labeling of the membrane protein is unexpected since the cells from which the protein was isolated are resistant to colchicine as well as vinblastine.

Resistance to vinblastine and other drugs can be produced in mammalian cells by the transfer of specific gene sequences that code for P-glycoprotein. This may be exemplified by studies demonstrating that introduction of a cDNA sequence obtained from drug-resistant carcinoma cells into a drug-sensitive cell line resulted in the expression of the multidrug resistance phenotype in the transfected cells (*20*). It bears emphasis that cellular resistance to vincristine and vinblastine can be dissociated from "classical" concomitants of the phenomenon of multiple drug resistance in some cell lines. Human lymphoblastic leukemia cells resistant to vinblastine and vincristine show marked increases in sensitivity to these drugs when treated with verapamil, while the sensitivities to drugs such as the anthracyclines are only slightly altered by verapamil treatment (*13*).

Rhabdomyosarcomas are common soft tissue sarcomas in children, and vincristine is a component of chemotherapeutic regimens used in the management of patients with these tumors. Rhabdomyosarcoma xenografts resistant to vincristine also exhibit some cross-resistance to the alkylating nitrogen mustard derivative L-phenylalanine mustard (L-PAM); such observations are of potential clinical pertinence in view of the relatively low degree of activity of L-PAM in pediatric patients unresponsive to drug regimens that include vincristine (*21*).

Horton and colleagues have found that the retention of vincristine in vincristine-resistant tumors was reduced substantially compared with retention of the drug observed in sensitive tumors; these investigators estimated an elimination half-time for vincristine in resistant tumors of 34 hr while that for sensitive tumors was estimated to be approximately 20-fold longer. Expression of the gene that codes for the P-glycoprotein associated with the multiple drug resistance phenotype in other tumor cells was not enhanced in vincristine-resistant rhabdomyosarcoma xenografts (*21*).

Vincristine resistance has been studied in Chinese hamster ovary cell lines; cells resistant to vincristine also are resistant to vinblastine and vindesine. Suggestions were made that, in cells with relatively low levels of drug resistance, at least two prominent mechanisms of resistance can occur (*22*). In the first instance, cellular resistance may be attributable to membrane alterations that are reversible, functionally, by treatment with verapamil. In the second, resistance has been postulated to be due to an altered sensitivity of tubulin to the effects of the drugs; the primary basis for postulating an altered interaction with tubulin was that a subgroup of cells resistant to vincristine showed enhanced sensitivity to taxol, a drug that can stabilize microtubules. It should be emphasized that differential sensitivities of tubulins from different tumor cells to the effects of vincristine or vinblastine has been proposed as a basis for the susceptibilities of cells to the cytotoxic effects of such drugs (*23*). Differences have been described in the electrophoretic patterns for tubulins obtained from vincristine-sensitive and vincristine-resistant tumor cells although no correlation has yet been established between these patterns and the capacities of the tubulins to bind vincristine (*24*).

Vincristine has been shown to enhance the accumulation of the folate antagonist methotrexate in murine leukemia cells, and the enhancement has been shown to involve inhibition of a specific efflux route for methotrexate (*25*); the suggestion has been made that the effect of vincristine on methotrexate efflux may be related to alterations of cell membrane electrical activity that appear to occur when cells are treated with vincristine. In this connection, it is worth mentioning that association of tubulin with membrane structures from bovine brain has been described (*25a*). Both vinblastine and vincristine have been reported to enhance the accumulation of the folate antagonist methotrexate in human leukemic cells (*8*); there is no evidence, however, to indicate that this interaction has significance in a clinical setting.

The majority of investigations of the cellular pharmacology of vinblastine and related alkaloids have involved aspects of the anticancer activity of the compounds. Among other types of pharmacological activity meriting consideration is the effect of vinblastine on the growth of the malarial parasite *Plasmodium falciparum* in cell culture. Vinblastine at a concentration of 100 n*M* completely inhibits plasmodial growth; mutants that are resistant to vinblastine also exhibit resistance to vincristine. These observations were interpreted as suggesting that cell division in the malarial parasite involves spindle formation and function comparable to that in mammalian cells (*26*).

IV. Preclinical Pharmacology

Early investigations of the antitumor activity of vinblastine and vincristine revealed that the compounds shared the ability to produce marked prolongations of life span in mice bearing intraperitoneally inoculated leukemic cells. Both compounds, at relatively low doses ($\leq$1 mg/kg/day for 10 days), could prolong the life span of leukemic animals by 100% and more (*1b*). Vincristine was particularly effective in treating mice with P1534 leukemia and could achieve "cures" of leukemic mice. It is interesting to note that the P1534 leukemia, initially "exquisitely" sensitive to the inhibitory effects of vinblastine and vincristine, over time developed substantial resistance to these agents (*27*).

An interesting difference in the experimental antitumor spectra of vinblastine and vincristine was noted in that vinblastine was inactive against the Ridgeway osteogenic sarcoma in mice whereas vincristine strongly inhibited the growth of this tumor. Vindesine inhibited the Ridgeway osteogenic sarcoma by 90–100% when administered intraperitoneally, daily for 8 days, at doses of 0.3–0.4 mg/kg; comparable effects were observed with vincristine treatment at doses of 0.15–0.20 mg/kg (*28*).

An interesting aspect of the preclinical pharmacological profiles of vinblastine and related compounds is the impressive activity characteristic of such compounds against the P388 murine leukemia *in vivo* and the lack of significant activity when the compounds are administered to mice inoculated with L1210 leukemia cells. Vinblastine and vindesine prolong the lives of mice with P388 by 200–300%, while treatment of mice with the L1210 leukemia using these agents results in little or no prolongation of life (*28*). The lack of efficacy of vinblastine and related compounds *in vivo* against the L1210 leukemia is particularly puzzling because these compounds exhibit a high potency when evaluated as inhibitors of the growth of L1210 cells *in vitro*. It should be recognized, however, that modest responses to vinblastine or vincristine treatment of mice with the L1210 leukemia have been reported by Goldin and colleagues, but the responses are much less pronounced than those noted for comparable treatment of mice with P388 leukemia (*29*).

Vinblastine, vincristine, and vindesine are active against the murine B16 melanoma *in vivo*, with many long-term survivors observed even when treatment was only moderately intensive; for example, treatment once every 4 days (on days 1, 5, and 9 following inoculation of the tumor) with vinblastine at 0.6 mg/kg resulted in four of 10 mice surviving for 45 days (*28*). In the murine tumor screening panel used until recently to assist in the selection of candidate drugs for clinical evaluation, vinblastine and vincristine exhibited significant activity in 70% of the tumor systems in which the drugs were tested; vinblastine was active in P388 leukemia, B16 melanoma, two colon tumors, and in a mammary tumor but was not effective against the Lewis lung tumor (Table II).

The profile of preclinical efficacy, in several standard systems, reported for vincristine was generally similar to that summarized for vinblastine (Table II). Houghton and colleagues have made important contributions to our understanding of the pharmacology of vincristine and vinblastine, using the rhabdomyosarcoma model. These investigators have shown that response of tumors to vincristine, but not vinblastine, can be explained on the basis of differential retention of the drugs in target tumor tissue and that responses to vincristine are schedule dependent (*30*). Houghton and colleagues also provided evidence that there is some selectivity in the retention of vincristine in tumor tissue and normal tissue of mice (*31*). Vindesine and deacetylvinblastine both show good activity (200–300% increase in life span) against P388 leukemia in mice and against B16 melanoma (*27*).

Conrad and colleagues have reported on the activities of a range of

TABLE II
PRECLINICAL EFFICACY OF VINBLASTINE AND VINCRISTINE

Tumor system	Response[a]	
	Vinblastine	Vincristine
L1210 leukemia	54	47
P388 leukemia	152	200
B16 melanoma	180	89
Lewis lung	11	16
Colon 26	88	30
Colon 38	(0)	(23)
CD8F1 mammary	(3)	(7)

[a] Percent prolongation of life, except for figures in parentheses which represent percentages of control tumor growth (data from Ref. *14*).

analogs related to vinblastine in experimental tumors (*32*). In terms of overall spectrum of antitumor activity, none of the analogs was judged to be superior to vindesine, which may be considered as a prototype vinblastine analog. A bisvindesine (Fig. 5), in which vindesine units are linked by a disulfide bridge, was reported to have antitumor activity comparable to that of vindesine in several standard evaluation systems but to differ

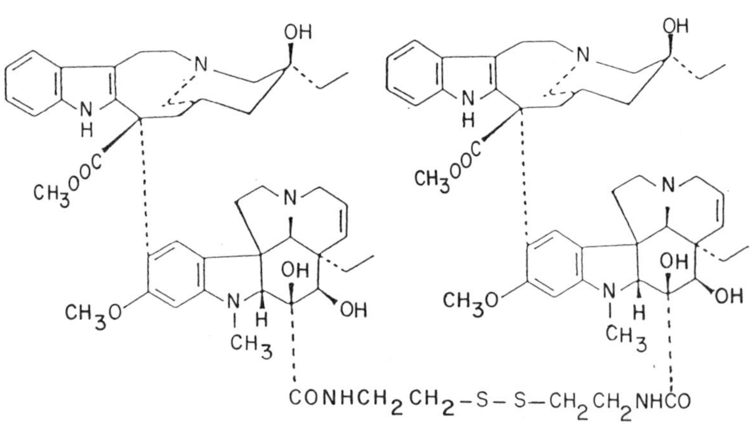

FIG. 5. Structure of a bisvindesine, bis(*N*-ethylidenevindesine) disulfide.

from vindesine in its ability to inhibit the growth of P388 leukemia resistant to vincristine and vinblastine.

Amino acid derivatives (Fig. 6) of vinblastine have been reported that show quite good activity against the P388 leukemia, with some treated mice surviving 60 days after inoculation of tumor cells; in these studies the average survival time for control mice was 10 days (*33*). Several of these amino acid derivatives also have modest activity against the L1210 leukemia. The amino acid derivatives show efficacy at doses considerably higher than those typically used experimentally for vinblastine and vincristine, and this is paralleled by a similarly higher dose required to produce gross toxicity with the amino acid derivatives.

The macrocyclic lactone antibiotic rhizoxin, obtained from a plant fungus, has been reported to have activity against vincristine-resistant cells, both *in vitro* and *in vivo* (*34*). Such observations are interesting in view of the similar effects of rhizoxin and vincristine on cell morphology and cell cycle distribution.

The conjugation of deacetylvinblastine to a monoclonal antibody that recognizes a tumor-associated antigen results in an agent with substantial antitumor activity in mice with relatively little toxicity (*1a*). Conjugates of this type are of obvious interest for future clinical trials as site-directed cancer chemotherapeutic agents.

There is substantial interest in drugs that have little or no antitumor activity when administered alone but that are capable of restoring sensitivity to vinblastine or vincristine in resistant tumor cell populations.

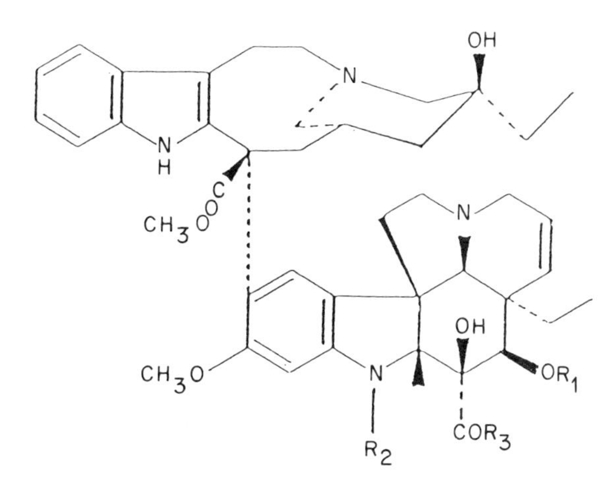

Fig. 6. Structure of a vinblastin-23-oyl amino acid derivative [R_1 = H, R_2 = CH_3, R_3 = $NHCHR_4$ ($COOR_5$)].

Verapamil is the prototype agent of this type, but a large number of apparently unrelated compounds are capable of "synergizing" with vincristine or vinblastine in a combination regimen that shows efficacy against tumors unresponsive to vincristine or vinblastine alone. An example of such a drug is the acridine derivative quinacrine; combinations of quinacrine and vincristine produce substantial increases in the life spans of mice inoculated with P388 leukemia despite the fact that neither drug alone shows marked activity (*35*) (Table III). The enhancement of the activity of vincristine by quinacrine was stated to bring the sensitivity of the P388 leukemia to a level comparable to that observed with fully sensitive tumor cells, but the incisiveness of these studies is diminished somewhat by the small but significant activity of vincristine and quinacrine alone in the "resistant" test system.

For a number of years following the discovery and initial clinical use of vinblastine and vincristine, there was relatively little definitive information about the pharmacokinetics of these compounds. Pharmacokinetic studies were accomplished typically using radiolabeled drugs and procedures that were of limited value in distinguishing parent drugs from putative metabolites.

The distribution and excretion of tritiated vincristine were studied by Castle and colleagues in rats and dogs (*36*). These investigators found that the drug was eliminated from the blood of both species in a biexponential manner, with half-lives of approximately 15 and 75 min. The persistence of low levels of radioactivity in the blood at late sampling times suggested that a third, very slow elimination phase contributes to the pharmacoki-

TABLE III
EFFECT OF QUINACRINE ON ACTIVITY OF VINCRISTINE AGAINST VINCRISTINE-RESISTANT P388 LEUKEMIA IN MICE

Treatment[a]	*T/C* (%)
None	100
Quinacrine (40 mg/kg)	129
Vincristine (0.1 mg/kg)	130
Vincristine (0.1 mg/kg) plus quinacrine (40 mg/kg)	185

[a] Vincristine was given once a day on days 1–5 following tumor inoculation. Quinacrine was given twice a day on the same days. (Table modified from Ref. *35*.)

netic profile of vincristine. After intravenous injection of tritiated vincristine to rats, radioactivity was found to be distributed uniformly to tissues such as liver, kidney, lung, and small intestine; levels of radioactivity in the brain were much lower than those in other tissues at 1 and 6 hr following drug administration, but the difference in radioactivity determined in brain and in other tissues was less pronounced at later times. For example, the concentration of vincristine equivalents in small intestine was 205 ng/g 1 hr after drug administration, while that in brain was 7.9 ng/g; however, by 24 hr the concentration of vincristine equivalents in brain was 5.4 ng/g, relatively close to that estimated for small intestine (12.4 ng/g). The clearance of vincristine and metabolites is predominantly by biliary excretion in the rat and dog. Relatively little metabolism of vincristine occurs in the rat. Creasey and colleagues investigated the pharmacokinetics of vinblastine in the dog (*37*). The elimination kinetics for vinblastine were comparable to those estimated for vincristine in the dog, but it appears that vinblastine is metabolized to a greater extent than is vincristine in this species.

Studies of the metabolism of vincristine and vinblastine have been complicated by chemical alterations of the drugs that occur during processing of samples for analysis. Extraction of these compounds under acidic conditions has been reported to minimize chemical degradation, and HPLC studies of the metabolism of vincristine indicate that microsomal preparations from mouse liver, but not those from human rhabdomyosarcoma tissue, convert vinblastine to 4-deacetylvinblastine *in vitro* (*38*).

The disappearance of tritiated vindesine from the blood of rats has been reported to be biphasic, with half-life estimates of 15 min (distribution) and 10 hr (elimination) (*39*); it is likely that the prolonged elimination phase represents a "hybrid" between the second elimination phase described above for vincristine and the prolonged third phase evident on inspection of log concentration–time plots for vincristine in the rat. Biliary excretion contributes heavily to the elimination of vindesine in the rat. The bioavailability of vindesine in the rat appears to be very poor. The distribution of vincristine to different tissues in the mouse has been correlated with the estimated concentration of tubulin in the tissues (*40*). Tubulin concentration was measured by the capacity of a tissue to bind colchicine (*40*); comparable relationships between tissue concentrations of vincristine and colchicine binding capacity were observed for the dog and the monkey, but it should be emphasized that the correlations were based on the assumption that tissue tubulin content is closely similar in the mouse, dog, and monkey.

Houghton and colleagues found that tumor tissue sensitive to the inhibitory effects of vincristine retains the drug much longer than some normal host tissues, and such selective retention may be critical in determining the efficacy of drugs in this class (*23*). In tumor tissue that is sensitive to vincristine but not to vinblastine, the elimination of vincristine is monophasic, with very little drug leaving the tissue over a period of 72 hr; in such tissue, elimination of vinblastine is biphasic, with elimination half-lives estimated to be 18 and 57–84 hr (*23*). The selective retention of vincristine in neoplastic tissue does not appear to be due to differential metabolism in neoplastic and nonneoplastic tissue (*23*). Complexes of vincristine with proteins from normal mouse tissues and tumor tissue (human rhabdosarcoma xenografts) have been found to differ substantially in terms of stability. For example, complexes formed with preparations from tumor tissue and from brain were stable for at least 2 hr at 37°C, while those from liver, kidney, and bone marrow degraded comparatively rapidly, with estimated half-lives of 30–40 min (*31*).

V. Preclinical Toxicology

Vinblastine and vincristine have a high toxic potential. Early toxicological studies provided estimates of LD_{50} values for vincristine of 2 mg/kg following a single intravenous injection to mice; the estimated LD_{50} for vinblastine was nearly 10-fold higher (17 mg/kg). The toxicity of vinblastine to mice was reported to be substantially higher when the drug was administered intraperitoneally (LD_{50} 3.2 mg/kg); no explanation was offered to account for the difference between the toxicities observed with different routes of drug administration (*1b*). Qualitatively similar differences have been reported for LD_{50} values for vincristine administered intravenously or intraperitoneally to mice; these differences may be due to localized tissue damage produced by the drug given intraperitoneally (*41*). Differences of more than two-fold have been reported for vincristine LD_{50} values for different strains of mice, and similar differences have been reported for LD_{50} values determined in the same strain (*41*). Mice treated with vincristine will show signs of neuromuscular toxicity, such as weakness and poor coordination of the limbs. The LD_{50} for vinblastine given intravenously to rats is 10 mg/kg, and that for vincristine is 2 mg/kg; the LD_{50} for vindesine in rats is 6.3 mg/kg (*28*). The LD_{50} for deacetylvinblastine in mice has been reported to be approximately half that of vinblastine (*27*).

In animals a prominent toxic effect is depression of white blood cell counts. This leukopenia is reversible and is dose related in rats treated with vinblastine. Early observations indicated that the death of animals treated with vinblastine was attributable to an overwhelming bacterial infection secondary to the leukopenia (*1b*). It is of interest to note that while leukopenia was a prominent toxicological consequence of vincristine administration to rats (intraperitoneal; 0.2 mg/kg daily for 10–21 days), there was an isolated observation of possible neurotoxicity (convulsions) in the study.

Johnson and colleagues have summarized the major toxicological observations made after administration of vinblastine or vincristine to other animals (*1b*). In rabbits and cats intravenous doses of 0.2 mg/kg vincristine were lethal after the second to the fifth dose on a schedule involving drug administration two to three times a week. Signs consistent with neurotoxicity were observed in rabbits and cats; for example, a head drop phenomenon, characteristic of neuromuscular dysfunction, was observed in rabbits, and clonic convulsions were observed in cats. Vincristine is lethal to dogs when administered at 0.05 mg/kg, intravenously, five times in a 1-week period. Vincristine is lethal to monkeys when administered five times on a schedule of 1 mg/kg weekly; comparable toxicity is observed when the drug is administered more frequently (daily) at lower doses (e.g., 0.2 g/kg).

In addition to leukopenia and manifestations of neurotoxicity (tremor, ataxia), monkeys treated with vincristine had degenerative changes in the liver and kidney. Vindesine at doses in the range of 0.1–0.3 mg/kg weekly produced leukopenia and reduced spermatogenesis in rats but apparently did not alter neural function (*42*). The acute intravenous LD_{50} for vindesine in mice is 6.3 mg/kg, and that for the congener in which two vindesine units are linked by a disulfide bridge is 6.9 mg/kg (*32*).

There is no currently accepted experimental model for the peripheral neurotoxicity produced by vincristine in humans. It has been suggested that chronic treatment of rabbits with vincristine might serve as a useful model for neurotoxicity (*43*). Rabbits were treated with vincristine (0.3 mg/kg) intravenously at weekly intervals for 5 weeks. The treated rabbits experienced little if any growth during the experimental period and had reduced appetite. Alopecia and anemia were observed, and some animals manifested signs of limb paralysis. Vincristine treatment resulted in the reduction of conduction velocity in the sciatic nerve.

Treatment of cells with vinblastine or vincristine can result in the formation of "paracrystals," complexes containing the alkaloid molecules and tubulin dimers in a 1 : 1 ratio. Paracrystal formation in neuronal tissue of a freshwater snail has been proposed as a model for the neurotoxic effects of *Catharanthus* alkaloids and derivatives (*44*). Vincristine is approximately 10-fold more active than vinblastine as an inducer of paracrystal formation when snail neuronal tissue is treated with high concentrations (150 μM) of the alkaloids.

Johnson and colleagues made a provocative observation in the course of exploratory preclinical toxicological studies of vincristine, namely, that folinic acid (Leucovorin; citrovorum factor; 5-formyl-5,6,7,8-tetrahydrofolic acid) was able to protect mice from the toxicity of high doses of vincristine (*1b*). Vincristine, at a dose of 2.5 mg/kg administered intravenously, resulted in a mortality of 90% over a period of 30 days, but treatment with folinic acid lowered the mortality to 25%. The protection against vincristine toxicity did not occur when folic acid was substituted for folinic acid. A report has appeared (*45*) indicating that there is no specific protective effect of folinic acid against vincristine toxicity in mice and that the "protection" can be observed by comparable treatment with isotonic saline solution. As discussed in Section VII, there is *not* conclusive evidence that folinic acid is able to ameliorate vincristine toxicity in humans (*46*).

VI. Clinical Pharmacology

Vinblastine is a component of the regimen of choice for the treatment of metastatic testicular cancer. In this regimen vinblastine is combined with the platinum complex cisplatin and the heterocyclic glycopeptide antibiotic bleomycin. Vinblastine is also used in combination with bleomycin, doxorubicin (an anthracycline antibiotic), and procarbazine (a hydrazine derivative) in the treatment of Hodgkin's disease. Other types of lymphomas may also respond to treatment with combination regimens that include vinblastine. Vinblastine also is used in the treatment of Kaposi's sarcoma, mycosis fungoides, and carcinoma of the breast. A regimen that includes vinblastine, bleomycin, and methotrexate has been reported to have a potential role as adjuvant therapy to radiation for patients with Hodgkin's disease (*47*).

Vincristine is a standard component of regimens used in the treatment of acute lymphocytic leukemia in children. A representative combination that results in a high percentage of favorable responses in this disease includes use of the corticosteroid analog prednisone and the anthracycline daunorubicin along with vincristine. Vincristine is also a component of regimens with established value in the management of Hodgkin's disease and other lymphomas and pediatric tumors such as Wilms' tumor and embryonal rhabdosarcoma. A treatment protocol that includes vincristine as part of the induction phase and the maintenance phase has been reported to be efficacious in adult patients with acute lymphoblastic leukemia (*48*).

A representative investigational regimen which includes vindesine is that employed in a study of the treatment of non-small cell lung cancer. The regimen consists of vindesine in combination with cisplatin; it produces a modest response rate (25%) but also produces substantial toxicity, including neurotoxicity (*49*).

Both vincristine and vinblastine are administered by intravenous injection, and this reflects the relatively poor bioavailability of the drugs indicated in preclinical studies. Oral administration of a vinblastine derivative, vinzolidine (Fig. 2), has been shown to produce some antitumor activity in clinical trials, but investigations of this oral agent have been discontinued because of unpredictable toxicity (*50*).

Clinical pharmacokinetic investigations with both vinblastine and vincristine have revealed a triexponential elimination pattern. As for preclinical pharmacokinetic studies, early information was obtained by analysis of samples from patients receiving radiolabeled drug, but more recent investigations make use of radioimmunoassays. It should be noted that radioimmunoassays, while very sensitive in terms of detecting drugs, may also measure structurally related drug metabolites, and some caution is needed in interpretation of pharmacokinetic results obtained from such studies.

Half-lives estimated after the administration of vinblastine to patients were 4 min, 1.6 hr, and 25 hr, indicating rapid distribution of the drug to most tissues, relatively rapid clearance, and a subsequent slow terminal elimination process. The distribution and initial clearance phase for vincristine are kinetically comparable to those observed for vinblastine; half-lives for these phases have been reported to be 4 min and 2.3 hr in studies with vincristine. The terminal elimination phase for vincristine has been reported to be three to four times longer than that estimated for vinblastine, and the slow elimination of vincristine from susceptible neuronal tissue has been suggested to play a role in the neurotoxicity commonly observed in clinical settings with vincristine but not with vinblastine (*51*).

Hepatic metabolism and excretion in the bile play major roles in the elimination of both vinblastine and vincristine in humans (*52*); small amounts of vincristine and vinblastine, of the order of 10% of the administered dose, are excreted unchanged in urine. Renal clearance of vinblastine has been reported to be less than 10% of total serum clearance (*53*). Vinblastine has been reported to inhibit a polymorphic cytochrome *P*-450 system in human hepatic microsomes, but the concentrations required were much higher than those observed in clinical settings (*54*).

Estimates for the apparent volumes of distribution for vinblastine, vincristine, and vindesine are extremely high; the mean value for vinblastine is approximately 2000 liters (for a 70-kg individual), and those for vincristine and vindesine are approximately 600 liters (*51*). Such high estimates for the apparent volumes of distribution indicate that extensive binding to various tissues is a prominent feature of the pharmacokinetics of these drugs.

Ohnuma and colleagues investigated the pharmacokinetics of vindesine given by rapid intravenous injection and by infusion (*55*). They found that the estimated volume (six times the serum volume) of the central compartment was considerably higher than that reported by other investigators but that the estimated mean apparent volume of distribution (400–500 liters) was comparable to previous estimates. Estimates of the terminal elimination half-lives for vindesine were 35 and 46 hr, respectively, for bolus injection and infusion. These estimates were also higher than those reported earlier by other investigators, and the differences could be ascribable to inclusion of data from patients with unusually low clearances in calculation of the parameters and also to a sampling schedule insufficient to define a slow elimination process.

There is substantial variability in the pharmacokinetics of vinblastine in patients. Evidence has been obtained that implicates altered liver function and dose-dependent elimination as contributing factors to the variable pharmacokinetics. When vinblastine was administered by a bolus injection, a mean terminal elimination half-life of 29.2 hr was estimated for a group of 24 patients, but the half-lives ranged from a low value of 16 hr to a high value of 65 hr (*53*). When vinblastine was administered by intravenous infusion, clearance of the drug appeared to decrease with time over a 4-month period; decreases in serum albumin values were found to be correlated with decreases in the clearance of vinblastine.

It is recommended that the dose of vinblastine and vincristine be reduced in patients with liver disease. Vincristine is conventionally administered intravenously, to adults, at a dose of 1.4 mg/m^2, with the total dose not to exceed 2 mg for a given administration. Sulkes and Collins have commented on adjustments that may be appropriate for conventional dosages of vincristine and other drugs (*56*). Of particular importance is the possibility that some patients may show good clinical responses and relatively little toxicity in dosage regimens that involve cautious use of larger amounts of vincristine. The initial dose of vinblastine for adults is 3.7 mg/m^2, with a typical dose later increasing to the 5.5–7.4 mg/m^2 range, administered weekly.

VII. Clinical Toxicology

The profiles of toxicities for vinblastine and vincristine observed during treatment of patients with malignant diseases differ. Vinblastine treatment commonly is associated with leukopenia that generally begins 5–10 days after completion of a treatment cycle and lasts for 1–2 weeks. Thrombocytopenia is less common with vinblastine than is leukopenia, although patients who have had prior treatment with radiotherapy or other drugs that are toxic to the bone marrow may show enhanced thrombocytopenia after vinblastine. Mouth sores (stomatitis) and gastrointestinal dysfunction, including nausea, vomiting, and constipation, also are observed in patients receiving vinblastine.

Although the incidence and severity of neurotoxicity is reduced substantially with vinblastine compared to vincristine, patients do suffer from neurotoxicity produced by vinblastine, and such toxicity may be manifested as paresthesias (numbness, tingling in the fingers or toes), loss of tendon reflexes, mental depression, headache, and convulsions (*57*). It should be noted that vinblastine also was observed to produce manifestations of neurotoxicity in preclinical studies. Alopecia (hair loss) is a commonly observed reversible effect in patients receiving vinblastine. Vinblastine shares with a number of basic natural products (including vincristine and vindesine) the capacity for producing severe local tissue damage if the agent leaks from the site of administration (extravasation) or if excessively high concentrations are given to vascular tissue.

While the dose-limiting toxicity for vinblastine usually is leukopenia, that for vincristine is most commonly neurotoxicity (*58*). Prominent manifestations of neurotoxicity are loss of the Achilles tendon reflex, paresthesias, loss of muscle strength (e.g., in the foot and wrist), and ataxia. Constipation and abdominal pain may occur and are thought to result, at least in part, from actions on the autonomic nervous system. Leukopenia and stomatitis are possible effects of vincristine treatment, but they occur relatively infrequently. Alopecia occurs with vincristine at a frequency comparable to that observed with vinblastine, and vincristine also is a potent tissue irritant. Vincristine may produce a syndrome of inappropriate secretion of antidiuretic hormone, and some manifestations of neurotoxicity, such as seizures, have been considered to be due to electrolyte disturbances associated with the relative excess of the antidiuretic hormone (*58*).

Incidents of vincristine overdosage have been reported relatively frequently in the medical literature. Some of these have involved inadvertent administration of the intravenous formulation into the central nervous system by the intrathecal route; this produces devastating results by a combination of chemical damage to sensitive neuronal tissue as well as biochemical perturbations. Two representative cases of vincristine overdose were described (*46*) involving administration of vincristine to patients scheduled to receive vinblastine. In one patient toxicity initially involved vomiting and diarrhea with subsequent constipation and paralytic ileus (inhibition of motor activity in the small intestine). Muscle pain and paresthesias in the extremities were observed for a period of 2 months. The patient was also observed to have mental confusion and hallucinations that persisted for 2 weeks. Myelosuppression was severe in this patient, but it is possible that this was exacerbated by previous treatment with other drugs. In a second patient the profile of neurotoxicity was similar except that the muscle pain and paresthesias lasted for only 3 weeks. The toxicity episode was generally more severe in the first patient, who received folinic acid, than in the second patient, who did not, and this observation, together with other reports, led to the conclusion that "apart from animal experimentation no reason exists for recommending this procedure [folinic acid rescue] as a treatment for vincristine overdose." As presented above, the animal data supporting the role of folinic acid in the management of vincristine toxicity are not convincing.

The toxicological profile for vindesine includes effects observed with both vinblastine and vincristine. Among the effects observed with vindesine are bone marrow depression, alopecia, and peripheral neurotoxicity.

References

1a. I. S. Johnson *Proc. Am. Assoc. Cancer Res.* **27**, 433 (1986).
1b. I. S. Johnson, J. G. Armstrong, M. Gorman, and J. P. Burnett, Jr., *Cancer Res.* **23**, 1390 (1963).
2. I. S. Johnson, H. F. Wright, and G. H. Svoboda, *J. Lab. Clin. Med.* **54**, 830 (1959).
3. M. A. Jordan, R. H. Himes, and L. Wilson, *Cancer Res.* **45**, 2741 (1985).
4. S. A. Lewis, D. Wang, and N. J. Cowan, *Science* **242** 936 (1988).
5. L. S. Borman, M. E. Kuehne, P. A. Matson, I. Marko, and T. C. Zebowitz, *J. Biol. Chem.* **263**, 6945 (1988).
6. G. C. Na and S. N. Timasheff, *Biochemistry* **25**, 6214 (1986).
7. G. C. Na and S. N. Timasheff, *Biochemistry* **25**, 6222 (1986).
8. R. A. Bender and B. A. Chabner, *in* "Pharmacological Principles of Cancer Treatment" (B. A. Chabner, ed.), p. 256. Saunders, Philadelphia, Pennsylvania, 1982.
9. S. Jaken and K. L. Leach, *Annu. Rep. Med. Chem.* **23**, 243 (1988).
10. S. T. Palayoor, J. M. Stein, and W. N. Hait, *Biochem. Biophys. Res. Commun.* **148**, 718 (1987).
11. B. T. Hill and R. D. H. Whelan, *Cancer Treatment Rev.* **7** (Suppl.), 5 (1980).
12. H. Madoc-Jones and F. Mauro, *J. Cell. Physiol.* **72**, 185 (1968).
13. W. T. Beck, M. A Cirtain, A. T. Look, and R. A. Ashmun, *Cancer Res.* **46**, 778 (1986).
14. R. W. Tucker, R. J. Owellen, and S. B. Harris, *Cancer Res.* **37**, 4346 (1977).
15. A. R. Safa, C. J. Glover, and R. L. Felsted, *Cancer Res.* **47**, 5149 (1987).
16. W. T. Beck, *Adv. Enz. Regul.* **22**, 207 (1984).
17. T. Tsuruo, H. Iida, M. Yamashiro, S. Tsukagoshi, and Y. Sakurai, *Biochem. Pharmacol.* **31**, 3138 (1982).
18. B. Vayuvegula, L. Slater, J. Meador, and S. Gupta, *Cancer Chemother. Pharmacol.* **22**, 163 (1988).
19. M. M. Cornwell, A. R. Safa, R. L. Felsted, M. M. Gottesman, and I. Pastan, *Proc. Natl. Acad. Sci. U.S.A.* **83**, 3847 (1986).

20. K. Ueda, C. Cardarelli, M. M. Gottesman, and I. Pastan, *Proc. Natl. Acad. Sci. U.S.A.* **84**, 3004 (1987).
21. J. K. Horton, P. J. Houghton, and J. A. Houghton, *Cancer Res.* **47**, 6288 (1987).
22. F. Brewer and J. R. Warr, *Cancer Treatment Rep.* **71**, 353 (1987).
23. J. A. Houghton, L. G. Williams, P. M. Torrance, and P. J. Houghton, *Cancer Res* **44**, 582 (1984).
24. J. A. Houghton, P. J. Houghton, B. J. Hazelton, and E. C. Douglass, *Cancer Res.* **45**, 2706 (1985).
25. G. B. Henderson and J. M. Tsuji, *Cancer Res.* **48**, 5995 (1988).
25a. C. S. Regula, P. R. Sager, and R. D. Berlin, *Ann. N.Y. Acad. Sci.* **466**, 832 (1986).
26. E. A. Usanga, E. O'Brien, and L. Luzzato, *FEBS Lett.* **209**, 23 (1986).
27. C. J. Barnett, G. C. Cullinan, K. Gerzon, R. C. Hoying, W. E. Jones, W. M. Newlon, G. E. Poore, R. L. Robison, M. J. Sweeney, G. C. Todd, R. W. Dyke, and R. L. Nelson, *J. Med. Chem.* **21** (1978).
28. M. J. Sweeney, G. B. Boder, G. J. Cullinan, H. W. Culp, W. D. Daniels, R. W. Dyke, K. Gerzon, R. E. McMahon, R. L. Nelson, G. A. Poore, and G. C. Todd, *Cancer Res.* **38**, 2886 (1978).
29. A. Goldin, J. M. Venditti, J. S. Macdonald, F. M. Muggia, J. E. Henney, and V. T. DeVita, *Eur. J. Cancer* **17**, 129 (1981).
30. J. A. Houghton, W. H. Meyer, and P. J. Houghton, *Cancer Treatment Rep.* **71**, 717 (1987).
31. J. A. Houghton, L. G. Williams, and P. J. Houghton, *Cancer Res.* **45**, 3761 (1985).
32. R. A. Conrad, G. J. Cullinan, K. Gerzon, and G. A. Poore, *J. Med. Chem.* **22**, 391 (1979).
33. K. S. P. Bhushana Rao, M.-P. M. Collard, J. P. C. Dejonghe, G. Atassi, J. A. Hannart, and A. Trouet, *J. Med. Chem.* **28**, 1079 (1985).
34. T. Tsuruo, T. Oh-hara, H. Iida, S. Tsukagoshi, Z. Sato, I. Matsuda, S. Iwasaki, S. Okuda, F. Shimizu, K. Sasagawa, M. Fukami, K. Fukuda, and M. Arakawa, *Cancer Res.* **46**, 381 (1986).
35. M. Inaba and E. Maruyama, *Cancer Res.* **48**, 2066 (1988).
36. M. C. Castle, D. A. Margileth, and V. T. Oliverio, *Cancer Res.* **36**, 3684 (1976).
37. W. A. Creasey, A. I. Scott, C. C. Wei, J. Kutcher, A. Schwartz, and J. Marsh, *Cancer Res.* **35**, 1116 (1975).
38. J. A. Houghton, P. M. Torrance, and P. J. Houghton, *Anal. Biochem.* **134**, 450 (1983).
39. H. W. Culp, W. D. Daniels, and R. E. McMahon, *Cancer Res.* **37**, 3053 (1977).
40. K. Wierzba, Y. Sugiyama, K. Okudaira, T. Iga, and M. Hanano, *J. Pharm. Sci.* **76**, 872 (1987).
41. A. M. Guarino, M. Rozencweig, I. Kline, J. S. Penta, J. M. Venditti, H. H. Lloyd, D. A. Holzworth, and F. M. Muggia, *Cancer Res.* **39**, 2204 (1979).
42. G. C. Todd, W. R. Gibson, and D. M. Morton, *J. Toxicol. Environ. Health* **1**, 843 (1976).
43. F. Norido, M. Finesso, C. Fiorito, P. Marini, G. Favaro, M. Fusco, F. Tessari, and M. Prosdocimi, *Toxicol. Appl. Pharmacol.* **93**, 433 (1988).
44. L. J. Muller, C. M. Moorer-van Delft, and E. W. Roubos, *Cancer Res.* **48**, 7184 (1988).
45. W. J. Thomas, M. T. Bailony, A. T. Lightsey, S. W. Lew, and W. M. Barnett, *Am. J. Ped. Hematol Oncol.* **8**, 266 (1986).
46. M. Beer, F. Cavalli, and G. Martz, *Cancer Treatment Rep.* **67**,746 (1983).
47. S. J. Horning, R. T. Hoppe, S. L. Hancock, and S. A. Rosenberg, *J. Clin. Oncol.* **6**, 1822 (1988).
48. J. E. Radford, Jr., C. P. Burns, M. P. Jones, R. D. Gingrich, J. D. Kemp, R. W. Edwards, D. B. McFadden, F. R. Dick, and B.-C. Wen, *J. Clin. Oncol.* **7** (1989).

49. E. Rapp, J. L. Pater, A. Willan, Y. Cormier, N. Murray, W. K. Evans, D. I. Hodson, D. A. Clark, R. Feld, A. M. Arnold, J. I. Ayoub, K. S. Wilson, J. Latreille, R. F. Wierzbicki, and D. P. Hill, *J. Clin. Oncol.* **6**, 633 (1988).
50. B. J. Takasugi, A. E. Robertone, S. E. Salmon, S. E. Jones, and D. S. Alberts, *Invest. New Drugs* **2**, 387 (1984).
51. R. L. Nelson, R. W. Dyke, and M. A. Root, *Cancer Treatment Rev.* **7** (Suppl.), 17 (1980).
52. R. J. Owellen and C. A. Hartke, *Cancer Res.* **35**, 975 (1975).
53. M. J. Ratain, N. J. Vogelsang, and J. A. Sinkule, *Clin. Pharmacol. Therap.* **41**, 61 (1987).
54. M. V. Relling, W. E. Evans, R. Fonne-Pfister, and U. A. Meyer, *Cancer Res.* **49**, 68 (1989).
55. T. Ohnuma, L. Norton, A. Andrejczuk, and J. F. Holland, *Cancer Res.* **45**, 464 (1985).
56. A. Sulkes and J. M. Collins, *Cancer Treatment Rep.* **71**, 229 (1987).
57. "AMA Drug Evaluations," 5th Ed., p. 1538. American Medical Association, Chicago, 1983.
58. H. D. Weiss, M. D. Walker, and P. H. Wiernik, *N. Eng. J. Med.* **291**, 127 (1974).

—CHAPTER 6—

THERAPEUTIC USE OF BISINDOLE ALKALOIDS FROM *CATHARANTHUS*

NORBERT NEUSS

Formerly of
Lilly Research Laboratories
Indianapolis, Indiana 46285

AND

MICHAEL N. NEUSS

Oncology, Hematology
Cincinnati, Ohio 45215

I. History of Therapeutic Use

The discovery of medicinal alkaloids from *Catharanthus roseus* G. Don (*Vinca rosea* L.) represents one of the most important introductions of plant products into the cancer chemotherapeutic armamentarium. The relatively unique effects and toxicities of these agents have allowed the design of multiagent chemotherapy programs that have demonstrated sufficient effectiveness to achieve cures even of advanced tumors in many instances. This great accomplishment is possible only because of the inclusion of many different drugs, including the binary "Vinca" alkaloids.

Drugs of plant origin and medicinal herbs have occupied an important place in therapeutics for centuries. Folklore, family tradition, and religious ceremony have allowed passage of knowledge of drug use from medieval times into the scientific era. Now, rigorous testing has confirmed the value of these agents.

The medicinal use of plant alkaloids, including, for example, morphine,

THE ALKALOIDS, VOL. 37
Copyright © 1990 by Academic Press, Inc.
All rights of reproduction in any form reserved.

quinine, strychnine, atropine, and colchicine, prompted the study of other alkaloid plant extracts. It was in the course of the systematic study of tropical flowering plants from the dogbane family (Apocynaceae) that the vinca alkaloids were isolated (*1*). While it was exceedingly difficult to obtain many species, one ornamental variety was plentiful, even in the midwestern United States. This plant, *Vinca rosea,* commonly and incorrectly called the periwinkle,* was sought for study because of its alleged value as a hypoglycemic agent. In the study of its proported action, rats were given crude plant extracts. Though no change in serum glucose was noted, leukopenia was regularly seen, prompting testing of these extracts for effect in a variety of animal tumor cell lines.

Concurrently, two groups became interested in these agents and first carried out research independently. Thus, while workers in the Collip Laboratories at the University of Toronto were isolating vincaleukoblastine from *Catharanthus* using white blood count as a criterion of activity (*2*), Svoboda, Johnson, and colleagues were testing plant extracts for effect on animal tumors at the Lilly Research Laboratories in Indianapolis (*3*). Interestingly, a cooperative effort between these two groups was later undertaken. Johnson *et al.* demonstrated that mice infected with P1534 leukemia enjoyed significant prolongation of survival if appropriately dosed with extracts of *Catharanthus*.

By careful fractionation followed by elution chromatography, three groups of alkaloids were identified (*1,4*). The first group includes "dimeric," highly active, oncolytic compounds, typified by vinblastine and vincristine.† These compounds contain vindoline or a vindoline derivative attached to a tetracyclic indole, carbomethoxyvelbanamine, derived from another major *Catharanthus* leaf alkaloid, catharanthine. Other alkaloids in this group are leurosine and leurosidine (*4,5*). The second group encompasses several binary alkaloids of miscellaneous structures different from the first group. The third group contains substances containing from 18 to 25 carbon and 2 nitrogen atoms in their ring system. Although the second and third groups contain some biologically active compounds, so far they have no value as therapeutic agents in contrast to the four oncolytic alkaloids in the first group.

Surprisingly, in P1534 leukemia-infected animals, four distinct fractions were shown to have tumoricidal and thereby life-prolonging activity. These contained the alkaloids vinblastine, leurosine, leurosidine, and vin-

* The correct botanical name of this pantropical flowering plant is *Catharanthus roseus* G. Don. A commonly used name is *Vinca rosea* L.

† Vinblastine and vincristine are the USAN council approved names of vinealeukoblastine and leurocristine. Vindesine is an approved, nonproprietary name for deacetylvinblastine amide.

cristine. All of these compounds come from the first mentioned group of *Catharanthus* alkaloids (*6*).

The P1534 leukemia used initially to study *Vinca* alkaloids is an acute lymphocytic leukemia line transplantable only in DBA/2 mice. In experiments, animals implanted with P1534 leukemia and treated with vinblastine survived 40–150% longer than saline-treated controls at doses ranging from 0.05 to 0.6 mg/kg. Similar results were obtained with leurosine therapy at doses ranging from 3 to 7.5 mg/kg. Survival prolongations ranging from 30 to 220% over control were obtained with vincristine at 0.05–0.35 mg/kg. Leurosidine achieved a 20–130% prolongation at doses from 2 to 10 mg/kg. At higher doses, 60–100% of the animals treated had indefinite survival and were thought to be "cured." This degree of specific activity against the P1534 lymphocytic leukemia line was remarkable and apparently unique (*3*).

With the exception of leurosine, all three alkaloids significantly prolonged the life of animals infected with the P1534 leukemic cell line. This effect was observed even when treatment was delayed until the animals were nearly moribund. However, and perhaps most interestingly, it was observed that "cured" animals were resistant to additional challenges with this tumor line (*3*). Sadly, these results could not be extrapolated to clinical situations.

Since the late 1960s, the oncolytic activity of binary vinca alkaloids is no longer seen in P1534 leukemia, undoubtedly because of a change in the tumor itself. Therefore, research testing new derivatives of *Catharanthus* alkaloids has relied on different models. Systems used have included the Ridgeway osteogenic sarcoma line, the Gardner sarcoma line, murine B16 melanoma, P388 leukemia, L1210 leukemia, and, of course, the (now differently behaving) P1534 leukemia line (*7*).

Two synthetic derivatives of "dimeric" *Catharanthus* alkaloids deserve mention. The first is navelbine or nor-5-anhydrovinblastine, which is being tested in France (*8*). Structurally, this compound differs from vinblastine by the presence of a eight-membered ring where a nine-membered ring is present in the latter. The second derivative of vinblastine was synthesized by Gerzon and colleagues (*7*). They prepared a series derived from functionalization of different substituents of vinblastine including deacetylvinblastine amide or vindesine. As the P1534 lymphocytic leukemia system could not be used, tumor inhibition by these derivatives was measured in the Ridgeway osteogenic sarcoma and Gardner lymposarcoma mouse tumors. Vindesine seems to resemble vincristine rather than vinblastine in its spectrum of activities against rodent tumor systems. Its neurotoxic potential, however, appears to be less than that of vincristine (*7*).

The pharmacologic activities of alkaloids from the second and third groups have been of little interest. Though they have some diuretic action, the activity is similar to chlorothiazide and dichlorothiazide, and, because of greater toxicity, these agents have not been developed further. A synthetic derivative of a "monomeric" vinca alkaloid, vinpocetine, is currently being investigated as a neuronal stimulant which may be useful in the treatment of cerebral dysfunction. This drug, while an alkaloid, is derived from *Vinca minor* and is chemically unrelated to the binary alkaloids derived from *Catharanthus roseus* (*9*).

II. Current Therapeutic Use

The vinca alkaloids are useful in the treatment of both malignant and nonmalignant disease. Though the former category accounts for a proportionally greater fraction of use of these compounds, their use in benign but often fatal platelet disorders, while evolving, is nonetheless established and important.

A. Use in Nonmalignant Disease

Sequential determination of platelet counts in patients receiving vincristine during early studies unexpectedly occasionally revealed thrombocytosis, which could not be accounted for by systemic response to treatment alone (*10,11*). Ultimately shown to most likely be the result of increased megakaryocytic endomitosis (*11*), the observation led to the use of vincristine, and later vinblastine, both alone and bound to platelets, in a variety of thrombocytopenic disorders. These include idiopathic thrombocytopenic purpura, thrombotic thrombocytopenic purpura, and chemotherapy-induced microangiopathic hemolytic anemia.

Idiopathic thrombocytopenic purpura is an immune-mediated disease in which immunoglobulin, either as antibody directed against platelet antigens or nonspecifically bound to platelets, is present in increased quantities on platelets. This leads to increased destruction of platelets and, in many instances, megakaryocytes. Standard treatment consists of corticosteroids and splenectomy (*12*). When these measures fail, treatment may include androgenic steroids, administration of intravenous γ-globulin, or injection of vinca alkaloids.

The best results of this treatment came in the first reported series of patients. As is often the case, this highly selected group of patients responded better than subsequently studied series. In the original report, 21 patients with idiopathic and 22 with secondary thrombocytopenic purpura were given 2 mg doses of intravenous vincristine (1 mg in children) every 7–10 days. Improvement in platelet count was seen in 16 patients (76%).

In patients with thrombocytopenia associated with a variety of conditions, 17 of 22 improved (*13*). Though these results have not been confirmed with this degree of success (*14*), vincristine remains an important agent in the treatment of refractory idiopathic thrombocytopenic purpura (ITP).

This same group of investigators in Miami later advocated the use of vinblastine-"loaded" platelets (*15*) to deliver the vinca alkaloid to the macrophages involved in the destruction of platelets in ITP patients (*12*). Vinblastine was chosen instead of vincristine because radiolabeled vincristine was not available and, therefore, the platelet–vincristine binding could not be measured. Results were again phenomenally successful, with 9 of 11 patients with "chronic refractory" ITP showing improvement (*15*). Other observers were unable to either repeat these results (*14*) or confirm sustained platelet–vinblastine binding, although others have suggested that vincristine–platelet binding may be effective (*16*).

The use of vincristine to treat thrombotic thrombocytopenic purpura has also been reported to be successful (*14,18*). This rare disease of sporadic thrombosis of small vessels with consequent intravascular hemolysis has been successfully treated recently with both plasma exchange and plasma infusion. However, there are obvious disadvantages to plasma infusion, including volume overload, transmission of infection, and cost and scarcity of plasma. Several patients have been successfully treated with vincristine alone or in association with plasma therapy. At present, however, this treatment should probably not be used alone because of the high success rate of plasma infusion and/or exchange.

Finally, the successful use of vincristine in the treatment of drug-induced microangiopathic hemolytic anemia, a disease resembling the hemolytic uremic syndrome rarely seen subsequent to administration of mitomycin C or other chemotherapy has been described in two patients with this disorder (*19*).

B. Use in Malignant Disease

Though vinca alkaloids are useful in platelet and platelet-associated disorders, it is in treating malignancy that they are truly an indispensable part of the pharmacopoeia. First available in the 1960s, vinca alkaloids now are included as part of virtually every successful combination chemotherapy program, both because of their unique action and because of their unique toxicities.

Vinca alkaloids are exceedingly important in both curative and palliative regimens. Curative regimens now exist that regularly allow long-term disease-free survival for a variety of adult and pediatric patients, even if treatment begins when the tumor is advanced. These allow treatment of lymphoproliferative disease, including Hodgkin's and non-Hodgkin's lymphoma and acute lymphoblastic leukemia, testicular cancer, Wilms' tumor, embryonal rhabdomyosarcoma, and Ewing's sarcoma. Less frequently, long-term survival is seen in small cell lung cancer patients (see Table I) (*20*). Real palliative benefit, even without cure, can be seen in patients with all of the diseases mentioned above as well as in breast, bladder, melanoma, and, perhaps, non-small cell lung cancer patients (see Table II).

Combination chemotherapeutic regimens are designed based on the belief that greater cumulative effects can be seen by using agents with differing sites (within the cell) and times (within the cell cycle) of action. Cumulative toxicities may be less than single toxicities because of varying toxicities of different agents (*20,21*). Because *Catharanthus* alkaloids uniquely inhibit microtubular assembly and mitosis and because of their relatively nonmyelosuppressive effects, they are part of the alphabet soup of important multiagent regimens, including COP, CAV, CHOP, ABVD, MOPP, PVB, PROMACE, m-BACOD, MACOP-B, and MVAC (Tables I and II).

The evolution of regimens in Hodgkin's lymphoma and testicular cancer is representative of the progress in treating malignancy over the 1970s and 1980s. Early studies with vinblastine and vincristine revealed the striking responsiveness of patients with Hodgkin's disease who were given vinca alkaloids as single agent therapy (*21–26*). The rate of response with vinblastine as a single agent in Hodgkin's disease is reported as high as 60% (*23*). As these tumors are, however, also responsive to alkylating agents, corticosteroids, and antimetabolites, a prototype regimen

TABLE I
CURATIVE REGIMENS CONTAINING *Catharanthus* ALKALOIDS

Acronym	Drugs	Disease
MOPP	Nitrogen mustard, vincristine, procarbazine, prednisone	Hodgkin's lymphoma
CHOP	Cyclophosphamide, daunomycin, vincristine, prednisone	Non-Hodgkin's lymphoma
ABVD	Daunomycin, bleomycin, vinblastine, dacarbazine	Hodgkin's lymphoma
PVB	Cisplatinum, vinblastine bleomycin	Testicular cancer
MACOP-B	Methotrexate, daunomycin, cyclophosphamide, vincristine, prednisone, bleomycin	Lymphoma
MVAC	Methotrexate, vinblastine daunomycin, cyclophosphamide	Bladder cancer

TABLE II
PALLIATIVE REGIMENS CONTAINING *Catharanthus* ALKALOIDS

Acronym	Drugs	Disease
COP	Cyclophosphamide, vincristine, prednisone	Lymphoma
POC	Procarbazine, vincristine, cyclophosphamide	Melanoma, small cell lung cancer
VATH	Vinblastine, daunomycin, Thio-TEPA, Halotestin	Breast cancer
VP	Vinblastine/vindesine, cisplatinum	Non-small cell lung cancer

of MOMP (nitrogen mustard, vincristine, methotrexate, and prednisone) was tried as treatment in Hodgkin's disease. Overall toxicity was acceptable, and complete responses were seen in 71% of previously untreated patients (*21*). With the substitution of procarbazine for methotrexate, the response rate climbed even higher.

As the frequency of curing patients has increased, efforts have been redirected toward decreasing the toxicity of multiagent regimens by decreasing the duration of treatment and, finally, by substituting other agents into the regimens. In an effort to reduce the delayed incidence of acute leukemia, ABVD has been substituted for MOPP with some success and is now, to many, considered the standard treatment of advanced Hodgkin's disease. Similarly, treatment for testicular cancer has evolved from observations of single agent efficacy to highly successful multiagent regimens, such as PVB or VAB-VI. Now, less aggressive but equally effective regimens with lesser numbers of cycles and the substitution of VP-16 for vinblastine have been tried and are considered standard therapy (*23*).

III. Toxicity

The potential toxicities of chemotherapeutic agents are, by the nature of their mechanism of introduction into practice, often best exposed in dose-defining and toxicity-defining early studies on the drugs. The spectrum of toxicity of *Catharanthus* agents is broad, and though there is overlap among the agents, each has a more characteristic pattern of toxicity.

The predominant effect of vinblastine is on bone marrow, and all cell lines are affected. The greatest effect is seen on white blood cells, specifically, the granulocyte. This toxicity is dose related. It is so regular that it may be defined as an expected effect and not a side effect. At usual therapeutic doses, neutrophilic depression is seen in approximately 35% of patients. Platelet depression is much less commonly encountered, but doses occur, as does depression of the erythron (*22–24*).

Gastrointestinal side effects of vinblastine are reported in up to 11% of patients who receive the drug. Nausea is overwhelmingly the most common gastrointestinal side effect, with 11% of patients experiencing this in standard dosing. Less than half of those who experience nausea actually vomit (*24*). The other predominant gastrointestinal side effect is paralytic ileus, which is seen in as high as two-thirds of patients given high-dose vinblastine (0.8 mg/kg). Very rarely, hemorrhagic colitis is also seen in patients given this agent (*23,24*).

Neurologic sequelae with vinblastine are much less common than those seen with vincristine and vindesine. Nonetheless, a causal relationship has been established for seizures, psychotic episodes, and confusional episodes. As common with other vinca agents, absence of reflexes and peripheral neuropathy are well described (*23,24*).

Cutaneous reactions to vinblastine are relatively rare. If the drug is, however, injected improperly and extravasates from a vein, necrosis at the site of injection is common (*22–24*). As opposed to cases with other chemotherapeutic agents, in this situation there is a treatment available, namely, subcutaneous injection of hyaluronidase, which is said to decrease the toxicity of this event. Local debridement is, unfortunately, also often required (*23,24,29*). A very unusual photosensivity has been well described and worked out in which a photosensitive dermatitis developed in a vinblastine-treated patient with Hodgkin's disease. As judged by both rechallenge and careful subcutaneous drug challenge, this appears to have been clearly caused by vinblastine (*30*). Other exceedingly rare side effects have been described, but the causal relationship of drug administration has not been completely proved.

Early studies of vincristine detailed a wide range of potential toxicities. This is, perhaps, not surprising when one considers that the drug was administered in dose ranges of as high as 7 mg total dose per week (as compared to current doses which rarely exceed 2 mg) (*25,26*). The most frequent toxicities observed in any dosage are neurologic. Paresthesia, with or without anesthesia or pain, is typically seen in the characteristic acral "stocking-glove" distribution of peripheral neuropathy. This is usually associated with loss of deep tendon reflexes, especially the Achilles reflex (*25,26,31,32*). Peripheral nerve involvement can include cranial nerves and visceral autonomic nerves. This leads to gastrointestinal neuropathy with atonic ileus and constipation, frequently associated with severe abdominal pain. Symptoms of impotence are also quite frequently reported when patients are questioned concerning this effect (*31*).

Because of sensory neuropathy, deep pain is often experienced by patients given vincristine. Through involvement of the glossopharyngeal nerve, throat pain may occur, as may deep pain of almost any other area of the body (*31,32*). Neuropathic changes are not always peripheral. Hallucinations and overall mental status changes, such as depression and/or psychosis, are also rarely reported (*31,32*). Another CNS effect is the syndrome of inappropriate antidiuretic hormone secretion, which is a well-characterized side effect of vincristine (*31,32*).

Some authors believe that neuropathic effects of vincristine are greatly exacerbated in patients with underlying neuropathy, such as in those with Marie-Charcot syndrome. This is based on experience with one patient who developed bulbar paralysis following vincristine injection. Some authors believe this reaction to have been idiosyncratic and report successful treatment of patients with underlying neuropathies with vincristine without untoward effects. At the least, extreme caution and careful consideration of alternate therapies seems reasonable in such patients (*31,32*). In either case, the neurologic symptoms are so common they should not limit the use of this agent, which in early studies was well tolerated at a weekly dose of 3.5 mg.

As opposed to vinblastine, vincristine's hematologic effects are minimal. In fact, thrombocytosis is often seen following vincristine administration (*10,25,26*). Alopecia occurs in from 13 to 22% of patients given full doses of vincristine. Cutaneous effects are, however, more unusual and are a practical problem only when the drug is extravasated (*21,22*). The spectrum of vindesine side effects is similar to that seen with vincristine.

The list of potential toxicities outlined above is by no means complete. It does, however, touch on the major toxicities regularly encountered in the administration of these agents.

IV. Mechanism of Action

The mechanism of action of *Catharanthus* alkaloids involves entry into the cell, binding to tubulin, and interference with cellular metabolic functions. The predominant observed effect is often mitotic arrest, but other effects on cellular organization and movement can also be demonstrated. It is not unequivocally clear that the mitotic effect is, *in vivo,* of greater importance than other tubulin-mediated effects (*33*).

In neoplastic disease, the consequences of this effect, namely, cell death and cessation of cell division, are obviously beneficial. In thrombocytopenic conditions, the mechanism is a little less clear. As mentioned above, the effect may simply be related to the promotion of endomitosis of megakaryocytes and consequent increased production of platelets (*11*). However, the effect may be more selective. Specifically, it is suggested that vincristine and vinblastine, especially when bound to platelets, are taken up by the reticuloendothelial system, as in benign platelet consumption disorders, and selectively destroy the phagocytic cell (*12*). This so-called selective chemotherapy of macrophages, however, seems less likely than when it was first hypothesized (*14*).

Once the drug is inside the cell, interference of tubulin aggregation results and accounts for the effects of the alkaloid. The direct mechanism of this effect is not completely clear, however, because of its complexity. Specifically, tubulin aggregation accounts for much of a cell's organized intracellular activity as well as its directional motility. Through interference with the former, mitosis is inhibited, as are many other cellular functions, including protein and DNA synthesis. Though it is often assumed that the cytostatic effect is due to mitotic inhibition because this process can be easily seen, the causal relationship of mitotic interruption to cell death is not unequivocally established (*33*). Furthermore, the *in vivo* tumoricidal effects of vinca alkaloids may result from observable effects on tumor cell motion, especially tumor invasion, and not from cytostatic activity (*33*).

V. Drug Resistance

The mechanism of action of *Catharanthus* alkaloids undoubtedly relates to their effect on tubulin aggregation and consequent microtubule assembly and function. As microtubular function is intracellular, the alkaloids have to enter cells and remain within them to be effective.

Drug resistance *in vitro* and probably *in vivo* results both from inhibition of influx of the vinca alkaloids and, perhaps more frequently, from promotion of their efflux out of cells (*34,35*). Until relatively recently, the former mechanism was thought to predominate, and, indeed, certain acquired drug-resistant states are clearly associated with the loss of membrane proteins which can be shown to bind and transport agents into cells (*34*). However, other resistant states have been shown to be associated with the acquisition of membrane transport proteins which remove toxins (and, therefore, chemotherapeutic agents) both from normal and malignant cells.

The phenomenon of multiple drug resistance has been well described.

In this situation, cell lines are shown to be resistant to colchicine, doxorubicin, vinblastine, and actinomycin D. This syndrome is accompanied by an increase in measurable membrane glycoprotein (the P-170 or permeability glycoprotein). It is believed that this protein transports hydrophobic chemicals out of cells and thereby prevents drug action. Current efforts to inhibit this efferent transport protein are currently underway but, sadly, have to date been largely unsuccessful (*35*).

In addition to efforts to overcome drug resistance by changing transport proteins, activity has been great in the field of modifying drugs so that they enter cells more readily and also enter only malignant cells. The former goal has been approached through chemical modification of *Catharanthus* alkaloids (*36*). Formation of amides and binding to larger protein molecules have promoted cell entry by pinocytosis. The most promising vinca–protein bonds have been those to hybridoma-derived monoclonal antibodies (*37,38*). Through these, tumor-directed therapy seems likely. Unfortunately, however, to data these complex molecules have been more effective *in vitro,* where immune clearance does not interfere.

The preceding discussion is obviously incomplete. Furthermore, chemotherapeutic and other indications for the use of *Catharanthus* alkaloids are always changing. Nonetheless, it seems apparent that their inclusion in therapeutic regimens is, at least in part, responsible for the success of these regimens in both malignant and nonmalignant disease. It is hoped that, through chemical modification, their effectiveness will only increase.

References

1. G. H. Svoboda, *J. Pharm. Sci.* **47**, 834 (1958).
2. R. L. Noble, C. T. Beer, and J. H. Cutts, *Ann. N.Y. Acad. Sci.* **76**, 882 (1958).
3. I. S. Johnson, G. H. Svoboda, and H. F. Wright, *Proc. Am. Assoc. Cancer Res.* **3**, 331 (1962).
4. G. H. Svoboda, *Lloydia* **24**, 173 (1961).
5. N. Neuss, M. Gorman, W. Hargrove, *et al., J. Am. Chem. Soc.* **86**, 1440 (1964).
6. N. Neuss, *in* "Indole and Biogenetically Related Alkaloids" (T. O. Phillipson and M. H. Zenk, eds.), p. 307. Academic Press, New York, 1980.
7. C. J. Barnett, G. J. Cullinan, K. Gerzon, *et al., J. Med. Chem.* **21**, 88 (1978).
8. P. Potier, *Pure Appl. Chem.* **58**, 737 (1986).
9. R. Balestreri, L. Fontana, and F. Astengo, *J. Am. Geriatr. Soc.* **35**, 425 (1987).
10. J. H. Robertson and G. M. McCarthy, *Lancet 2*, 353 (1969).
11. R. J. Ratzean, S. Tucker, and L. R. Weintraub, *Blood* **40**, 965 (1972).
12. W. F. Rosse, *N. Engl. J. Med.* **298**, 1139 (1978).
13. Y. S. Ahn, W. J. Harrington, R. C. Seelman, and C. S. Eytel, *N. Engl. J. Med.* **291**, 376 (1974).

14. J. G. Kelton, J. W. D. McDonald, R. M. Barr, *et al., Blood* **57**, 431 (1981).
15. Y. S. Ahn, J. J. Byrnes, W. J. Harrington, *et al., N. Engl. J. Med.* **298**, 1101 (1978).
16. G. Agnelli, M. DeCunto, P. Gresele, and G. G. Nenci, *Blood* **60**, 1235 (1982).
17. L. A. Gutterman and T. D. Stevenson, *J. Am. Med. Assoc.* **247**, 1433 (1982).
18. M. L. Sennett and M. E. Conrad, *Arch. Int. Med.* **146**, 266 (1986).
19. J. L. Grem, J. A. Merritt, and P. P. Carbone, *Arch. Int. Med.* **146**, 566 (1986).
20. E. Frei, *Cancer Res.* **45**, 6523 (1985).
21. E. Frei, V. T. DeVita, J. H. Moxley, and P. P. Carbone, *Cancer Res.* **26**, 1284 (1966).
22. E. Frei, A. Franzino, B. I. Schnider, *et al., Cancer Chemother. Rep.* **12**, 175 (1961).
23. J. G. Armstrong, R. W. Dyke, P. J. Fouts, and J. E. Gahimer, *Cancer Chemother. Rep.* **18**, 49 (1962).
24. T. L. Wright, J. Hurley, D. R. Korst, *et al., Cancer Res.* **23**, 169 (1963).
25. P. P. Carbone, V. Bono, E. Frei, and C. O. Brindley, *Blood* **21**, 640 (1963).
26. J. F. Holland, C. Scharlau, S. Gailani, *et al., Cancer Res.* **33**, 1258 (1973).
27. A. Santoro, V. Bonfante, and G. Bonadonna, *Ann. Int. Med.* **96**, 139, (1982).
28. S. D. Williams, R. Birch, L. H. Einhorn, *et al., N. Engl. J. Med.* **316**, 1435 (1987).
29. R. Rudolph and D. L. Larson, *J. Clin. Oncol.* **5**, 1116 (1987).
30. T. S. Breza, K. M. Halprin, and R. Taylor, *Arch. Dematol.* **111**, 1168 (1975).
31. H. D. Weiss, M. D. Walker, and P. H. Wiernik, *N. Engl. J. Med.* **291**, 127 (1974).
32. S. Rosenthal and S. Kaufman, *Ann. Int. Med.* **80**, 733 (1974).
33. M. Mareel and M. DeMets, *Int. Rev. Cytol.* **90**, 125 (1984).
34. B. T. Hill, *J. Antimicrob. Chemother.* **18**, 61 (1986).
35. I. Pastan and M. Gottesman, *N. Engl. J. Med.* **316**, 1388 (1987).
36. K. S. Rao, M. Collard, and A. Trouet, *Anticancer Res.* **5**, 379 (1985).
37. J. R. F. Corvalan, W. Smith, V. A. Gore, *et al., Cancer Immunol. Immunother.* **24**, 138 (1987).
38. A. G. Casson, C. H. Ford, C. H. Marsden, *et al., NCI Monogr.* **3**, 117 (1987).

CUMULATIVE INDEX OF TITLES

INDEX

X

Y